Optoelectronics: Devices, Techniques and Applications

Optoelectronics: Devices, Techniques and Applications

Edited by Julian Silva

Clanrye International
www.clanryeinternational.com

Clanrye International,
750 Third Avenue, 9th Floor,
New York, NY 10017, USA

ISBN: 978-1-63240-845-7

Cataloging-in-Publication Data

Optoelectronics : devices, techniques and applications / edited by Julian Silva.
p. cm.
Includes bibliographical references and index.
ISBN 978-1-63240-845-7
1. Optoelectronics. 2. Optoelectronic devices. 3. Infrared technology.
4. Photonics. I. Silva, Julian.

TA1750 .O68 2019
621.36--dc23

For information on all Clanrye International publications
visit our website at www.clanryeinternational.com

Clanrye International

Printed in China.

Contents

Preface

As a sub-field of photonics, optoelectronics refers to the practice of using and studying electronic devices and systems, which control, source and detect light. Optoelectronics is based on the fundamentals of quantum mechanical effects of light. It focuses on light's effect on semiconductors, and other electronic materials. The different optoelectronic devices are optoisolators, photoresistors, phototransistors, quantum cascade lasers, OLEDs, photomultipliers, light-emitting diodes, etc. Research in this field has enabled scientists to apply principles of optoelectronics in optocoupler and optical fiber communications. This book presents the complex subject of optoelectronics in the most comprehensible and easy to understand language. Such selected concepts that redefine the subject have been presented in it. This extensive book is a vital tool for all researching and studying this field.

This book has been the outcome of endless efforts put in by authors and researchers on various issues and topics within the field. The book is a comprehensive collection of significant researches that are addressed in a variety of chapters. It will surely enhance the knowledge of the field among readers across the globe.

It gives us an immense pleasure to thank our researchers and authors for their efforts to submit their piece of writing before the deadlines. Finally in the end, I would like to thank my family and colleagues who have been a great source of inspiration and support.

Editor

1

Hybrid Optoelectronic Router for Future Optical Packet-Switched Networks

Salah Ibrahim and Ryo Takahashi

Additional information is available at the end of the chapter

Abstract

With the growing demand for bandwidth and the need to support new services, several challenges are awaiting future photonic networks. In particular, the performance of current network nodes dominated by electrical routers/switches is seen as a bottleneck that is accentuated by the pressing demand for reducing the network power consumption. With the concept of performing more node functions with optics/optoelectronics, optical packet switching (OPS) provides a promising solution. We have developed a hybrid optoelectronic router (HOPR) prototype that exhibits low power consumption and low latency together with high functionality. The router is enabled by key optical/optoelectronic devices and subsystem technologies that are combined with CMOS electronics in a novel architecture to leverage the strengths of both optics/optoelectronics and electronics. In this chapter, we review our recent HOPR prototype developed for realizing a new photonic intra data center (DC) network. After briefly explaining about the HOPR-based DC network, we highlight the underlying technologies of the new prototype that enables label processing, switching, and buffering of asynchronous arbitrary-length 100-Gbps (25-Gbps × 4λs) burst-mode optical packets with enhanced power efficiency and reduced latency.

Keywords: optical packet switching, optical signal processing, optoelectronic devices

1. Introduction

Optical networks have been playing a pivotal role in achieving the current unprecedented capability of communications that has transformed human experience. In addition to their conventional role in the infrastructure core, metro, and access networks, optical networks are also indispensable in enabling other vital applications such as large-scale data centers (DCs) and supercomputers. With such diversity of application domains, the traffic handled by the network varies between stable and bursty traffic. The network should thus utilize the optical transmission scheme that matches its traffic nature.

Among different schemes of optical transmission, optical circuit switching (OCS) and optical packet switching (OPS) are two basic schemes that possess complementary features. The OCS scheme allows uninterrupted data transmission where a link is established between two network nodes before starting to transmit data in between. The optical link is realized in a way similar to reserving a closed circuit and thus the scheme is entitled circuit switching. The OCS scheme is suitable for transmission of stable traffic and hence it is widely utilized in core networks. Differently, no link establishment is required with the OPS scheme as data is transmitted as individual packets in a connectionless manner. Each packet is equipped with a given label and based on that label, the packet is forwarded along network nodes until arriving at its destination. More importantly, the packets are forwarded without going through optical-electrical-optical (OEO) conversion in a real photonic network. This feature together with the elimination of the link establishment time makes the OPS scheme very suitable for handling bursty traffic.

On the other hand, the capacity of optical links has been significantly boosted [1] by using higher data rates and complex data formats, which increases the burden on current network nodes that are mostly relying on electrical packet switching (EPS). The resulting extensive dependence on electronic processing is the reason for several shortcomings that are difficult to overlook such as the high power consumption and end-to-end latency. The EPS approach can be also identified as a limiting factor for the network scalability which is the problem faced by current large-scale DC networks. This condition can be overcome by realizing a photonic network instead, and in this regard, we have proposed a new photonic intra DC network based on the hybrid optoelectronic router (HOPR).

In this chapter, we review our recent HOPR prototype developed for realizing the DC network. We briefly explain about the HOPR-based photonic DC network, and highlight the underlying technologies of the new prototype that enable label processing, switching, and buffering of asynchronous arbitrary-length 100-Gbps (25-Gbps × 4λs) burst-mode optical packets with enhanced power efficiency and reduced latency.

2. New photonic data center network

DCs have considerably evolved to become main players in the big-data era, providing a diversity of services with unprecedented volumes of data traffic [2–4]. However, as mentioned earlier, current intra DC networks that are mainly based on EPS have been facing increasing difficulties to cope with the growing demand. The advancement of the EPS-based DC networks has continuously relied on the progress of the CMOS and transceiver technologies. But these technologies have already reached an advanced level after which it becomes difficult to achieve further improvement at the quick pace required for fulfilling the ongoing demands [5]. Realizing a photonic DC network is a radical solution that can let DCs surpass their current difficulties. To serve this end, we have proposed the photonic intra DC network

[6] illustrated in **Figure 1**. The network has a torus topology and it depends on the deployment of HOPRs and a centralized network controller. More about the key network aspects can be found as follows.

- **Basic operation:** A HOPR unit is located at each node of the torus DC network and connected to neighboring HOPRs via optical links. High-speed burst-mode optical packets are transmitted over these links. Each HOPR unit is also connected to a group of Top-of-Rack (ToR) switches that handle Ethernet packets from the servers connected to them. An Ethernet packet is transformed into a burst-mode optical packet at the corresponding HOPR unit and from there on it is transmitted in the optical domain throughout the torus network until it reaches the HOPR unit attached to the destination server. Thus, each HOPR unit has a twofold role as an optical-packet switch that forwards the optical packets of the torus network and an aggregation switch that handles the servers' packets.
- **Topology:** A highly dimensional torus topology is considered for realizing the network. This topology has been widely adopted in supercomputers such as the CRAY (3D), Blue Gene (3–5D), and K Computer (6D). It enables a highly scalable network that can strongly support fault tolerance with redundancy of links. The torus topology also exhibits several features that are advantageous for the OPS scheme; as (1) it is a direct topology in which a low-radix switch is sufficient for deployment at each network node, (2) the uniform arrangement of the network nodes allows the utilization of a simple deterministic algorithm for forwarding incoming packets, and (3) the presence of multiple equidistant routes between the source and destination nodes with the same latency can efficiently help in resolving packet contention when it occurs.
- **Transmission schemes:** Three data transmission schemes are supported by the photonic DC network; namely the OPS, OCS, and virtual V-OCS schemes. Unlike current DC networks, the OPS scheme is used to allow latency-sensitive applications over a wide network scale without being limited to nodes in close proximity.

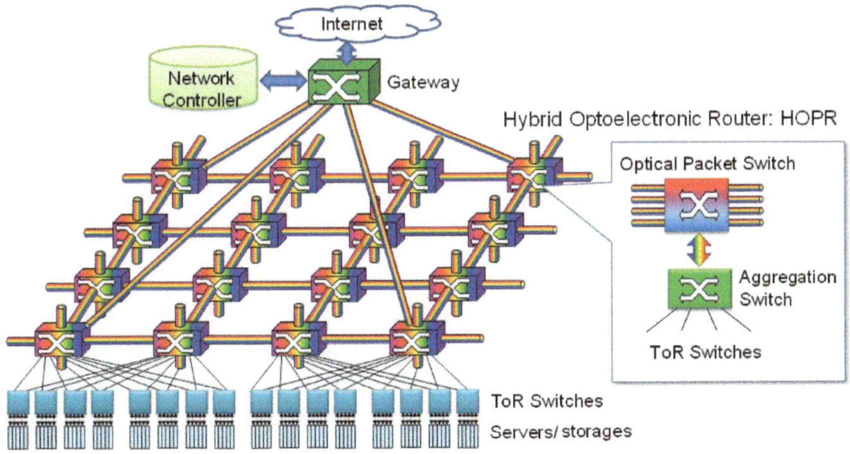

Figure 1. Illustration for the torus DC network based on HOPR and NW controller.

However, an OPS packet can reach its destination via different routes and might encounter collision with other packets and thus the probability of packet loss cannot be completely avoided. For the OCS case, an optical path is reserved before transmission is started to avoid packet loss and the sent packets arrive with the same order and same latency. However, as the number of OCS paths increases, the normal OPS packets are strongly counteracted due to the increasing difficulty of finding a vacant route to reach their destination. Hence, the virtual OCS scheme has been devised [7, 8] to enable the coexistence of the OPS and OCS packets almost without counter effects. To realize such DC network, we have developed a new HOPR prototype that can meet the demand of handling burst-mode optical packets without OEO conversion unless necessary.

3. Hybrid optoelectronic router

In this section, we highlight HOPR's operation, architecture, and implementing approach. **Figure 2a** shows HOPR unit for a 3-D torus network, and **Figure 2b** shows a 6-D network where a 16 × 16 optical switch is used and the shared buffer interface is upgraded to 200 Gbps allowing the aggregation of more servers' traffic. Using the WDM configuration as shown in **Figure 2c**, the link capacity can be increased as the product of the number of wavelength layers and data rate of generated packets to allow achieving 0.4–1 Tbps.

When an optical packet arrives at HOPR, the destination address is recognized by the label processor (LP) while keeping the packet in the optical domain. The optical switch is then configured to forward the packet via the desired output switch-port. If the port is occupied, a contention resolution plan is followed to resolve the condition. At the torus DC network, the packet can be forwarded to another output switch-port and still goes through one of the shortest routes toward its destination, that is, deflection routing. HOPR is also equipped with optical and optoelectronic buffers to help with resolving contention, where (1) the packet can have a fixed time delay by going through a fiber delay line (FDL) without OEO conversion, or (2) it can be electronically stored in the optoelectronic buffer for an arbitrary storage time.

Unlike most conventional electrical routers that adopt the store-and-forward mechanism for packet forwarding, HOPR relies on the cut-through mechanism where the optical switch is configured once the packet label is recognized without demanding the whole packet to be received first. This matches well with forwarding the packets without OEO conversion. The packet path via HOPR only includes the LP and optical switch, whereas the shared buffer is located in parallel without obstructing the normally forwarded packets. The role of these three basic functional units is summarized in the following.

- **Label processor:** The LP's role is divided into: (1) extraction and recognition of the incoming packet label, (2) deciding the output switch-port accordingly, (3) performing arbitration for colliding packets, and then (4) configuring the switch with the corresponding control signals. On the other hand, the LP should be realized with (1) low power consumption to enable a widely scalable network, (2) low latency as the packet remains in the optical domain and the label processing time is compensated for just by delaying it, (3) a sufficiently high dynamic range to exhibit tolerance for power variation among the incoming packets.

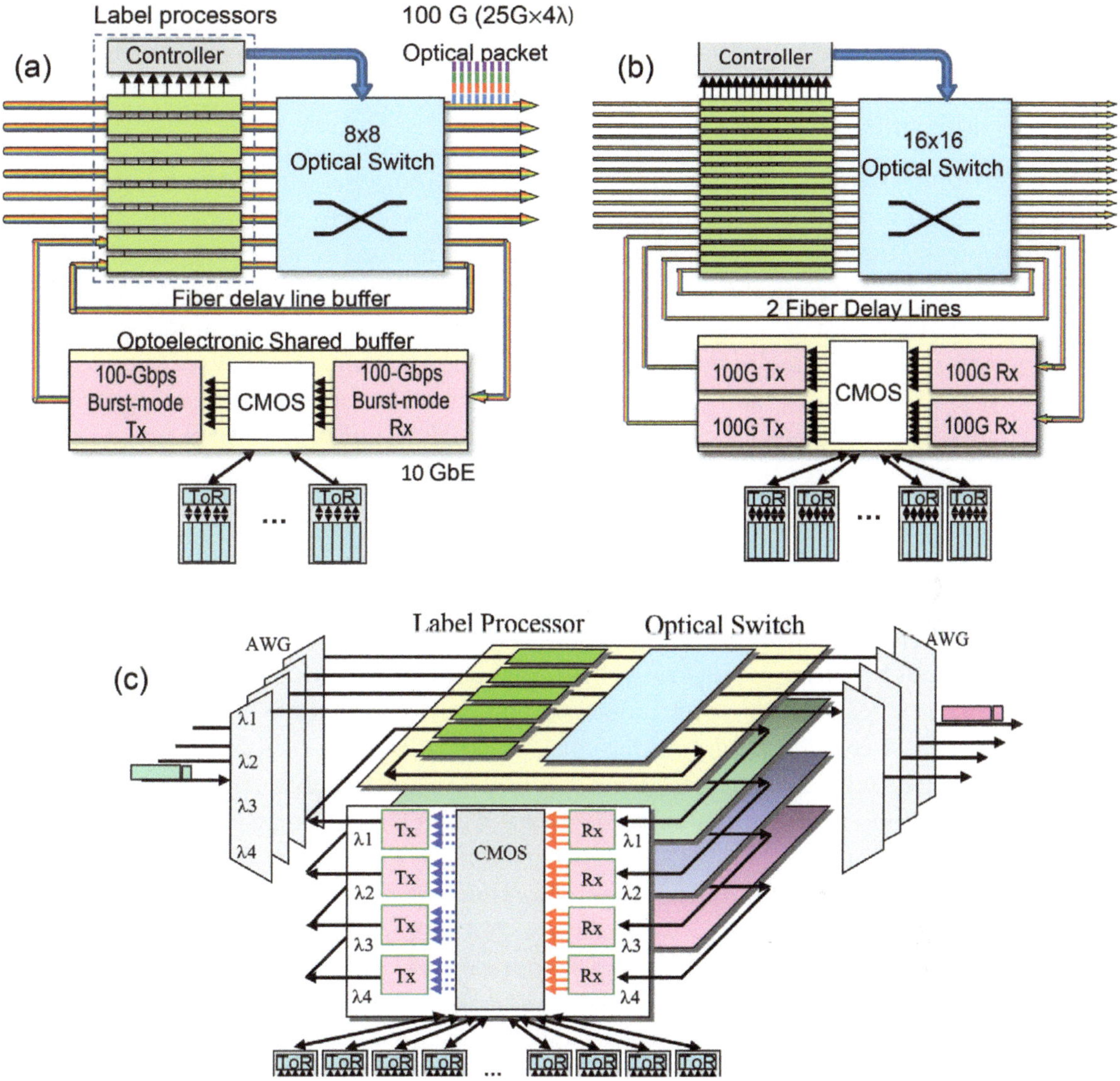

Figure 2. HOPR's architecture and basic subsystems in different configurations.

- **Optical switch**: There is a general trade-off between the switch speed and its maximum port count. For example, mechanical switches such as MEMS-based switches [9] can exhibit large port-counts, that is, over 1000 ports, but their reconfiguration time lies in the millisecond order. Differently, the OPS switch should operate in the nanosecond order; and with current technologies, only low port-count switches, for example, 16 × 16 ports, are available. A fast switch with low port-count can still enable realizing a highly scalable network with good performance if a suitable network topology is selected. It is beneficial to realize a switch with independent characteristics in terms of bit-rate, packet format, wavelength, and polarization. The switch is also demanded to possess low power consumption, high extinction ratio, low crosstalk, and ease of controllability. To meet these demands, several switch types have been considered by researchers [10–12]. Examples include the matrix switch that consists of cascaded smaller switches, phased-array switch, wavelength routing switch, and broadcast-and-select (B&S) switch. Among them, the B&S switch is selected as it meets most of the requirements to a good extent. We have also considered wavelength-routing switches to achieve a switch with higher port-counts.

- **Optoelectronic shared buffer:** Realizing an optical memory has attracted a lot of research interest [13, 14] but a reliable solution is still missing. Available solutions are either passive all-optical buffering via an FDL [15] for a fixed duration or electrical buffering where the optical packets are interfaced to an electronic memory. The role of shared buffer exceeds the demand for packet storage as it is essential for resolving contention and for enabling higher network functionalities such as packet regeneration, Quality of Service (QoS) control, and format conversion. Another important role for the optoelectronic buffer is interfacing the photonic network to other transmission domains with different data formats and data rates, where for the DC network case, it should act as an aggregation switch for the servers Ethernet packets.

Unlike conventional electrical routers handling continuous data streams to prevent clock loss at the receiver side, HOPR should operate in a burst-mode fashion where a packet suddenly arrives after a period of no received signal. Knowing the bit timing of the incoming packet is essential, but there is no prior synchronization between the incoming packets and HOPR. To enable clock recovery, burst-mode routers rely on preamble bits [16] that precede the packet, where a conventional way for clock recovery is used such as the phase locked loop (PLL) method that demands a long locking time [17], or the phase picking method [18] that is widely selected for handling lower data rates (~10 Gbps). The over-sampling method [19] is another alternative, where a clock recovery time of 31 ns has been recently reported [20] by combining this method and an approximation algorithm. However, the continuous increase of data rates reduces the packet duration, and keeping the dependence on preamble bits will degrade the efficiency of utilizing the optical link. Thus to allow HOPR to handle preamble-free optical packets, special burst-mode optoelectronic devices are developed to interface optical packets to electronic circuits with a novel optical clocking (triggering) method.

HOPR prototype (**Figure 2a**) is developed with (1) total throughput of 1.28 Tbps, where six input/output ports handle 100-Gbps optical packets and four input/output ports handle the 10-GbE connections, (2) total power consumption of 110 W, with ~40 W for the optical packet switch part that includes 8 LPs and an 8 × 8 optical switch, and ~70 W for the aggregation switch (optoelectronic shared buffer), including the contribution of the control plane, cooling, GUI, 10-GbE transceiver modules, and so on, and (3) latency of 140 ns, with ~60 ns for the transmission delay via the optical switch and EDFA, and ~80 ns for the LP dominated by the arbitration time for resolving contention. Electrical switches have also been significantly enhanced, but still they require a Network Interface Card (NIC) and optical transceiver module at each port. The CFP 100-GbE optical transceiver, for example, demands 6–20 W for different transmission ranges. HOPR's optical packet switch part consumes 5 W/100-G-port which is lower than a single CFP module. The future increase in packets data-rates, the power of optical packet switch would remain almost unchanged. The power consumed by the shared buffer can also be further reduced if instead of combining a Field Programmable Gate Array (FPGA) and discrete electronic components, the state-of-the-art Application Specific Integrated Circuit (ASIC) technologies are used in a way similar to the electrical switches.

As the name implies, the hybrid optoelectronic router is based on a hybrid implementation approach, where both electrical and optical technologies are used, each where it fits most. This approach has been vital for fulfilling the basic demand of reducing HOPR's power consumption and latency without sacrificing its performance. The hybrid approach can be seen at HOPR's underlying device level, where, for example, the operation principal of some essential devices is based on using optical timing pulses, that is, clock pulses, for triggering electronic circuits to enable handling the packets' ultrafast bits in an efficient way. Moreover, the hybrid approach can also be seen at the subsystem level as for instance in the presence of different packet buffering options, represented by the all-optical FDL-based buffer and the optoelectronic buffer. These alternatives provide the flexibility demanded to efficiently cope with diversified conditions of operation.

4. Optical label processing technologies

In this section, we explain about our latest LP [21] that handles 25-Gbps burst-mode optical labels without preamble bits. **Figure 3** shows the structure of the LP that consists of a set of label extractors (LE) connected to a shared controller.

At each input port of HOPR, a split from the incoming packet is directed to the attached LE, whereas the main part goes toward the optical switch but first passes through an FDL to compensate for the label processing time. Unlike traditional Serializer/Deserializer (SerDes)

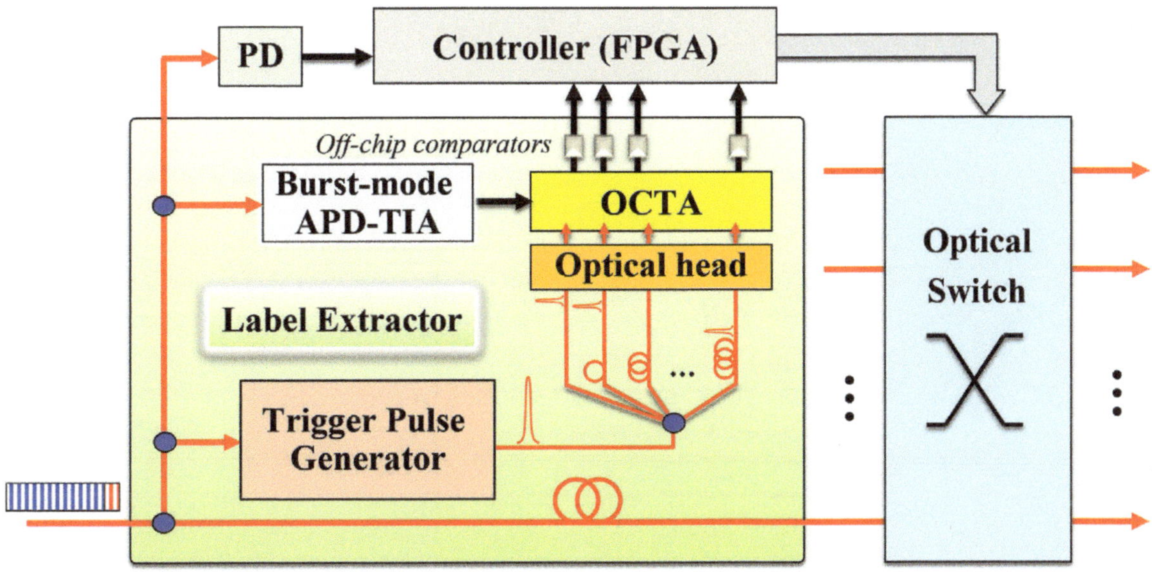

Figure 3. The overall structure of the burst-mode label processor.

that converts the bits of the whole packet, the LE handles only the packet label. The label's ultrafast bits are separated in parallel and their speed is sufficiently slowed down by undergoing serial-to-parallel conversion to allow their direct latching into the CMOS based LP's shared controller. At the LP controller, the forwarding table is looked up with the extracted label and when a match is found and the corresponding output port of the switch comes out as the result. The occupancy of the resulting output port is examined; and if the port is free, the control signals necessary for configuring the optical switch are generated and applied, whereas if the port is occupied, the second priority output is selected and so on until finding an available port. To enable label processing of preamble-free packets, the LE adopts an operating mechanism that relies on two basic elements: (1) a burst-mode serial-to-parallel converter (SPC) that operates once supplied by optical triggers and it is referred to as the optically clocked transistor array (OCTA), and (2) a burst-mode optical trigger pulse generator (TPG) that selectively utilizes the first bit of incoming packet to produce a synchronized optical trigger pulse for OCTA.

4.1. Optical trigger pulse generator

The TPG produces an optical trigger pulse by selectively amplifying the first bit of incoming packet with a semiconductor optical amplifier (SOA). Being originally a part of the incoming cket, the resulting optical pulse is a synchronized trigger that allows jitter-free serial-to-parallel conversion when applied to OCTA. **Figure 4a** shows the SOA's driver integrated circuit developed to drive a narrow current pulse (~1 ns) of high peak current (>600 mA).

A split from the incoming packet as shown in **Figure 4b** is applied to the metal-semiconductor-metal (MSM)-PD at the driver's discharge-based (DB) circuit and produces a single electrical pulse that undergoes reshaping before turning on a set of integrated high electron mobility transistors (HEMTs) shortly to enable the flow of high electric current through the SOA. The red curve in **Figure 4c** shows the normalized amplified spontaneous emission (ASE) of the

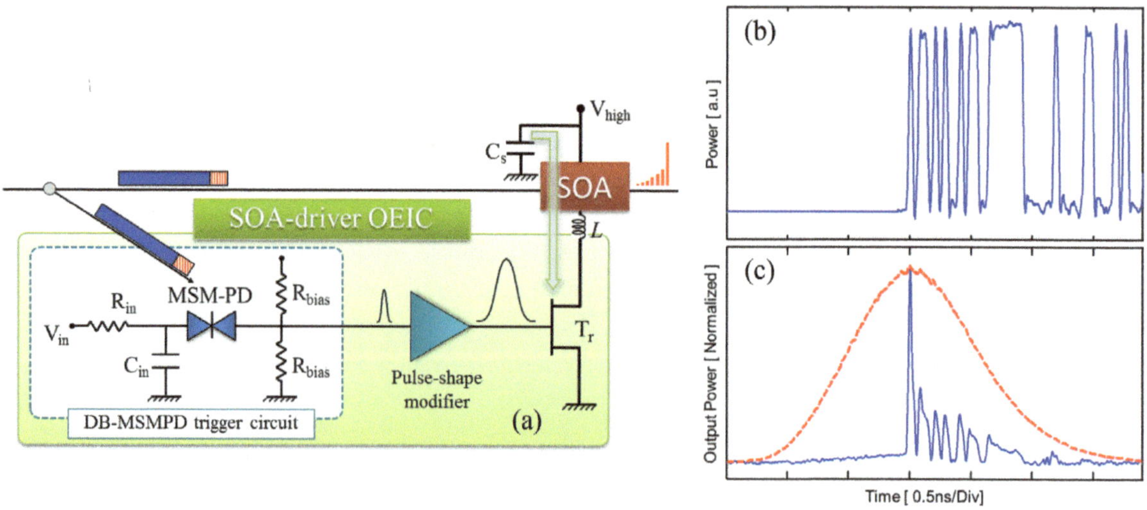

Figure 4. (a) The TPG circuit, (b) incoming packet, and (c) SOA output.

SOA that corresponds to its gain profile when the current pulse of the driver is applied to the SOA without having it supplied with any optical input. The packet arrival at the SOA is roughly adjusted to be around the gain peak. The packet's first optical pulse experiences strong amplification and quenches the accumulated carriers at the SOA, and the following pulses fade out due to lack of gain as shown in **Figure 4c**. As the SOA's power dissipation is limited only to a short instant after packet arrival, the power consumption of the TPG has been significantly reduced to 120 mW compared to an older TPG [21] that consumed 3 W.

The DB circuit [22] is a fundamental element in our burst-mode devices, used for converting an optical trigger signal into a narrow electrical pulse that can control the gate of a HEMT transistor. It utilizes an MSM-PD that has a fabrication advantage due to its surface structure. Unlike PIN photodetectors whose response is limited by an RC time constant, the MSM-PD has a very quick rise-up time only limited by the electron transit time due to its ultralow capacitance. However, the MSM-PD response suffers from a long tail due to the low mobility of holes. The discharge-based configuration is thus employed to allow the MSM-PD to generate a sufficiently narrow electrical pulse. The input capacitor C_{in} of the DB circuit is initially charged maintaining a high bias voltage. When an optical pulse is then applied to the MSM-PD, it causes the flow of photocurrent. A corresponding electric current cannot be injected by V_{in} due to the high resistance R_{in}, and thus C_{in} is discharged resulting in the reduction of the bias voltage. One important feature of the MSM-PD is that in the absence of bias voltage, the current cannot flow even when the carriers are still present. The first pulse of incoming packet almost depletes the carriers of C_{in} leaving the following packet bits ineffective. Thus even with the MSM-PD direct irradiation with the whole packet, the DB circuit can produce a single short electrical pulse.

4.2. Optically clocked transistor array

Figure 5 illustrates the structure of OCTA which is a monolithically integrated circuit that consists of 16 serial-to-parallel conversion channels attached to a common transmission line (TL) each via a separate HEMT (T_m).

An avalanche photodetector and trans-impedance amplifier (APD-TIA) burst-mode module is used to convert the packet split at the LE unit (**Figure 3**) into an electrical signal that is then coupled to OCTA and propagated along its TL. The channels are used in turn, each for converting a different label bit. The timing of a separate optical trigger pulse is adjusted to match the presence of a given bit at the TL, and the bit is converted by applying that trigger pulse to the respective channel. OCTA's correct operation demands T_m to be turned on shortly to convert only a single bit, and hence the DB-MSMPD trigger circuit is used to produce a single narrow electrical pulse to control T_m. OCTA-based serial-to-parallel conversion was initially done by the sample-and-hold (S&H) scheme as shown in **Figure 5a**, where an electric charge corresponding to the considered bit level is sampled into the capacitor C_{hold} through T_m. The voltage-change induced at C_{hold} is then amplified to produce the channel's final output. This SPC scheme suffers the limited difference between the sampled charge that corresponds to the "1" and "0" bits, respectively. The reason is that charging C_{hold}, that is, in case of a "1" bit, increases the voltage at T_m source terminal and hence forces it to get turned-off. Moreover,

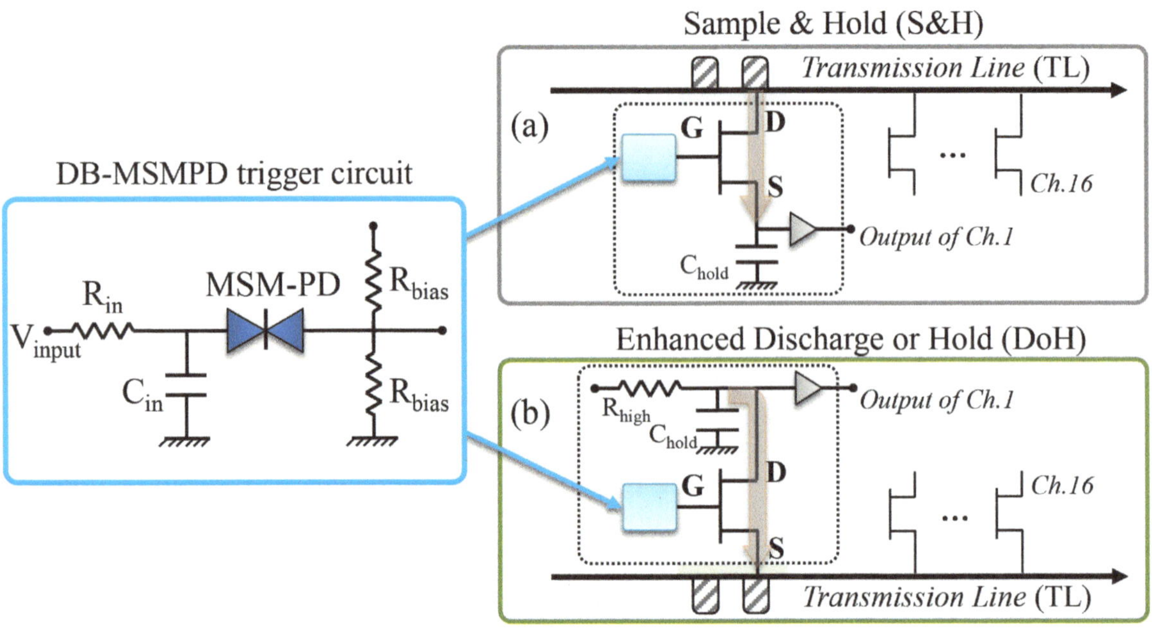

Figure 5. The structure of OCTA and illustration of its operation schemes (a), and (b).

charging C_{hold} cannot be done efficiently as a part of the bit voltage is unavoidably dissipated in the TL's characteristic impedance that is present in parallel to the turned-on conversion channel. The S&H scheme was then replaced by the discharge-or-hold (DoH) scheme [21] illustrated in **Figure 5b**, where the charge initially present at C_{hold} is either discharged into the TL or kept unchanged. Discharging C_{hold} is done more efficiently than charging it, and a much higher difference of charge is produced at C_{hold}. The benefit of the DoH scheme has been further elevated by using a label signal with negative voltage span, that is, negative voltage for a "1" bit and zero voltage for a "0" bit.

Figure 6 shows a comparison for using the DoH scheme with the TL signal having either a positive or a negative polarity. The gate voltage signal generated by the optical trigger pulse is shown in solid line. When the transistor T_m is turned on as V_{GS} exceeds the threshold voltage V_{th}, the bias voltage between its drain and source terminals, that is, ΔV_{DS}, is obviously higher in case of negative polarity. This allows a more efficient discharge for C_{hold} with the higher electrical current enabled by the higher ΔV_{DS}. Then if the energy of optical trigger pulse is reduced, a corresponding reduction in the gate pulse amplitude takes place as highlighted by the dotted line. Even with this reduction in ΔV_{GS}, the initially higher value of ΔV_{DS} enables conversion as efficient as in the case of a positive polarity signal with unreduced optical trigger energy.

4.3. Enhanced packaging

Optoelectronic integrated circuits (OEICs) are attractive for their high-speed operation and low power consumption. To make the best use of these key features, the optical trigger pulses

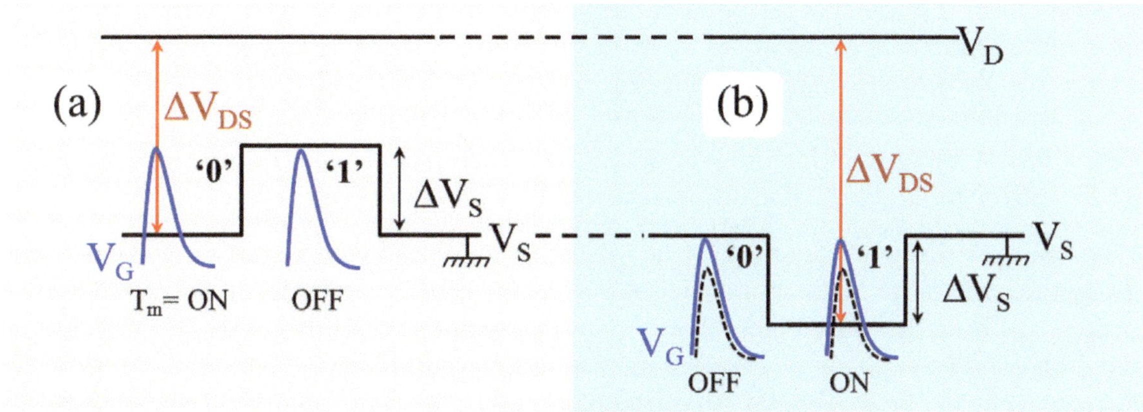

Figure 6. The DoH scheme with TL signals of (a) positive and (b) negative polarities.

necessary for OEIC operation should be provided with suitable device packaging. The older way of device packaging was complicated and costly, and thus it had to be revised.

Figure 7a shows a photo for the old packaged OCTA, where a set of lens array is used to focus the optical triggers on the MSM-PDs located at the front side of the chip. The old packaging method was wasting the trigger energy due to the shadowing effect that occurs for front-illuminated light by the MSM-PD's interdigitated metal electrodes [23]. This method also had other shortcomings such as the need for active alignment and an expensive lens array. To overcome these issues, a new method was developed as shown in **Figure 7b**. The optical head employed is a commercially available fiber-array block, where SMF fibers are placed in grooves with the same pitch as the chip's MSM-PDs. The chip is directly attached to the optical head after reducing its thickness to ~130 μm. The high refractive index of the chip's InP substrate limits the divergence of the optical beams launched from the SMF fibers. The $1/e^2$

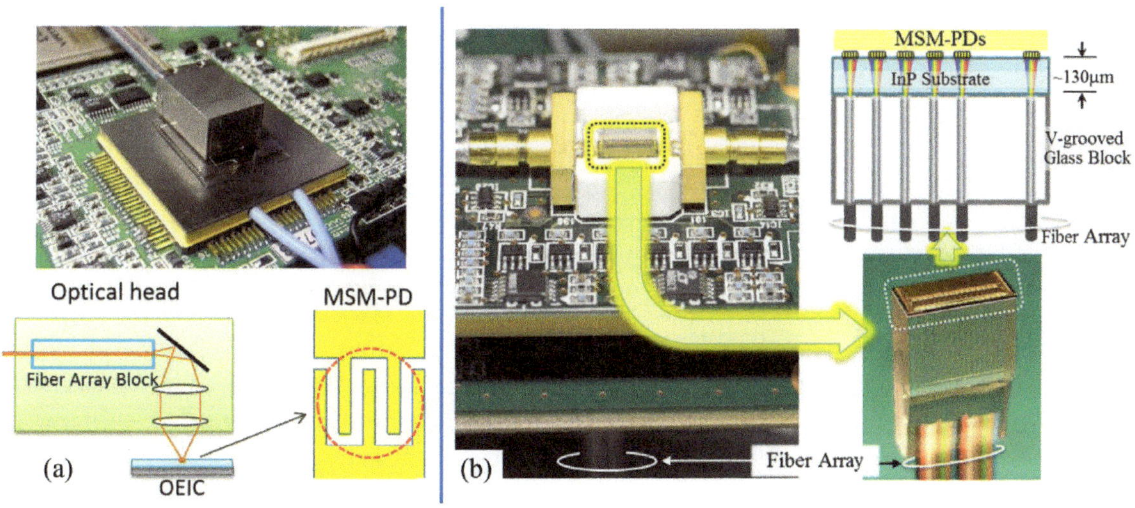

Figure 7. (a) Old packaging of OCTA versus (b) new packaging.

diameter of the optical spot resulting at the MSM-PD is ~13 μm, which is narrow enough to fit into the 20-μm active area of the MSM-PD without using lenses and while still exhibiting a large misalignment tolerance. Another key feature of this method is that optical triggers are supplied via the backside of the chip. With back illumination, the responsivity of the MSM-PDs is almost doubled and the trigger pulse energy necessary for performing serial-to-parallel conversion for 25-Gbps label bits has been reduced to 0.35 pJ/bit.

When an optical signal is applied to the MSM-PD via its backside, a part of the signal goes though the MSM-PD and can still be seen from the chip's front side. A simple alignment process has been developed by making use of this feature, where the optical beams coming out of the MSM-PDs of all the conversion channels are simultaneously observed by using an infrared camera. The best alignment position is achieved simply when a full set of clear optical spots is visually observed.

5. Optical switching technologies

In this section, we present two types of fast optical switches that we have been considering, namely the wavelength-routing switch and the broadcast-and-select switch.

5.1. Wavelength routing switch

Figure 8a illustrates the basic architecture of an N × N optical switch that operates with the wavelength routing mechanism. The switch consists of a cyclic arrayed waveguide grating (AWG) equipped with a tunable wavelength converter (TWC) at each input port and a fixed wavelength converter (FWC) at each output port. By changing the wavelength of the optical signal input to the AWG with the TWC, the signal can be directed to a different output port, whereas the FWC is used to return the signal back into its original wavelength. Based on the well-established AWG technology, it is feasible to realize such switch with a medium port-count, for example, 64 × 64 ports, provided that the used TWC can cover the whole C-band. The full exploitation of this switching method demands a TWC with independent characteristics in terms of modulation format and data-rate, but it is still hard to realize such TWC.

Currently, the most reliable TWCs are based on signal regeneration with OEO conversion where the wavelength of a TL is changed to match the desired output port, and the data of incoming packet is converted into an electrical signal that is then used to modulate the new wavelength. The OEO-based TWC does not allow the switch usage for handling WDM packets or coherent packets.

The components required for realizing the OEO-based TWC are (1) a fast tunable laser for generating an optical carrier with desired wavelength, (2) a burst-mode photodetector and an electrical amplifier for converting the incoming optical packet into electrical signal with sufficient amplitude, and (3) a modulator for modulating the desired wavelength with the electrical packet signal. We have developed a parallel-ring resonator-based tunable laser (PRR-TL) that is integrated with an electro-absorption modulator (EAM) on the same InP chip. The

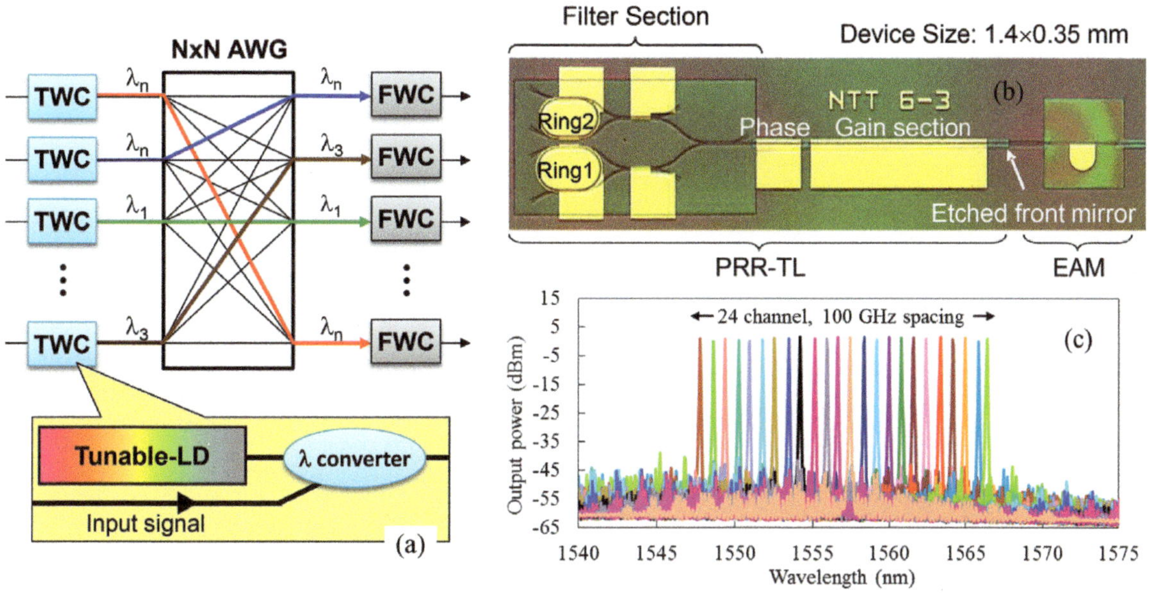

Figure 8. (a) Illustration for wavelength-routing switch, (b) photo of tunable transmitter chip, and (c) superposed spectrum of the PRR-TL.

PRR-TL relies on the Vernier effect for providing a wide tuning range [24]. The chip's photo is shown in **Figure 8b** where the TL consists of a phase section, gain section, and parallel-ring resonator (PRR). The TL has a partially reflective output port, whereas the PRR section is the other end of its laser cavity that selectively reflects the optical signal with desired wavelength. The wavelength is tuned with low current injection, that is, mA order. This enables the PRR-TL to exhibit a very low wavelength drift of less than 5 GHz. The low current injection and parallel ring design results in a laser cavity with low optical loss. This allows the PRR-TL to exhibit an output power variation of less than 1 dB over a wide tuning range of 35 nm as shown in **Figure 6c**. Moreover, it also allows the TL to generate a high output power while exhibiting a high-speed wavelength tuning of less than 6 ns.

5.2. Broadcast-and-select switch

As the wavelength-routing switch is not suitable for handling WDM packets (25-Gbps × 4λs), we have considered the B&S switch instead. **Figure 9a** illustrates the basic structure of the B&S switch, where an optical 1 × N passive splitter is located at each input port, with an optical gating unit placed at each of the splitter outputs. On the other hand, an optical N × 1 coupler is placed at each output port, with each coupler input connected to one output of the splitter that serves for a different input switch port. The packet undergoes switching by turning on the optical gate that can lead it to the desired output port.

- **Optical gate:** SOAs are widely selected as optical gating units in the B&S switches as they can provide optical gain to compensate for the optical loss while unavoidably adding some ASE noise. SOAs exhibit pattern-dependent effects for high bit-rate packets and inter-channel crosstalk for DWDM-based packets due to four-wave mixing and cross-gain modulation,

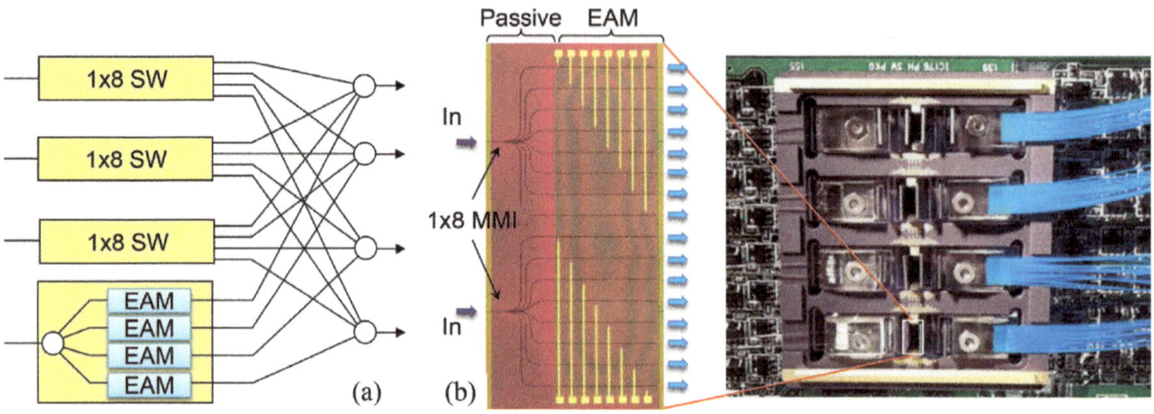

Figure 9. (a) Illustration for the B&S switch and (b) realized switch with inset of fabricated chip.

and they also require high-speed drivers with large current modulation which leads to high power consumption. Instead of SOAs, we have considered EAMs as optical gates, and the full switch system requires an EDFA for loss compensation at each output port. But yet still the total power consumption is lower than that of the SOA-based switch as the power consumption of an EDFA can be reduced (~1 W) by optimizing it for a given gain range (~25 dB). In addition, the power consumption of the reverse-biased EAM and its electrical driver is very low compared to the SOA.

- **Device and module**: **Figure 9b** shows the device photo for two monolithically integrated 1 × 8 EAM gate switches. The EAM section has a shallow-ridge waveguide structure and the passive waveguide section including the 1 × 8 multimode interference splitter is formed as a deep-ridge to reduce the required bending radius. The bulk InGaAsP (λ_{gap} = 1.4 μm) used for the absorption layer of the EAM section operates based on the Franz-Keldysh (FK) effect. To simplify the fabrication process, the EAM and passive sections share the same core/cladding layers, in which EAMs are not electrically isolated from each other. Thus when voltage is applied to a given EAM, an undesirable change is caused in the output power of other switch ports due to the electro absorption by FK effect induced in the passive waveguide section. One way to eliminate such electrical crosstalk is to replace the p-InP cladding layer by a high-impedance material such as Fe-doped InP. However, this would demand an additional step of regrowth process. To avoid that, a surface-ground electrode is used to cover the MMIs of the passive waveguide section. The simple addition of the surface ground electrode has been successful in suppressing the electrical crosstalk [25].

A set of micro-lenses is used for the optical coupling between an optical fiber array and the switch waveguide array. High coupling loss and loss variation among different waveguides were observed due to the large and asymmetric numerical aperture of ridge waveguide (NA ~ 0.8). To improve coupling loss and alignment tolerance, spot size converters (SSCs) were then added at both switch sides to decrease the waveguide's NA by expanding its optical mode field. The core size is set to 0.4 μm × 0.4 μm to achieve a symmetric NA of 0.5. The fabricated module exhibits low WDL and PDL of typically ±0.5 dB in the wavelength range of

1540–1560 nm. The module is attached to an electrical control board including an FPGA and EAM drivers that enable a switching rise/fall time of less than 10 ns. With a reverse bias of ~7 V, all the output ports exhibited extremely high extinction ratios of more than 50 dB, which is sufficient to avoid any signal degradation caused by inter-symbol interference.

6. Optical buffering technologies

HOPR comprises an FDL-based all-optical and optoelectronic shared buffer. This section is devoted to the optoelectronic shared buffer where packet processing is performed to enable higher network functions such as packet regeneration, QoS control, and format conversion.

6.1. Operation

Figure 10 illustrates the optoelectronic buffer used to handle 100-Gbps WDM optical packets composed of four wavelengths each modulated at 25 Gbps. A CMOS processor is located at the buffer core, and it can independently handle the servers' Ethernet packets delivered through the ToR switches, and the network OPS packets. The OPS packets are fed into and out of the processor by using a set of burst-mode serial-to-parallel converters (SPCs) and parallel-to-serial converters (PSCs), respectively. The incoming packet four wavelengths are de-multiplexed with an AWG at the buffer input. Each packet part belonging to a different wavelength is converted into fast electrical signal by using an APD-TIA module. Each two

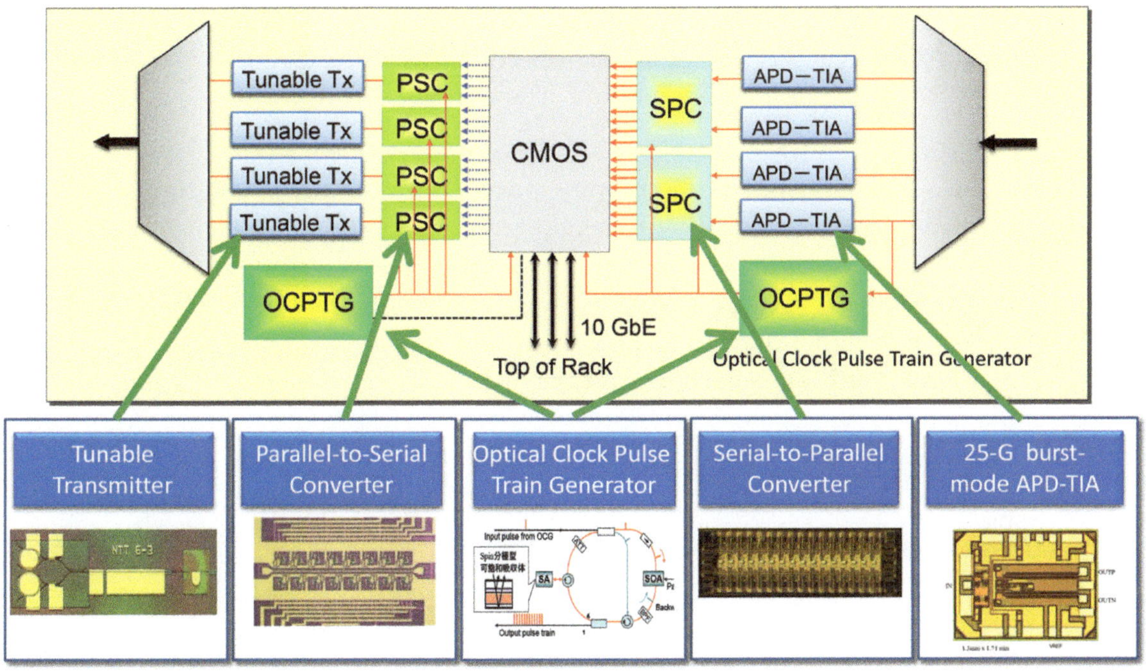

Figure 10. The optoelectronic shared buffer that can handle burst-mode WDM packets.

packet parts are then simultaneously fed into the same SPC device, where the speed of their ultrafast bits is reduced by the ratio of 1:16.

The two SPC devices are supplied with optical clock pulses to keep them operating until the serial-to-parallel conversion is done for the whole packet. The optical clock pulse train generator (OCPTG) module present at the buffer input side provides the optical clock pulses necessary for operating the SPCs and also the optical clock pulses required to latch the SPCs output to the CMOS processor.

To generate a 100-Gbps optical packet, four different PSCs separately receive the electrical data for each wavelength from the processor, and operated in parallel to reduce the speed of the data bits by the ratio of 16:1. Each PSC output is used at a separate tunable transmitter that generates an optical carrier adjusted to a desired wavelength. After modulation, the four wavelengths are multiplexed by an AWG to form the output 100-Gbps optical packet. The OCPTG module at the buffer output side provides the optical clock pulses required for reading out data from the processor and for operating the PSCs.

6.2. Enabling devices

The devices necessary for realizing the shared optoelectronic buffer are briefly highlighted here, and more details can be found in the corresponding references.

- **OCPTG:Figure 11** illustrates the OCPTG structure that has an optical clock generator (OCG) at the input followed by the pulse train generator (PTG)'s optical loop that includes an SOA and spin-polarized saturable absorber (SA) with high-extinction ratio. The OCG generates a single optical pulse in response to the first bit of incoming packet. This pulse is then fed into the PTG module to produce a train of optical pulses with fixed separating intervals between pulses throughput the whole length of the packet. The OCPTG [26] can thus handle preamble-free asynchronous optical packets with variable lengths.

The optical pulse train is generated by tapping out a portion of the circulating seed pulse initially provided by the OCG. The loop is made to have a round-trip gain function with

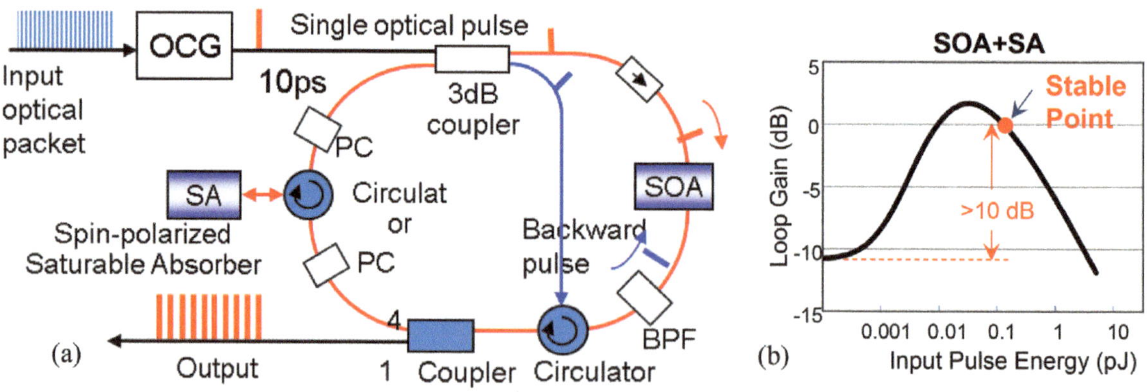

Figure 11. (a) The structure of OCPTG and (b) loop transfer function of PTG.

negative slope based on the SOA gain saturation (**Figure 11b**). This allows the circulating pulse energy to converge even if some fluctuation occurs, where the gain and absorption of the SOA and SA, respectively, finally balance out. The ASE of the SOA is suppressed by the SA. The SA is characterized by fast absorption recovery due to spin relaxation, but still after the pulse comes out of the SA, a backward optical pulse is launched into the SOA to prevent the increase of the ASE inside the loop with the recovery of SOA gain. Without any external control for the loop gain, a stable optical pulse train with low jitter is generated from a single 10-ps seed optical pulse even with a 10-dB variation in the seed pulse intensity.

- **APD-TIA module:** The high-speed avalanche photodetector (APD) and burst-mode transimpedance amplifier (TIA) module have been newly developed [27]. The APD has a new p-down inverted structure [28] to prevent undesirable edge breakdown and to suppress surface leakage current. A back-illuminated InAlAs/InGaAs APD based on this structure exhibited a 3-dB bandwidth of 27 GHz. On the other hand, the TIA was designed and fabricated by using 0.13-μm SiGe BiCMOS technology with a cut-off frequency (f_t) and maximum frequency (f_{max}) of 200 and 270 GHz, respectively. Both chips are DC coupled to enable burst-mode operation with a record of high sensitivity.
- **SPC:** The basic structure of the SPC used at the optoelectronic buffer resembles that of the SPC used at the LP, and operates based on the same discharge-or-hold scheme. But unlike the LP's SPC that is used in a single shot fashion every several 10's of nanoseconds when a new label comes in, each conversion channel of the buffer SPC is operated repeatedly every 640 ps, that is, 16 bits at 25 Gbps, until the whole packet is converted, and thus the implementation circuits are different. More details about the buffer's SPC can be found in [29].
- **PSC:** The parallel-to-serial converter (PSC) is an optoelectronic integrated circuit (OEIC) that resembles the SPC in consisting of several conversion channels attached to a common TL. The input to each channel of the PSC is a low speed electrical signal provided by the CMOS processor, whereas the TL carries the final device output which is the high-speed electrical pulses that form a packet part used in a following step to modulate a given wavelength. Each channel comprises a customized MSM-PD discharge-based circuit operated with an optical trigger pulse to produce a positive electrical pulse and a negative electrical output from the opposite sides of the MSM-PD. Both pulses are used to generate a non-return-to-zero (NRZ)-like output electrical pulse. More details about the buffer's PSC can be found in [30].
- **Tunable transmitter**: A fast tunable transmitter is realized by monolithically integrating an EAM section composed of InGaAlAs MQWs [31] and a PRR-based tunable laser (**Figure 8b**). The InGaAlAs MQWs allow for a steep extinction curve and large E/O frequency bandwidth [32], and hence can support operation at 25-Gbit/s with a sufficient extinction ratio and over a wide wavelength range. **Figure 12a** shows the NRZ eye diagram for a 25-Gbps pseudorandom bit stream (PRBS) of 2^{31}-1. A dynamic ER larger than 10 dB was achieved for wavelengths up to 1570 nm with DC bias levels ranging from -0.8 to -1.5 V, and with a constant voltage swing of 2.0 V maintained at all wavelengths (**Figure 12b**).

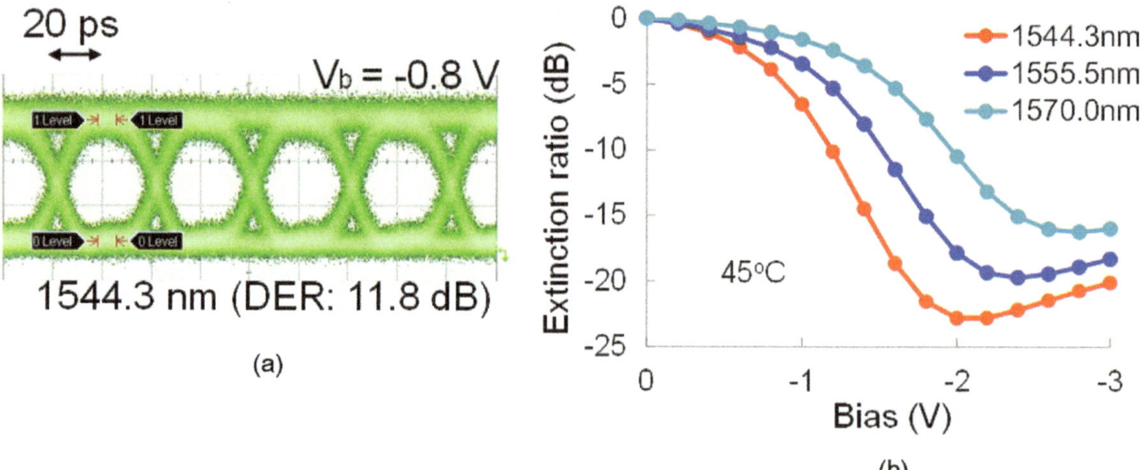

Figure 12. (a) The eye diagram of tunable transmitter and (b) extinction ratio.

7. Conclusion and outlook

In this chapter, we have reviewed our recent HOPR prototype developed to realize a new photonic intra DC network. The basic aspects of the network have been briefly introduced, followed by highlighting the enabling technologies of the new prototype. HOPR's role at the DC network has two folds as an optical-packet switch for forwarding high-speed burst-mode optical packets in a torus-topology network, and as an aggregation switch for handling the traffic of the DC servers. HOPR's operation relies on three functional units namely the label processor, optical switch, and shared buffer. HOPR is realized based on a hybrid implementation approach with an attempt to make optimal use of optics and electronic for enabling low power consumption and latency without sacrificing performance. The use of both all-optical and optoelectronic shared buffers shows the hybrid approach adopted at the subsystem level. The hybrid approach can also be seen at the device level where, for example, special optoelectronic devices are developed to efficiently interface preamble-free optical packets to electronic circuits with a novel optical clocking mechanism.

To meet the pressing demand for power reduction, HOPR is developed to have the power dissipation limited to incoming packet duration. The dissipation time has been even further reduced as, for example, in the label processor case where it occurs only during the short label duration. HOPR's latency is dominated by performing arbitration for resolving packet contention where a fast arbitration algorithm is demanded. On the other hand, to maintain a high-quality signal after multi-hop transmission, EDFAs and not SOAs are used for packet amplification, whereas other than amplification, the SOA attractive features of high-speed operation and compactness are employed for signal processing. New device packaging has been also considered as for instance in the label processor to enhance the responsivity of used photodetectors and to enable an easy lens-free and low-cost alignment.

OPS-based photonic DC networks provide a radical solution that can take data centers into new frontiers, and let them surpass their current difficulties caused by excessive electronic processing. Performing more node functions by combining optics/electronics is thus pivotal, and to enable that the demand for new innovative concepts and implementation strongly exists. HOPR's new prototype is a step toward fulfilling this objective which together with other community achievements reveals the high potential of this networking approach.

Acknowledgements

This work has been funded by the National Institute of Information and Communications Technology (NICT) R&D program Basic Technologies for High-Performance Hybrid Optoelectronic Router (2011–2016).

Author details

Salah Ibrahim* and Ryo Takahashi

*Address all correspondence to: ibrahim.salah@lab.ntt.co.jp

NTT Device Technology Laboratories, NTT Corporation, Kanagawa, Japan

References

[1] D.J. Richardson, J.M. Fini, and L.E. Nelson. Space division multiplexing in optical fibers. Nature Photonics 7: 353-362, 2013.

[2] Cisco visual networking index: forecast and methodology, 2013-2018. White paper c11-481360.pdf, 10 June 2014.

[3] M. AI-Fares, A. Loukissas, and A. Vahdat. A scalable, commodity data center network architecture. In Proceedings of the ACM SIGCOMM, USA, pp. 1-12, 2008.

[4] C. Kachris, K. Kanonakis, and T. Tomkos. Optical interconnection networks in data centers: recent trends and future challenges. IEEE Communications Magazine 51(9), pp. 39-45, 2013.

[5] K. Bernstein, R.K. Cavin, W. Porod, A. Seabaugh, and J. Welser. Device and architecture outlook for beyond CMOS switches. Proceedings of the IEEE 98(12), pp. 2169-2184, 2010.

[6] R. Takahashi, T. Segawa, S. Ibrahim, T. Nakahara, H. Ishikawa, A. Hiramatsu, Y-C. Huang, and K-I. Kitayama. Torus data center network with smart flow control enabled by hybrid optoelectronic routers. Journal of Optical Communications and Networking 7: B141–B152, 2015.

[7] Y-C. Huang, Y. Yoshida, S. Ibrahim, R. Takahashi, A. Hiramatsu, and K-I. Kitayama. Novel virtual OCS in OPS data center networks. Photonic Network Communications 31(3): 448-456, 2016.

[8] Y-C. Huang, Y. Yoshida, S. Ibrahim, R. Takahashi, A. Hiramatsu, and K-I. Kitayama. Bypassing route strategy for optical circuits in OPS-based data center networks. IEEE Photonics Journal 8(2): 1-10, 2016.

[9] W. Mellette, G. Schuster, G. Porter, G. Papen, and J. Ford, "A scalable, partially configurable optical switch for data center networks," J. Lightwave Tech., 35(2), pp. 136-144, Jan.15, 2017.

[10] S. J. Ben Yoo. Optical packet and burst switching technologies for the future photonic internet. Journal of Lightwave Technology 24, pp. 4468-4492, 2006.

[11] G. I. Papadimitriou, C. Papazoglou, and A. S. Pomportsis. Optical switching. Hoboken, NJ: Wiley-Interscience, 2006.

[12] M.-J. Kwack, T. Tanemura, A. Higo, and Y. Nakano. Monolithic InP strictly non-blocking 8×8 switch for high-speed WDM optical interconnection. Paper presented at the European Conference and Exhibition on Optical Communication, Amsterdam, The Netherlands, 2012, Paper Th.3.B.3.

[13] R. S. Tucker, P.-C. Ku, and C. J. Chang-Hasnian. "Slow-light optical buffers: capabilities and fundamental limitations," J. Lightwave Tech., 23(12), 2005.

[14] K. Nosaki, A. Shinya, S. Matsuo, Y. Suzaki, T. Segawa, T. Sato, Y. Kawaguchi, R. Takahashi, and M. Notomi. Ultralow power all-optical RAM based on nanocavities. Nature Photonics 6: 248-252, 2012.

[15] T. Zhang, K. Lu, and J. R. Jue. Shared fiber delay line buffers in asynchronous optical packet switches. IEEE Journal on Selected Areas in Communications 24(4): 118-127, 2006.

[16] S. Porto, C. Antony, A. Jain, D. Kelly, D. Carey, G. Talli, P. Ossieur, and P. D. Townsend, "Demonstration of 10 Gbit/s burst-mode transmission using a linear burst-mode receiver and burst-mode electronic equalization," J. Opt. Comm. and Net. 7(1), A118–A125 (2015).

[17] Roland E. Best. Phase locked loop: design, simulation and applications, 4th ed. New York: McGraw-Hill, 1999.

[18] R. Yu, R. Proietti, S. Yin, J. Kurumida, and S. J. Ben Yoo. "IO-Gb/s BM-CDR circuit with synchronous data output for optical networks," IEEE Photon. Technol. Lett., 25, pp. 508-511, 2013.

[19] N. Suzuki, K. Nakura, T. Suehiro, M. Nogami, S. Kosaki, and J. Nakagawa. Over-sampling based burst-mode CDR technology for high-speed TDM-PON systems. In Optical Fiber Communication Conference and Exposition (OFCINFOEC), 2011 and the National Fiber Optic Engineers Conference, Los Angeles, CA, 2011, pp. 1-3.

[20] A. Rylyakov, J. Proesel, S. Rylov, B. G. Lee, J. Bulzacchelli, A. Ardey, C. Schow, and M. Meghelli. A 25 Gb/s burst-mode receiver for low latency photonic switch network. IEEE Journal of Solid-State Circuits 50(12), pp. 3120-3132, 2015.

[21] S. Ibrahim, T. Nakahara, H. Ishikawa, and R. Takahashi. Burst-mode optical label processor with ultralow power consumption. Optics Express 24: 6985-6995, 2016.

[22] K. Takahata, R. Takahashi, T. Nakahara, H. Takenouchi, and H. Suzuki. 3.3 ps electrical pulse generation from discharge-based metal-semiconductor-metal photodetector. Electronics Letters 41(1): 38-40, 2005.

[23] H. Kim Jae, H. T. Griem, R. A. Friedman, E. Y. Chan, and S. Ray. High-performance back-illuminated InGaAs/InAlAs MSM photodetector with a record responsivity of 0.96 A/W. IEEE Photonics Technology Letters 4(11): 1241-1244, 1992.

[24] T. Segawa, S. Matsuo, T. Kakitsuka, T. Sato, Y. Kondo, and R. Takahashi. Semiconductor double-ring-resonator-coupled tunable laser for wavelength routing. IEEE Journal of Quantum Electronics 45(7): 892-899, 2009.

[25] Y. Muranaka, T. Segawa, Y. Ogiso, T. Fujii, and R. Takahashi. Performance improvement of an EAM-based broadcast-and-select optical switch module. In The Proceedings of the International Conference on Photonics in Switching (PS), Florence, 2015.

[26] T. Nakahara, and R. Takahashi. Self-stabilizing optical clock pulse-train generator using SOA and saturable absorber for asynchronous optical packet processing. Optics Express 21, pp. 10712-10719, 2013.

[27] M. Nada, M. Nakamura, and H. Matsuzaki. 25-Gbit/s burst-mode optical receiver using high-speed avalanche photodiode for 100-Gbit/s optical packet switching. Optics Express 22, pp. 443-449, 2014.

[28] M. Nada, Y. Muramoto, H. Yokoyama, N. Shigekawa, T. Ishibashi, and S. Kodama. Inverted InAlAs/InGaAs avalanche photodiode with low–high–low electric field profile. Japanese Journal of Applied Physics 51, p. 02BG03-1, 2012.

[29] S. Ibrahim, H. Ishikawa, T. Nakahara, and R. Takahashi. A novel optoelectronic serial-to-parallel converter for 25-Gbps burst-mode optical packets. Optics Express 22, pp. 157-165, 2014.

[30] H. Ishikawa, T. Nakahara, H. Sugiyama, and R. Takahashi. A parallel-to-serial converter based on a differentially-operated optically clocked transistor array. IEICE Electronics Express 10(20); pp. 20130709, 2013.

[31] T. Segawa, W. Kobayashi, T. Sato, S. Matsuo, R. Iga, and R. Takahashi. A flat-output widely tunable laser based on parallel-ring resonator integrated with electroabsorption modulator. Optics Express 20: B485-B492, 2012.

[32] W. Kobayashi, M. Arai, N. Fujiwara, T. Fujisawa, T. Tadokoro, K. Tsuzuki, Y. Kondo, and F. Kano. Design and fabrication of 10-/40-Gb/s, uncooled electroabsorption modulator integrated DFB laser with butt-joint structure. Journal of Lightwave Technology 28(1): 164-171, 2010.

2

Hybrid Silicon-Organic Heterojunction Structures for Photovoltaic Applications

Andrey Kosarev, Ismael Cosme,

Svetlana Mansurova, Antonio J. Olivares and

Hiram E. Martinez

Additional information is available at the end of the chapter

Abstract

The concept for inorganic-organic device is an attractive technology to develop devices with better characteristics and functionality due to the complementary advantages of inorganic and organic materials. This chapter provides an overview of the principal requirements for organic and inorganic semiconductor properties and their fabrication processes and focus on the compatibility between low temperature plasma enhanced chemical vapor deposition (PECVD) and polymer organic materials deposition. The concept for inorganic-organic device was validated with the fabrication of three hybrid thin film photovoltaic structures, based on hydrogenated silicon (Si:H), organic poly(3-hexythiophene): methano-fullerenephenyl-C61-butyric-acid-methyl-ester (P3HT:PCBM), and poly(3,4 ethylenedioxythiophene): poly(4-styrenesulfonate) (PEDOT:PSS) films. Optoelectronic characteristics, performance characteristics, and interfaces of the different configurations aspects are discussed. Hybrid ITO/PEDOT:PSS/(i)Si:H/(n)Si:H structure results in a remarkably high short circuit current density as large as 17.74 mA/cm^2, which is higher than the values in organic or inorganic reference samples. Although some hybrid structures demonstrated substantial improvement of performance, other hybrid structures showed poor performance, further R&D efforts seem to be promising, and should be focused on deeper study of organic materials and related interface properties.

Keywords: organic semiconductors, plasma deposited materials, thin film devices, hybrid devices, solar cells

1. Introduction

Polymer-based organic semiconductors have attracted considerable attention due to their excellent light absorption properties and compatibility with the solution processing technologies enabling new approaches to fabricate cost-effective, organic thin-film solar cells by printable process. On the other hand, hydrogenated silicon (Si:H) prepared by plasma enhanced chemical vapor deposition (PECVD) is a widely known inorganic semiconductor with rather mature deposition technology, which has been successfully implemented in several commercial optoelectronic devices, including photovoltaics. Important advantages of Si:H and related semiconductors are their high photocarrier generation, good transport properties of photocarriers, possibility to vary optical gap in the range of 1.2–3 eV, and acceptable stability under sun irradiation. Organic solar cells based on conjugated polymers as electron-donor materials blended with small molecule sensitizer as an electron acceptor have achieved 7–9% conversion efficiency in small area single bulk heterojunction devices. The efficiency of single-junction Si:H solar cells is around 10% even with optimal optical design. A few attempts have been undertaken to merge the beneficial properties of both materials in the so-called hybrid silicon-organic heterojunction solar cells. The main idea behind this concept is based on the possibilities to enhance photovoltaic performance characteristics by engineering suitable material and improving interface properties and design. Rather developed deposition and doping process of amorphous silicon and related thin films (intrinsic and p-n-doped) allow to tune the important structural and electronic parameters (such as degree of order, optical gap, mobility, Fermi level position, etc.) of the material which determines the device performance characteristics. Thus, the combination of organic and inorganic materials would be a promising approach for developing new types of large area, lightweight, thin film efficient solar cells.

We present a concept for silicon-organic heterojunction structures for photovoltaic devices, based on mainly PECVD noncrystalline silicon (can be also extended and involve crystalline semiconductors) describe fabrication of both plasma deposited films and organic materials and their relevant electronic properties, fabrication of hybrid device structures, and their characteristics comparing with those reported in literature.

2. Fabrication and electronic properties of plasma deposited and organic materials

2.1. Inorganic materials based on silicon deposited by PECVD

Technology based on amorphous silicon semiconductors has become an important part in the multibillion business in electronics in the last decades. There was a time, when the theory of the physical properties of solids seems to be impossible without considering a stable and periodic structure, Ziman [1]. However, nowadays, the theory of amorphous materials has not only been developed but also technology based on these materials has become omnipresent in wide range of electronic applications. The breakthrough of amorphous silicon materials was directly related to the development of a new method called GD that allowed for the first time to deposit intrinsic hydrogenated amorphous silicon (a-Si:H) films [2, 3] with high photoconductivity and low concentration of ESR centers, p-type and n-type doped amorphous silicon films, Spear and Le Comber [4], with sufficient quality (low DOS) to be used in functional p-n junctions [5].

The modern approach of the GD system used for first time is now well-known (PECVD). The actual PECVD technique has reached the versatility to deposit not only amorphous silicon but also silicon materials in their entire range of structure: nanostructured [6], microcrystalline [7, 8], epitaxial crystalline silicon [9], and alloys such as SiGe:H, SiC:H, SiO:H, SiN:H, etc. The optoelectronic properties of these materials depend on the deposition parameters such as pressure, flow rate, substrate temperature, power, and excitation frequency. Good electronic properties for interesting high Ge-content nanostructured Si_xGe_y:H films deposited at 300°C, high H-dilution ratio, and low frequency have been systematically reported in Ref. [10]. On the other hand, one of the main characteristics of the PECVD technique to take advantage for actual applications is the relative low temperature process in the range of 100–300°C. Low temperature process is a requirement for new applications such as large area, lightweight, flexible devices on low-cost plastic substrates, and, more recently, hybrid organic-inorganic devices based on polymers. It was general to think that reduction of temperature below 200°C leads to the increase of dangling bonds (defects) and results in films with electronic properties that are generally poor [11]. However, it has been found that deterioration of electronic properties in materials deposited at temperature lower than $T < 300$°C can be corrected by other fabrication parameters. Low-temperature (LT) a-Si:H and LT nanocrystalline (nc) Si:H have been optimized at $T = 120$ and 75°C as reported in Ref. [12].

A-Si:H films deposited at reference temperature ($T = 300$°C) show a dark conductivity in the range of 10^{-10} Ω^{-1} cm^{-1} in comparison to LT a-Si:H films that show values of 4×10^{-11} Ω^{-1} cm^{-1} for 120°C and 9×10^{-11} Ω^{-1} cm^{-1} for 75°C, and optical gap parameters were 1.75, 1.92, and 1.90 eV for 300, 120, and 75°C, respectively [12]. The higher optical gap of LT films could be explained for changes in the way that hydrogen is incorporated to the films at low temperatures: 10% for 300°C, 10.9% for 120°C, and 9.5% for 75°C, this effect should be considered for optimization at low temperatures. Intrinsic LT (75°C) nc-Si:H films show a dark conductivity 3×10^{-7} Ω^{-1} cm^{-1}, a crystallinity of 75% and a grain size of ~20 nm, and a doped n^+-doped nc-Si:H film shows a dark conductivity of 0.3 Ω^{-1} cm^{-1}, a crystallinity of 72%, and a grain size of ~15 nm. As reference, a nc-Si:H deposited at 300°C shows a dark conductivity of 10^{-6} Ω^{-1} cm^{-1}, crystallinity of 82%, and a grain size of ~30 nm [12]. For solar cell development, the use of μc-Si:H and a-SiGe:H layers is critical to improve IR long wavelength sensitivity [13]. However, only few systematic studies of the role of temperature on the film growth process and its effect on the film optoelectronic properties have been reported. Modeling of properties for microcrystalline silicon films deposited by PECVD at low temperature (175°C) on flexible low-cost substrate is reported in Ref. [14]. Films under microcrystalline growth conditions with different deposition temperatures for solar cell devices were studied [15]. In Ref. [16], amorphous and microcrystalline silicon films were deposited by RF PECVD at deposition temperatures of 25 and 100°C. Values of properties for μc-Si:H films deposited by RF PECVD were at 25°C: deposition rate $V_d = 0.09$ A°/s, optical gap $E_{opt} = 2.10$ eV, dark conductivity $\sigma_d = 8.6 \times 10^{-9}$ Ω^{-1} cm^{-1}, photoconductivity $\sigma_{ph} = 5.9 \times 10^{-8}$ Ω^{-1} cm^{-1}, and activation energy

E_a = 0.63 eV, and at 100°C: V_d = 0.12 A°/s, E_{opt} = 2.14 eV, $\sigma_d = 3 \times 10^{-5}\ \Omega^{-1}\ cm^{-1}$, $\sigma_{ph} = 1.7 \times 10^{-5}\ \Omega^{-1}\ cm^{-1}$, and E_a = 0.31 eV. On the other hand, a-SiGe films could be an option for application on flexible hybrid solar cells. Deposition of silicon-germanium alloys at low temperatures has been studied since the 1970s [17–20]. The Ge atom is slightly larger and remarkably heavier compared to Si atom, thus changing growth, structure, and consequently electronic properties of Si-Ge:H and Ge:H films. To compensate this, several approaches have been made, such as ion bombardment [21], thickness optimization [22], and hydrogen dilution [23]. Nanostructured high Ge-content GeSi:H films have been studied in [10] (**Figure 1**). These films demonstrated good electronic properties such as low defect and tail state-related absorption (**Figure 2**). Nanostructured $Ge_{0.96}Si_{0.4}$:H films deposited by LF PECVD at deposition temperature T_d = 300°C and high H-dilution ratio R_H = 75 showed a E_{opt} = 0.94 eV, $\sigma_d = 2.5 \times 10^{-4}\ \Omega^{-1}\ cm^{-1}$, and E_a = 0.28 eV. A comparison of the nanostructured $Ge_{0.96}Si_{0.4}$:H film deposited at 300°C and a LT GeSi:H film deposited at 160°C is presented in Ref. [24]. The LT GeSi:H shows a $\sigma_d \sim 10^{-4}\ \Omega^{-1}\ cm^{-1}$ and E_a = 0.32 eV, the absorption coefficient spectra also showed good electronic properties as low defect. **Table 1** shows a summary of electronic properties of silicon-based films deposited by PECVD at high and low temperature. Thus, these results show that high quality silicon materials can be deposited at low temperatures compatible with organic polymer semiconductors deposited on plastic substrates.

2.2. Polymer organic materials

Organic materials based on polymer semiconductors have been in focus during the last decade. The most studied materials such as the bulk heterojunction P3HT:PCBM and the conductive semiconductor PEDOT:PSS have been widely implemented in new organic thin-film solar cells [26]. One important advantage of these materials is their solution type processing techniques for fabrication that allows printable deposition of the films at room temperature and at atmospheric pressure in comparison to inorganic material deposition that usually

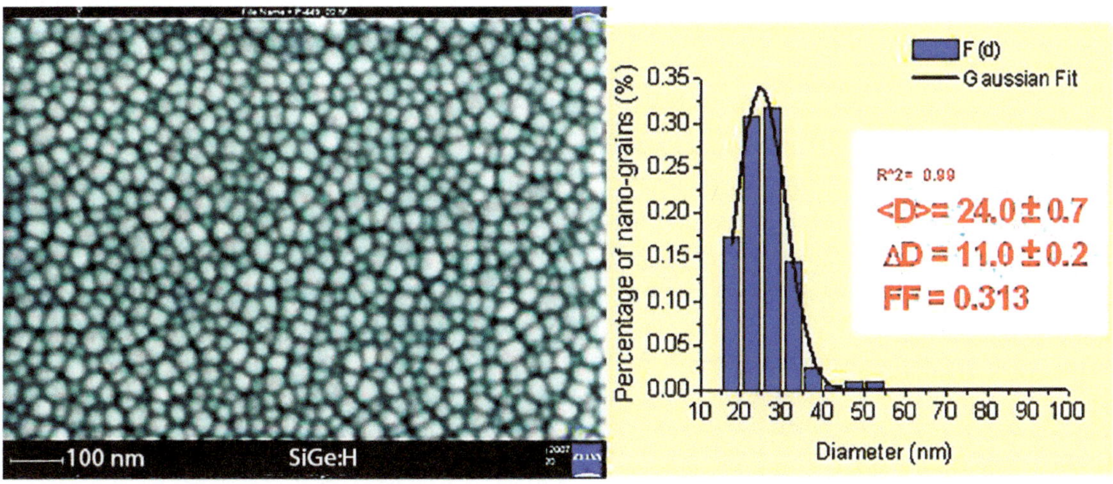

Figure 1. SEM images (left) of nanostructured $Ge_{0.96}Si_{0.4}$:H films deposited by LF PECVD at high H-dilution ratio R_H = 75. Diameter distribution function of grains (right).

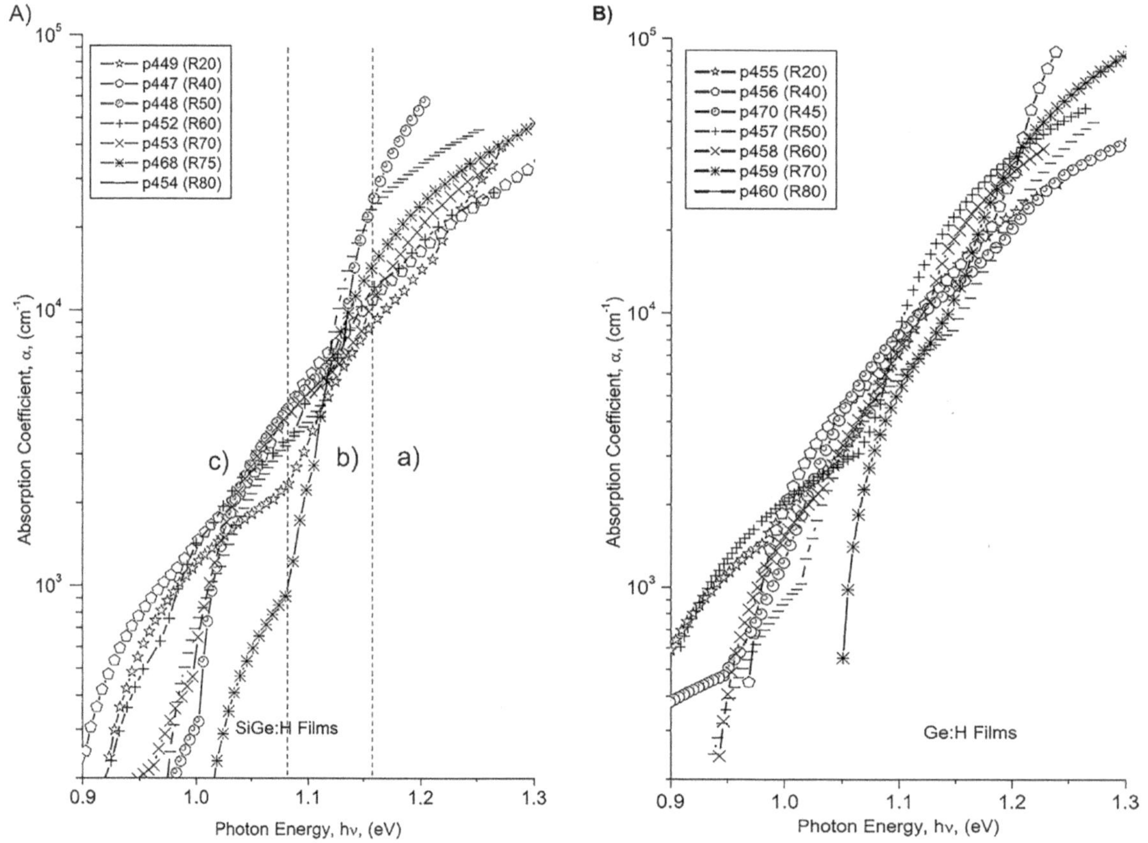

Figure 2. Spectral dependence of optical absorption coefficient $\alpha(h\nu)$for different hydrogen dilution, R, (a) SiGe:H and (b) Ge:H films.

Film	Temperature (°C)	V_d (A°/s)	E_{opt} (eV)	σ_d (Ω^{-1} cm^{-1})	σ_{ph} (Ω^{-1} cm^{-1})	E_a (eV)	Other	Ref.
a-Si:H	300	–	1.8	8×10^{-10}	–	0.60	–	[24]
		–	1.75	$\sim10^{-10}$	–	–	–	[12]
	220	0.93	1.72	1.1×10^{-9}	1.8×10^{-4}	0.93	RF	[16]
	120	–	1.92	4×10^{-11}	–	–	–	[12]
	75	–	1.90	9×10^{-11}	–	–	–	[12]
nc-Si:H	300	–	–	$\sim10^{-6}$	–	–	–	[12]
	75	–	–	3×10^{-7}	–	–	–	[12]
μc-Si:H	100	0.12	2.14	3×10^{-5}	1.7×10^{-5}	0.31	RF	[16]
	25	0.09	2.10	8.6×10^{-9}	5.9×10^{-8}	0.63	RF	[16]
GeSi:H	300	2.2	0.94	2.5×10^{-4}	–	0.28	LF	[10]
	160	1.3	0.95	1.3×10^{-4}	7.5×10^{-6}	0.32	LF	[25]

Table 1. Optoelectronic characteristics comparison for silicon-based films deposited at different temperatures.

requires high substrate temperature and complex high vacuum process. The film-forming processes for organic materials can be classified in (1) coating process methods that include spin coating, blade coating, spray coating, and painting and (2) printing process methods thus include screen printing, offset printing, and inkjet printing [27]. Spin coating method has been one of the important techniques for the deposition of polymer materials. The final characteristics of the films depend on the specific material and its solvent at a given concentration, see, for example, **Figure 3**, it is shown that the thickness dependence as a function of the rotation velocity for PEDOT:PSS films deposited by spin coating at different IPA solvent concentrations. Spin coating deposition is a complex process that also depends among others properties of viscosity, molecular weight, diffusivity, and volatility of the material; however, it is a highly reproducible process [27]. Nevertheless, there are some aspects that limit the application of the technique for industrial production such as no possibility for deposition on large areas, serial production, and problems with patterning. A review of alternatives to spin coating method for the deposition of polymer materials is presented in Ref. [29], just to recognize that the "ideal" process for the deposition of organic materials is still unresolved and an urgent task for the future of this technology.

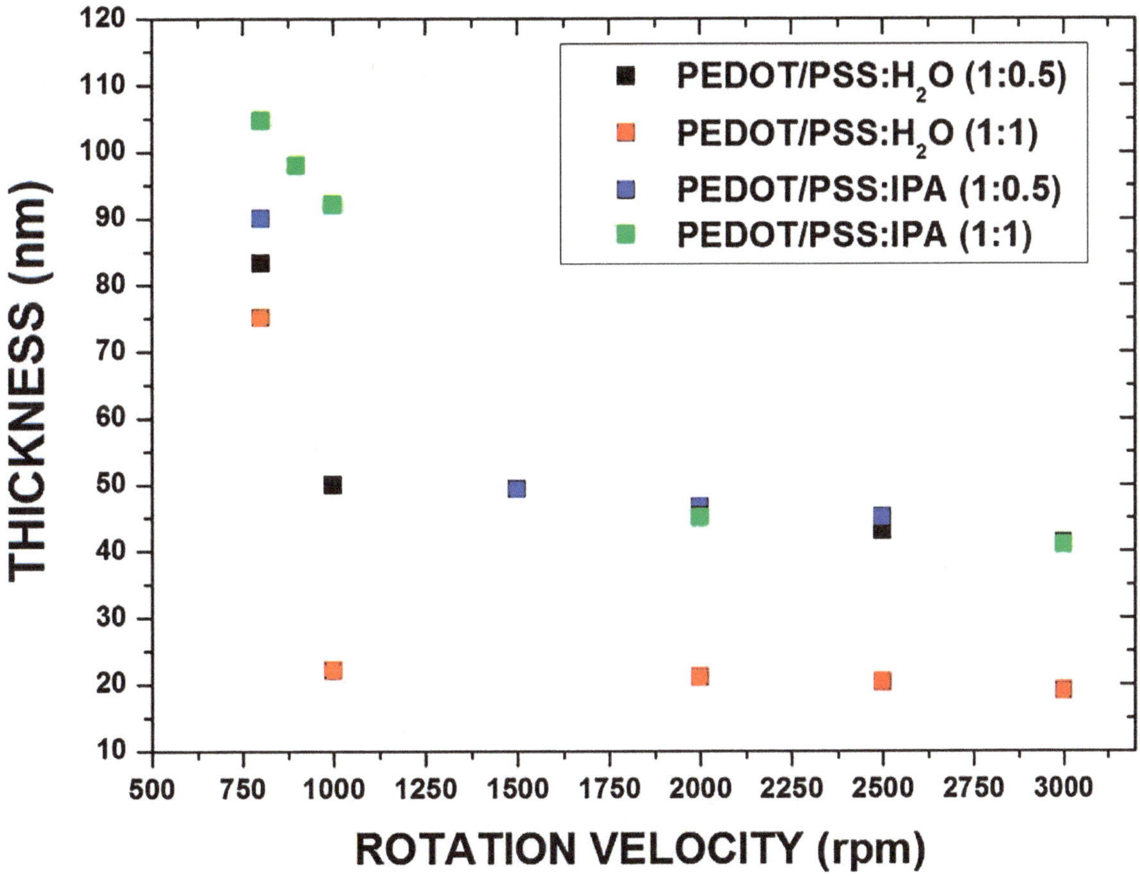

Figure 3. Deposited thickness of PEDOT:PSS films deposited by spin coating as function of frequency rotation for different dilution preparation.

Despite the large number of developed organic materials, there are a few materials that have been usefully applied in promising optoelectronic devices, i.e., PEDOT:PSS is a very attractive conductive polymer semiconductor for electronic and photovoltaic applications (not only for organic electronics) due to its optical transparency, high conductivity, and processing techniques. The use of PEDOT:PSS in hybrid organic-inorganic solar cells and devices has attracted the attention in recent years specially as a candidate for substitution of p-type a-SiC:H:B layer in silicon-based photovoltaic devices [28–30]. The Si/PEDOT interface has shown interesting properties such as effective formation of blocking barrier for electrons, acting similar to an "ideal" barrier [29]. Other advantage of PEDOT:PSS versus traditional inorganic p-type a-Si:H is the flexibility of processing; **Figure 4** shows the DC conductivity, $\sigma_{DC}(T)$, for the spin-coated PEDOT:PSS films deposited under different dilutions. A change in room temperature conductivity of one order of magnitude is obtained only with a change in concentration of dilution with isopropyl alcohol (IPA) [32]. This process seems to be simple in comparison to the required doping process for inorganic semiconductors. However, despite the application of PEDOT:PSS in structures and devices, only a few studies of the electrical properties has been performed. Organic materials are following similar history to amorphous semiconductors, where the practical applications were growing faster than the development of theoretical physics of the materials. In [31],

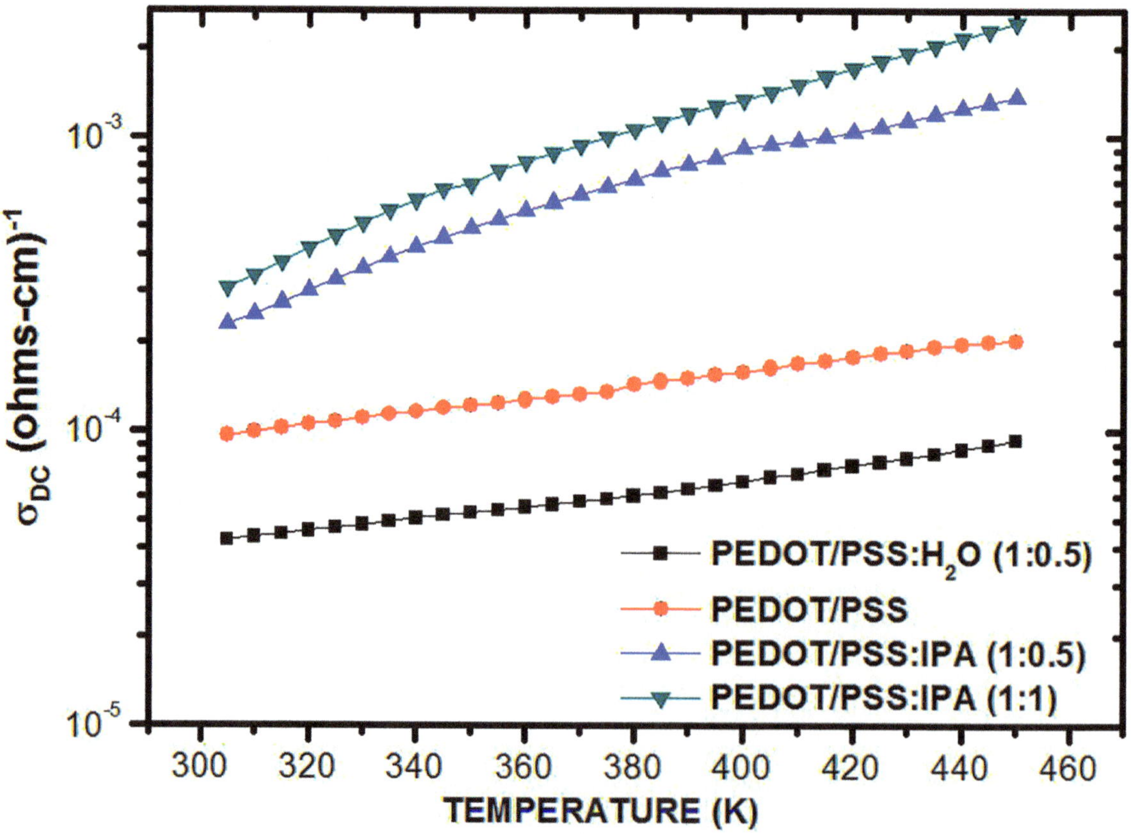

Figure 4. DC conductivity, $\sigma_{DC}(T)$, for the spin-coated films: PEDOT/PSS: H_2O (1:0.5), PEDOT:PSS (without dilution), PEDOT/PSS: isopropyl alcohol (IPA) (1:0.5), and PEDOT/PSS:IPA (1:1) from 300 to 440 K.

the temperature dependence of DC conductivity of spin-coated poly PEDOT:PSS films has been investigated, $\sigma_{DC}(T)$ was measured in films deposited from different mixtures of PEDOT:PSS (**Figure 4**). Experimental data was fitted with two different transport models: thermal activated conduction and variable range hopping (VRH) model. It was found that $\sigma_{DC}(T)$ of PEDOT:PSS/IPA (1:1) sample can be described by the one-dimensional (1D) VRH model; however, at least at below room temperature it is not possible to distinguish unambiguously a concrete model from these two. The versatility of organic materials allows different functions to realize with the same material. PEDOT:PSS has been studied also as replacement of transparent contact oxides (TCO) as it will be discussed in the next section.

2.3. Transparent contact issues

For several years, TCOs were a unique option used as transparent contacts in optoelectronic applications due to their transparency in visible range ($0.4 < \lambda < 0.7$ μ) and low resistivity ($R_{sq} < 100$ Ω/sq). TCOs can be deposited as thin films by techniques as chemical vapor deposition, pulsed laser deposition, spray pyrolysis, and sputtering. There are many studies with a wide background about the development of this technology and film properties [32]. However, to fabricate transparent contacts compatible with the new hybrid inorganic-organic technology and/or compatible with flexible substrates, it is necessary to achieve good optoelectronic properties at low process temperatures (<200°C). To reach this goal, the following approaches have been proposed: (1) to modify the standard deposition methods, (2) to propose new techniques of deposition, or most recently (3) to propose new materials with fabrication methods at room temperature, i.e., organic-based materials. Sputtering method has been modified to achieve TCOs with high optical transmittance and low resistivity at low temperature deposition. ITO (In_2O_3) films deposited by roll-to-roll sputtering at low substrate temperature <50°C show sheet resistance of 47.4 Ω/sq. and a transmittance of 83% [33]. On the other hand, AZO films have been extensively studied as alternative to ITO films due to their stability, high transmission, high conductivity, lower cost, and lower deposition temperature in comparison to ITO films (AZO films, R_{sq} ~ 8.5 Ω/sq and T_{avg} = 84.44% @ T_d = 200°C on glass) [34]. In [35], ZnO/Al (AZO) films were codeposited using pulsed-direct current (DC)-magnetron reactive sputtering processes at 188°C using "soft power." The obtained ZnO:Al films show a sheet resistance of about R_{sq} ~ 13 Ω/sq and a transmittance of ~85%.

Recently with the introduction of organic materials and new technologies, new options have been studied for substitution of traditional TCOs to overcome some disadvantages as low transmission in blue region, high cost, complex high vacuum process, mechanical characteristics, and scarcity. To achieve this, here are some proposals: thin film grids [36], high carbon-based materials [37], and organic polymers [38–40]. PEDOT:PSS has been one of the most investigated organic materials as alternative to TCOs, the polymer has shown higher transparency and conductivity than traditional ITO contacts. Additionally, PEDOT:PSS can be deposited by dispersion in water which is compatible with large area, flexible, and low-cost substrates [41]. The electrical conductivity of PEDOT:PSS can be increased by orders of magnitude using secondary "dopants" as solvents, polyols, surfactants, among others [42]. By solvent dilution and post-treatment method the conductivity of PEDOT:PSS films has reached values of 1418 Ω^{-1} cm^{-1}, low sheet resistance below 65 Ω/□, and transmission in visible range >80%, which have been reported in [38]. In Ref. [43], silver nanostructures were used to enhance conductivity of PEDOT:PSS contacts using a codeposition of spray and inject printing techniques, and the lower sheet resistance achieved by this method was 26.2 Ω^{-1} cm^{-1} with an average transmission of 80.7%.

2.4. Flexible substrates issues

Hybrid inorganic-organic devices seem to be an attractive approach for large area, lightweight flexible solar cells. Flexible electronics also opens the landscape to new applications such as flexible displays, X-ray sensors, electronic textiles, and electronic skin [42]. Many materials have been investigated as flexible substrates: thin glass substrates [44], metal foils [45], and different polymeric films (PET, PEN, Polyimide – KAPTON®, etc.) [41]. The substrate-specific requirements depend strongly of the application, process fabrication, and technology: optical properties, surface roughness, thermal properties, chemical properties, mechanical properties, and electrical and magnetic properties [12]. In the case of applications in organic hybrid solar cells, special attention should be paid to thermal properties, optical properties, and surface roughness. **Table 2** presents a comparison of the main properties of flexible substrates. One of the most attractive materials for their flexibility, low cost, and transparency are the plastic substrates. **Figure 5** shows the spectral dependence of the optical transmittance in the visible range of plastic substrates. PEN and PET are most desirable materials for solar cells application due to the optical transmittance and relative low cost. Another interesting material is PTFE or TEFLON®; the comparison of transmittance spectra presented in **Figure 5** shows that TEFLON transparent thin film is an attractive candidate for application in solar cell application with transparency from 250 to 900 nm and high transmittance (~96%) of above 350 nm. However, TEFLON has a nonpolar surface or a relative low surface energy to the polymer and inorganic materials, which result in low adhesion [46]. Polymer substrates require relatively low processing temperature, below T = 250°C, in the case of both technologies, organic and PECVD Si-based materials, for fabrication of hybrid solar cells. In both technologies, several approaches have been developed for fabrication of flexible devices like PECVD roll-to-roll fabrication. Inorganic PECVD materials have a long history in this field [12]. However, flexible plastic substrates despite having a scope of challenges are very attractive for large-area scale, lightweight, and relatively low cost hybrid organic-inorganic devices.

Substrate	Temperature process (°C)	Transparency (%)	Flexibility	Planarization
Thin glass	600	Yes	Some	No
Metal foil	1000	No	Some	Yes
Polyamide	<300	Some	Yes	No
PEN	<180	Yes	Yes	No
PET	<120	Yes	Yes	Maybe
TEFLON	<240	Yes	Yes	Maybe

Table 2. Comparison of substrate properties for solar cell applications [12, 25].

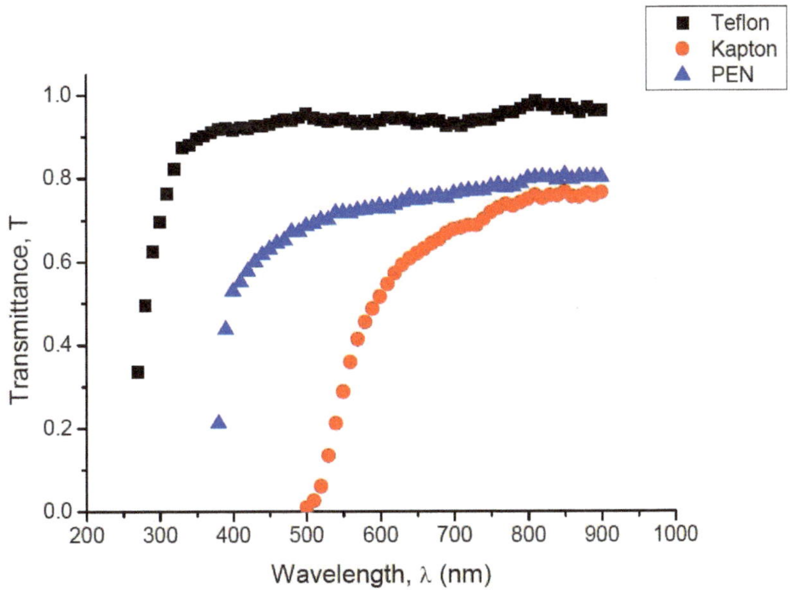

Figure 5. Spectral dependence of optical transmittance for TEFLON, KAPTON, and PEN.

3. Concept for inorganic-organic hybrid device structure

3.1. Concept

A concept for inorganic-organic device (CIOD) structures can be developed taking into account the following material aspects: (a) compatibility of deposition technologies for inorganic-organic semiconductors and electrode materials, (b) complementary of electronic properties of these materials resulting in improvement of device performance, and (c) some technological advantages in device fabrication, e.g., organic materials can be deposited by printable PC controlled technique resulting in simplification of patterning. Inorganic semiconductors can be presented by both crystalline materials and PECVD noncrystalline materials. The latter is applicable for large area flexible devices. In general, CIOD includes various combinations of crystalline and noncrystalline materials together with organic materials. In our further consideration, we shall be focused on PV applications based on PECVD films and organic materials because of their promise for large area flexible devices, although for achieving very high efficiency more complex (crystalline inorganic)-(noncrystalline inorganic)-(organic) material combinations would be of much interest.

3.2. Strategies for CIOD structures aimed at performance improvement

Performance of PV device is determined by: (a) generation of photo-induced carriers controlled by light absorption, (b) transportation of photo-induced carriers controlled by built-in electric field and diffusion, (c) life time of photocarriers that determine both concentration and transport of photocarriers, and (d) properties of interfaces, which control generation-recombination near interface and charge transport through the interface. We select p-i-n configuration as basic PV device structure, where p-layer is p-doped semiconductor, i-layer is intrinsic (nondoped) semiconductor, and n-layer is n-type semiconductor. Internal built-in electric field in this structure is created because of difference in work functions for p- and n-layer. PECVD silicon is conventionally used as intrinsic semiconductor (nondoped), boron-doped as p-layer and phosphor-doped as n-layer. Optical gap for Si:H is typical E_{opt} = 1.7 eV. PECVD materials allow changes in optical gap and consequently spectral absorption in a wide range. To improve optical absorption in short wave range, PECVD silicon-carbon alloys SiC:H are used with optical gap E_{opt} up to 2.5 eV. For long wave range, PECVD silicon-germanium alloys are used with optical gap as low as E_{opt} = 1.2 eV. Interface properties in the structures even prepared in PECVD cluster tool systems depend strongly on process fabrication and deposition regimes used. Let us analyze the above-mentioned principal factors one by one and start with the increase of effective optical absorption in device structure. The latter can be obtained by means of bi-layer (inorganic-organic) absorber of nondoped materials. In this case, organic material typically with wider band gap than Si:H could improve absorption in short wave range. For improvement of absorption in long wave range, SiGe:H layer could be added. To improve transport of photocarrier in organic nondoped materials, built-in electric field can be formed by means of p- and n-doped PECVD films. It is well known, that device performance in shortwave range depends critically on frontal interface (usually it is p-i interface). Improved photocarrier collection for shortwave have been reported for interfaces with organic p-layer [28, 29]. On the other hand, organic semiconductor with narrow band gap have not sufficiently developed yet, therefore, long wave absorption can be improved by incorporation of SiGe:H films. Some of these ideas have been realized and their fabrication and characterization will be described further.

Most of mature solar cell technologies are being developed with silicon-based materials. The stability with sufficient efficiency of these solar cells has been demonstrated for several years. However, in comparison to new thin film organic materials, they have complex fabrication process, high relative cost, and do not suit well for large area flexible PV devices. New technologies such as organic polymer materials bring new approaches of fabrication techniques and materials that can be molecular designed and tailored with specific optoelectronic properties but efficiency and stability are relatively low in comparison to inorganic counterpart. Thus, the combination of both technologies is looking for devices that can take advantage of the strong optical absorption of organic polymer materials, high charge carrier mobility, and stability of inorganic materials while looking for simplification of process fabrication. Different approaches have been proposed to combine both technologies, the versatility of organic materials results in many possibilities for forming hybrid devices structures [47]. Silicon remains a good candidate for hybrid devices due to favorable and well-known electronic properties. Different ways have been reported to combine silicon with polymers in hybrid solar cell: hybrid crystalline silicon-PEDOT:PSS solar cell [29], nanowire-Si/polymer solar cells [48], and a-Si:H/PEDOT:PSS pin solar cell [28]. **Figure 6** shows examples of some ideas for application of organic and inorganic materials in hybrid solar cells. In the structure labeled H1 with configuration (ITO)(p) SiC:H/P3HT:PCBM/(n) Si:H), P3HT:PCBM bulk heterojunction film is used as active absorber film and conventional a-Si:H and a-SiC:H are used as n-type and p-type layers, respectively. The structure labeled H2 (ITO/PEDOT:PSS/(i) Si:H/(n) Si:H) is an a-Si:H p-i-n structure, where the p-type SiC:H layer was substituted by an organic conductive film (PEDOT:PSS). Finally, the structure labeled H3 is a double-absorber layer structure (ITO/PEDOT:PSS/P3HT:PCBM/(i) Si:H/(n) Si:H).

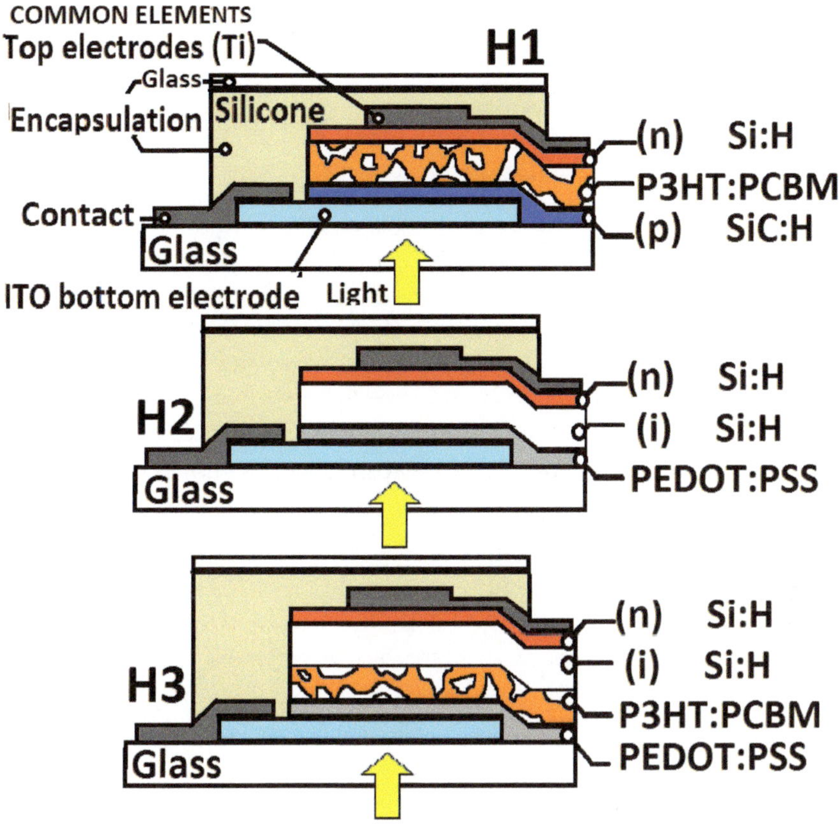

Figure 6. Hybrid photovoltaic structures based on Si:H and organic semiconductors (H1) ITO/(p) SiC:H/P3HT:PCBM/(n) Si:H, (H2) ITO/PEDOT:PSS/(i) Si:H/(n) Si:H, and (H3) ITO/PEDOT:PSS/P3HT:PCBM/(i) Si:H/(n) Si:H.

3.3. On electron energetical diagrams for hybrid device structures

Electron energetic diagram is a conventional and convenient instrument applied to the analysis of device structures and their performance. Such diagram can be created if electronic properties (optical gap, work function, electron affinity, etc.) are known (well measured) that does not exist even in crystalline semiconductors. Moreover, knowledge of bulk electronic characteristics is not sufficient because information on interfaces is required as well. As far as organic semiconductors are concerned, data reported on bulk electronic properties are very poor and those for interfaces have not been reported. In many publications, "energetic diagrams" look as a set of energetic sketches for different materials with different Fermi levels and some arrows denoted hypothetical electron transitions. This is misleading because in multilayered device structure without illumination a thermal equilibrium is suggested (i.e., exchange of particles and their energies between materials) meaning the same Fermi level in the system, rather than different ones in isolated materials. Under illumination, electronic diagram becomes significantly more complex and can only be created based on special experimental study of device structure. Nevertheless, design of energetic diagram based on available experimental data (even with some shortages and errors in details) is strongly motivated. In **Figure 7**, such hypothetical electron energetic diagrams in thermal equilibrium for H1,

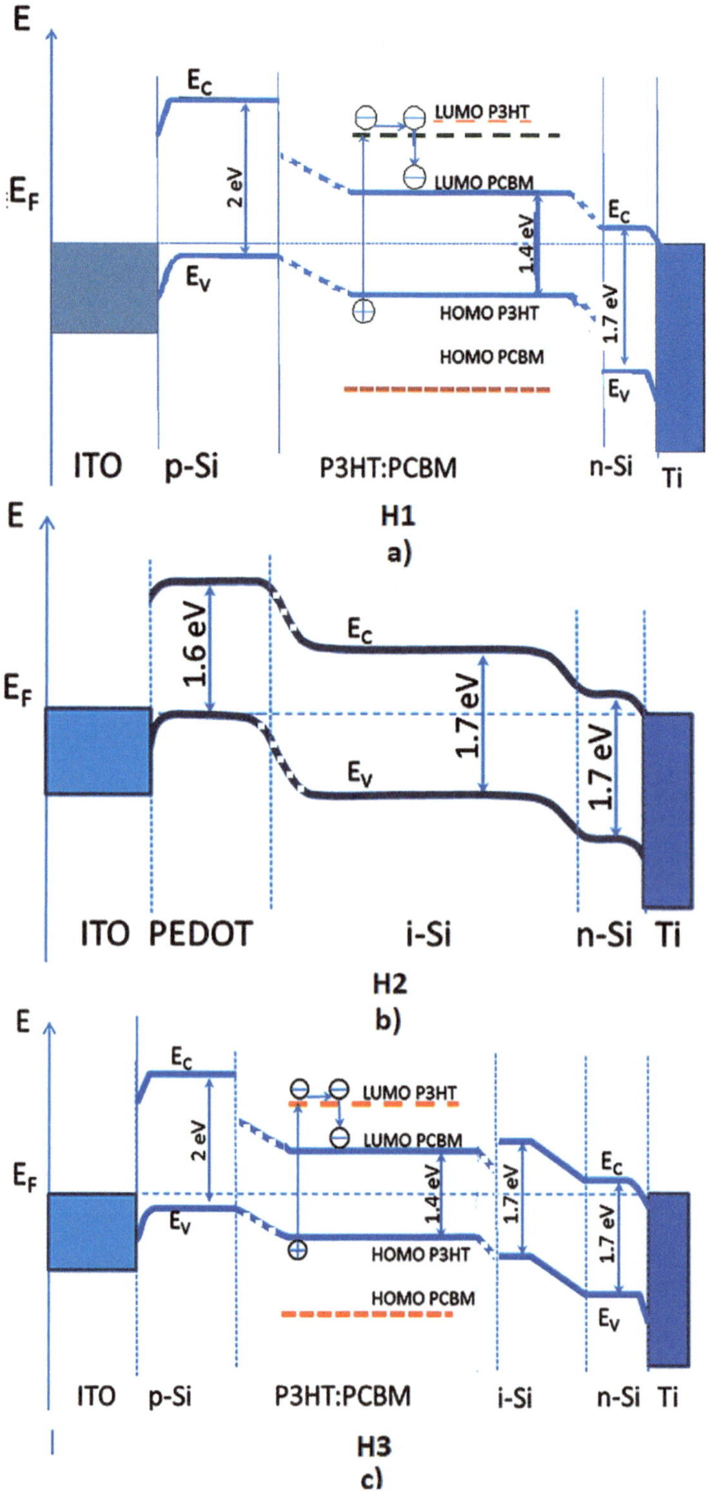

Figure 7. Electron energetic diagrams in thermal equilibrium for the studied structures: (a) ITO/(p) SiC:H/P3HT:PCBM/(n) Si:H (H1), (b) ITO/PEDOT:PSS/(i) Si:H/(n) Si:H (H2), and (c) ITO/PEDOT:PSS/P3HT:PCBM/(i) Si:H/(n) Si:H (H3). Dotted lines represent hypothetical interfaces and offsets in the energy levels due to lack of information on the values reported in literature.

H2, and H3 structures are represented, more details about them can be found in Ref. [28]. In H1, energetic diagram (**Figure 7a**) shows that the light absorption occurs in the organic bulk heterojunction formed by P3HT and PCBM. The creation of an exciton (bounded electron-hole pair) in P3HT is followed by separation of the charge at the internal interfaces of the heterojunction P3HT/PCBM. The inorganic p- and n-type silicon-based layers are used to create an internal electrical field to transport the "free" charge to electrodes. In H2 structure the PEDOT:PSS layer acts as a heavily doped p-type layer and its comparable optical gap (~1.6 eV) to amorphous silicon (~1.7 eV) forms an optimal internal build-in-field and prevents back diffusion of electrons at the frontal interface. In H3 structure improvement of light absorption is expected by using double active absorber layers (a-Si:H and P3HT:PCBM), the energetic diagram shows that one obstacle of this is that the alignment of energetic band between organic and inorganic materials is complex and may create barriers, which oppose to electron and hole transport.

4. Experimental results on fabrication and characterization of different configurations of silicon-organic p-i-n heterojunction structures

4.1. Fabrication

The fabrication flow chart for the H2 structure (ITO/PEDOT:PSS/(i) Si:H/(n) Si:H) is shown in **Figure 8**. The process begins with a coated ITO substrate. The bottom electrode was defined by etching using hydrochloric acid and then the substrate is cleaned by ultrasonic cleaning in acetone and then isopropyl alcohol. The PEDOT:PSS solution was prepared with a 1:6 weight ratio, filtered using a pore size filter of 0.45 μm and deposited by a spin-coating in N_2 ambient with a rotation speed of 2500 rpm that results in a thickness of the film of $d = 45$ nm. The inorganic layers were deposited by RF PECVD process in a multichamber system. The silicon films were grown by a 10% SiH_4 + 90% H_2 gas mixture for intrinsic film and 0.01% PH_3 + 9.9% SiH_4 + 90.09 % H_2 gas mixtures for n-type layers, at pressures of 550 mTorr and substrate temperature of 160°C. All the gases used were semiconductor purity from "Matheson Inc." The compatibility of PECVD process and organic materials for this flow process has been reported in [25].

4.2. Characterization

Figure 9 shows the DC current density—voltage characteristics J(U) under AM1.5 standard illumination for nonoptimized hybrid structures presented in **Figure 6**. As seen, the structure H1 with organic semiconductor as nondoped intrinsic material demonstrate low U_{OC} = 185 mV, while expected values based on difference of Fermi levels for p- and n-layers could be close to $\Delta E_F \approx 900$ mV. The reduced U_{OC} value may be related to low shunt resistance due to leakage transversal current resulting in low J_{SC} current. Mechanism of leakage current in such structures has not been studied yet and could be rather complex because of the complex structure of the organic material (e.g., see SEM image in **Figure 13**). The highest value of J_{SC} = 5.87 mA/cm^2 is observed in H2 structure with PEDOT/PSS as p-layer and i-Si:H as nondoped absorber and Si:H(P) as n-layer. U_{OC} = 400 mV in this structure is better than in H1 but still

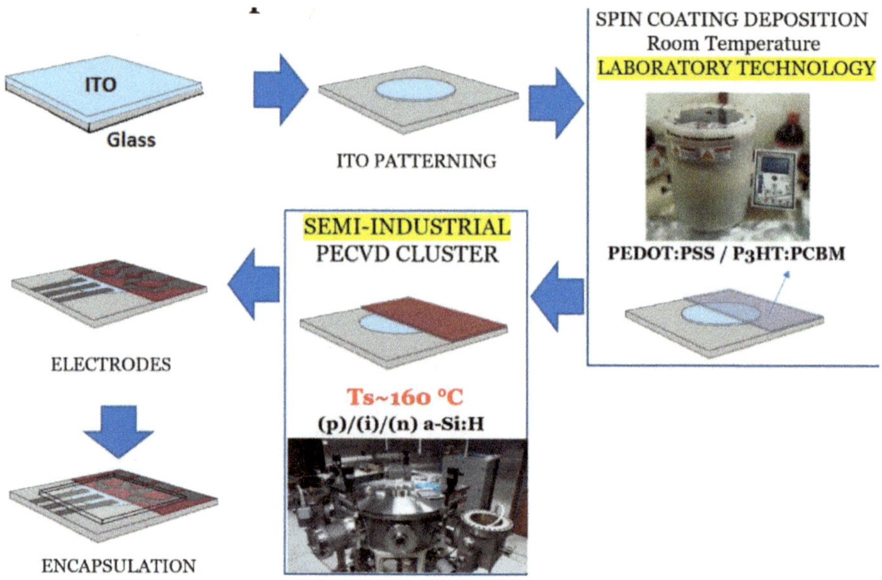

Figure 8. Flow fabrication for thin film silicon/organic p-i-n structure involving PECVD process and spin on coating deposition for inorganic and organic films, respectively.

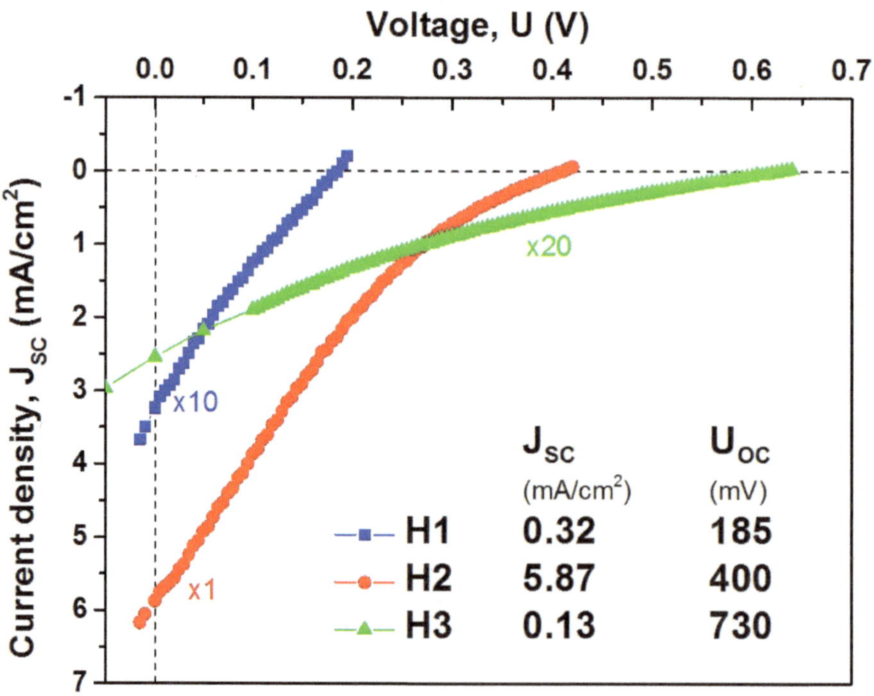

Figure 9. Current-voltage characteristics and extracted *Uoc*, J_{SC}, *FF* values of hybrid photovoltaic structures based on Si:H and polymer organic semiconductors (H1) ITO/(p) SiC:H/P3HT:PCBM/(n) Si:H, (H2) ITO/PEDOT:PSS/(i) Si:H/(n) Si:H, and (H3) ITO/PEDOT:PSS/P3HT:PCBM/(i) Si:H/(n) Si:H.

less than that expected. The H2 structure clearly shows current suppression at low internal field, when the potential is close to U_{OC} suggesting still high value of transversal leakage current. The structure with double absorber layer H3 has the highest U_{OC} = 700 mV, but very low current density J_{SC} = 0.13 mA/cm^2. It means that the structure has sufficient internal electric field but very low current. For speculative purpose, the following reasons of low current can be mentioned: (a) too high resistivity of nondoped organic and inorganic layers, (b) not optimal thicknesses for nondoped layers, and (c) blocking barrier between two nondoped layers. However, special study is necessary to determine unambiguously the mechanism of current reduction. In terms of device performance characteristics, the main results of comparison of three types of hybrid structures can be summarized as follows: the H2 structure shows the highest short circuit current density J_{SC} = 5.87 mA/cm^2, whereas the H3 structure has the highest open circuit voltage U_{OC} = 730 mV, while both J_{SC} and U_{OC} of the H1 structure are rather low. It is worth to note that this comparison provides only some guideline for further study.

Figure 10 presents the spectral dependence of short circuit current density Jsc (λ) for the structures H1, H2, and H3 in comparison to the solar spectral irradiance. For Jsc (λ) form comparison, the Jsc (λ) curves were normalized to Jsc (λ) response of an a-Si:H inorganic solar cell. An interesting relative higher photo current than that for inorganic reference sample is found in the region of short wavelength from λ = 410 to 354 nm associated to the frontal interface of the structures. It suggests better electronic properties of organic-inorganic frontal interfaces due to lower surface back diffusion coefficient of carriers generated close to this interface or/and lower surface recombination rate and probably more stable chemical interface [28]. Some confirmation on quality of undoped absorber material can be obtained from subgap for $h\nu < E_{opt}$ photocurrent. Slope of $J_{SC}(h\nu)$ characterized by E_0 in the vicinity of E_{opt} is approximately the same in the all structures meaning that the organic semiconductor has narrow band tail density of states. As far as deep states are concerned, the sample H1 shows the highest value, sample H2 higher value than that in the reference sample, but less than that in H1 sample. In the sample H3 was not able to measure signal for $h\nu$ < 1.6 eV. In other words, quality of nondoped absorber with organic semiconductor is still less than that for Si:H films that motivates further efforts in improvement of electronic properties of organic semiconductors.

Some sort of "optimization" (see details in [28]) has been performed for H2 type of structure. **Figure 11** shows the current voltage characteristics of an improved hybrid H2* structure in comparison to those for organic and inorganic solar cells. It is most interesting that J_{SC} = 17.7 mA/cm^2 value exceeds values for both reference samples (organic and inorganic Si:H p-i-n structure). This high value may result from: (a) higher electric field at the frontal interface PEDOT/PSS-i-Si:H, and consequently reducing back diffusion and increasing current, (b) reducing surface recombination rate at the interface, and (c) better electronic quality of Si:H film grown on PEDOT/PSS. Though the latter cannot be excluded, no experimental evidence on changes electronic properties of Si:H film in this structure. U_{OC} = 645 mV for this H2* structure is higher than that for organic and less than that for Si:H p-i-n structure. At low voltage in the vicinity of U_{OC}, photocurrent is suppressed indicating high transversal leakage current. Finally, the comparison of performance characteristics for H2* structure reported for PEDOT/(i) Si:H/(n) Si:H devices is performed (see **Table 3**). H2 structure demonstrates a large high value of short circuit current density J_{SC} = 17.7 mA/cm^2 in H2* structures, this makes promising further study of this kind of photovoltaic cell, especially if the transversal leakage current is significantly reduced.

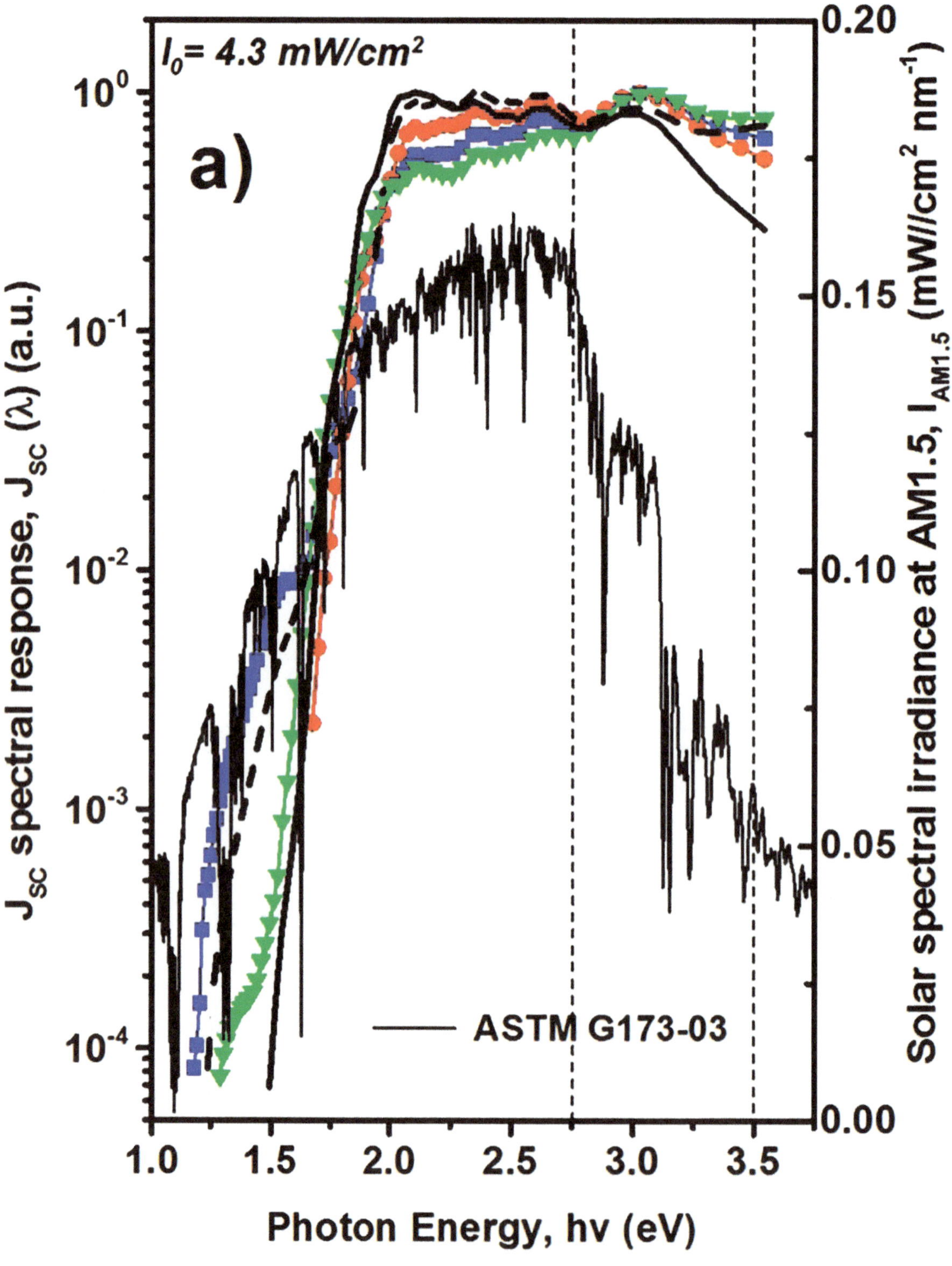

Figure 10. Spectral dependence of short circuit current density in the structures H1, H2, and H3 in comparison with solar spectral irradiance.

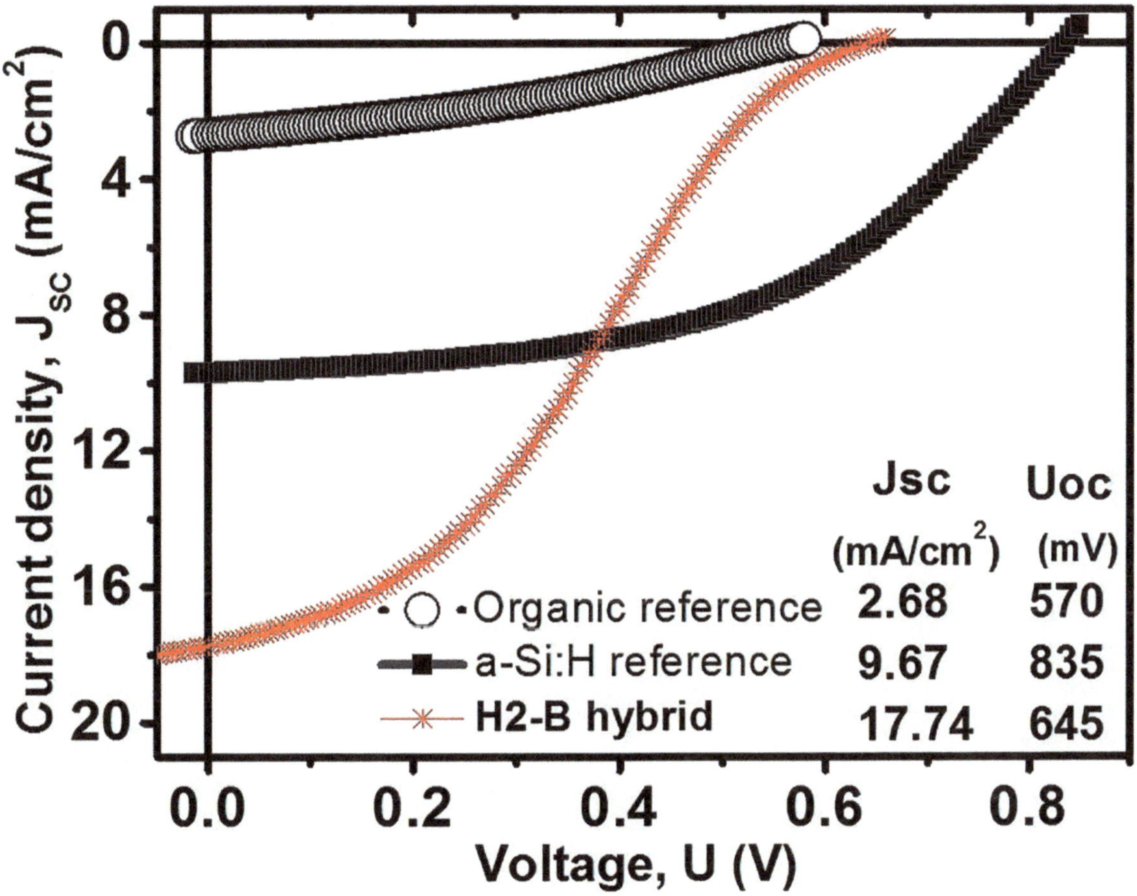

Figure 11. Current-voltage characteristics $J(U)$ and extracted U_{OC}, J_{SC}, FF and conversion efficiency values for hybrid photovoltaic structures in H2* configuration, organic, and inorganic reference solar cells.

Frontal interface	J_{sc} (mA/cm²)	U_{oc} (V)	FF	PCE (%)
PEDOT:PSS/(i)a-Si:H	4.55	0.88	51	2.1
	16.0	0.54	50	4.7
	17.74	0.64	32	3.75
a-SiC:H:B/(i) a-Si:H efficiency record	16.36	0.89	71	10.2

Table 3. Comparison of PV characteristics reported in literature with the best structure (H2) PEDOT:PSS/Si:H [28].

5. Hybrid structures on flexible substrates

Flexible solar cells on plastic substrates have several advantages such as lightweight and relative low cost. Inorganic and organic semiconductor technologies have demonstrated to be compatible with flexible substrates [12, 33–35]. However, the new concept of hybrid PECVD-polymer solar cell on flexible substrates has not been reported until now. A hybrid solar cell ((AZO/PEDOT:PSS/

(i) Si:H/(n) Si:H//Ag) on PEN and polyimide substrate shown in **Figure 12** have been fabricated and studied. A transparent electrode of AZO was deposited on the flexible substrate by sputtering method. The inorganic silicon layers were deposited by RF PECVD process at substrate temperature of $T_d = 160°C$ and Ag top contact was deposited by sputtering at the same temperature and the PEDOT:PSS film was deposited by spin coating. A reference structure was fabricated on Corning glass substrate and demonstrated a $Jsc = 10.4\ mA/cm^2$ and $Voc = 565\ mV$.

Similar structures on PEN with different AZO-film thickness were compared; it was found that the performance characteristics of the cell were strongly dependent on the thickness of the AZO film. The short circuit current density increased from $Jsc = 3.21$ to $9.79\ mA/cm^2$ and the open circuit voltage from $Voc = 405$ to $565\ mV$ for a AZO thickness change from $d = 250$ to $500\ nm$, respectively. Average roughness measured on AZO film deposited on plastic substrate as references shows a reduction from $\langle r \rangle = 1.11\ nm$ for 250 nm thick AZO layer to $\langle r \rangle = 0.87\ nm$ for 500 nm thick AZO layer, both on PEN substrate. Polyimide KAPTON® substrates from Du Pont™ are known for some applications such as circuits and space applications. The average roughness for 250 nm thick AZO films on polyimide show a small difference in value (1.11) in comparison to 250 nm thick AZO film on PEN substrate (1.4). The difference of average roughness between

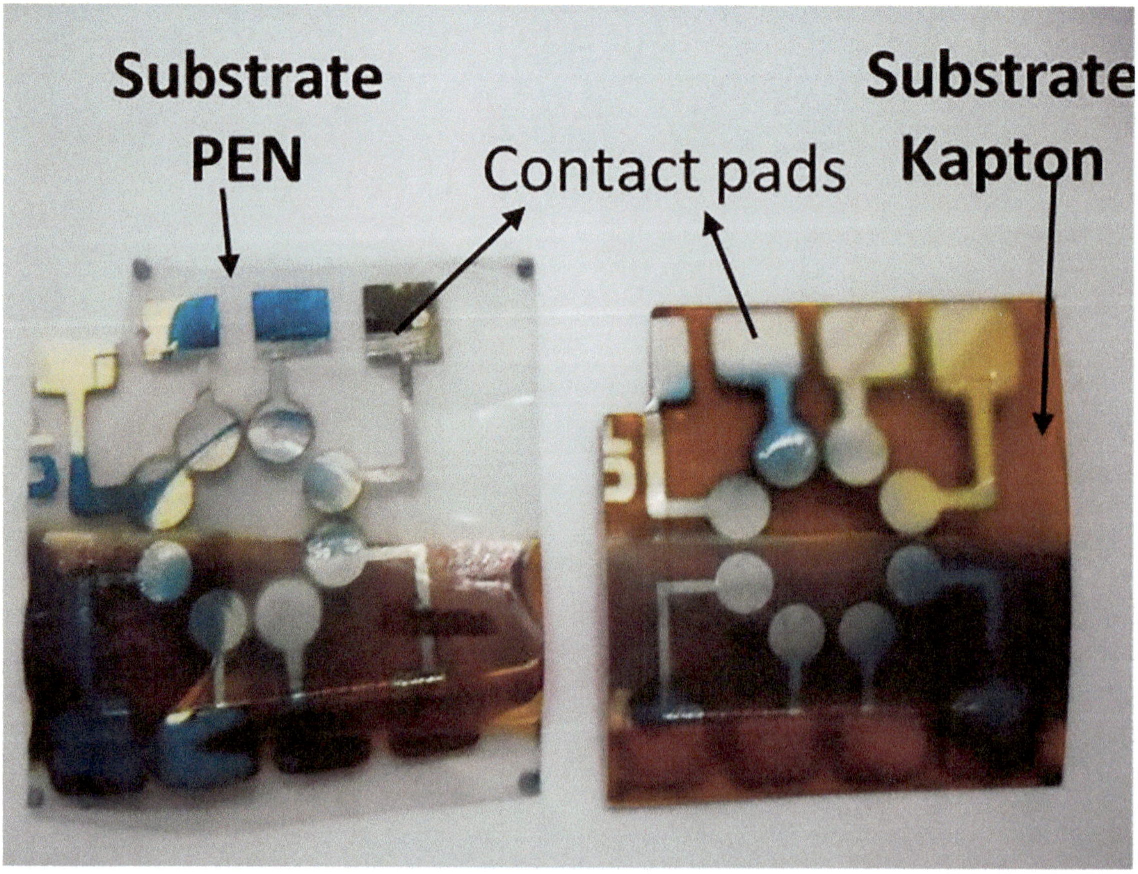

Figure 12. Hybrid photovoltaic structure on flexible substrate (PEN, left and right, KAPTON) based on Si:H and organic semiconductors PEDOT:PSS as p-type layer.

AZO on polyimide and PEN disappears for those with AZO film thickness of d = 500 nm. Polyimide has some advantages over other plastics such as high performance in electrical, chemical, and mechanical properties at relatively high temperature process of up to T = 450°C. However, its low transmission in visible range (**Figure 5**) limits implementation in solar cells. For similar solar cell structures and similar 500 nm thick AZO layer, the Jsc reduces from 3.21 mA/cm^2 on PEN to 1.9 mA/cm^2 on polyimide because of the short wavelength limit (T < 50%) at 590 nm. The cross-sectional scanning electron microscopy (SEM) image of the solar cell on PEN is shown in **Figure 13**. The 500 nm thick AZO, 240 nm PEDOT:PSS, 300 nm thick (i) a-Si:H and 200 nm thick Ag layers and PEN substrate are well defined in the images. A large difference between the expected thickness (~45 nm) on glass substrate and PEN substrate (250 nm) was observed for PEDOT:PSS. This may be attributed to different surface properties for glass and PEN. As it can be seen, PEDOT layer looks as dispersive nonuniform material. Still it is not clear if this nonuniformity and its characteristics are intrinsic (inherent) material properties or they may depend on substrate (or previous layer) surface properties. Deviations of expected thickness and real thickness on plastic substrate were not observed for the PECVD films; however, the deformation (curvature, see **Figure 12**) of the substrates after device fabrication is an issue to be considered due to stress caused by deposition temperature and other factors in PECVD process. These curvatures are used to be the cause of misalignment effects on flexible devices. SEM images also illustrate how quality of substrate and defects of substrate are important. As we can see a defect in the substrate is transferred through all the thickness of AZO film and is

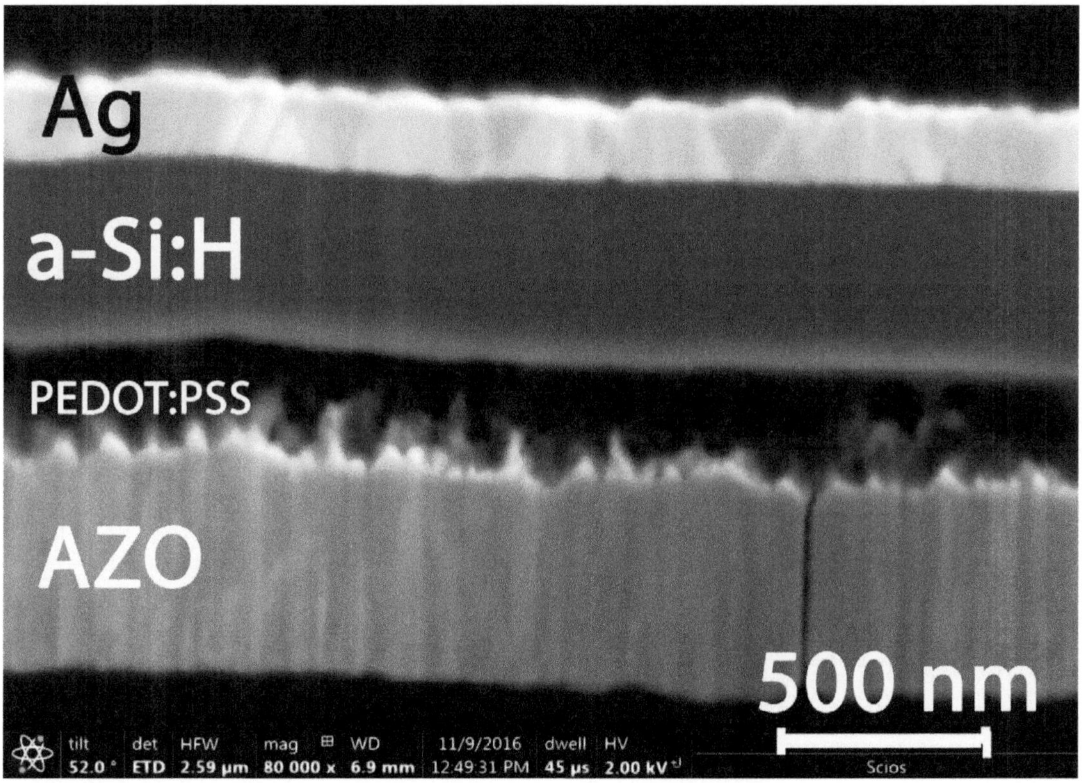

Figure 13. Hybrid solar cell cross-sectional SEM image on flexible substrate PEN/ITO/PEDOT:PSS/(i) Si:H/(n) Si:H (H2).

interrupted by the organic PEDOT:PSS layer. The surface quality is an important parameter that influences the final quality of the devices.

6. Summary and conclusion

In the beginning, a brief overview of polymer organic and PECVD inorganic material semiconductor technologies has been made to focus on optoelectronic characteristics and fabrication process at low substrate temperature for application on hybrid photovoltaic devices. Inorganic semiconductor materials deposited by PECVD have been reported in literature to be a mature technology compatible with large area flexible devices; while organic semiconductors and their technology are significantly less studied. On other hand, organic materials have shown potential as PV materials with several advantages in the fabrication process such as room temperature deposition and reduction of complex vacuum stages for deposition but the "ideal" process for the deposition of organic materials is still unresolved.

A concept of inorganic-organic photovoltaic devices mainly for PECVD inorganic materials has been considered. The concept is based on compatibility of both technologies and complementary electronic properties. Three types of hybrid structures have been fabricated and studied: (1) with OS as intrinsic absorber and p-, n-Si:H layers creating built-in electric field (H1 structures), (2) ITO/PEDOT:PSS/(i) Si:H/(n) Si:H structures (H2 type), and (3) structures with double intrinsic absorber such as ITO/PEDOT:PSS/P3HT:PCBM/(i) Si:H/(n) Si:H (H3). All the structures showed PV performance with different characteristics discussed and compared with those reported in the literature. Performance of the structures including those fabricated on flexible plastic substrates confirmed clear compatibility OS and PECVD technologies. The device diagnostics of the studied structures has been performed by measurements of photocurrent spectral dependencies. The measurements of subgap photocurrent revealed, that electronic quality of intrinsic OS materials, was sufficient, but worse comparing to that of Si:H films. The structures with OS frontal interfaces showed better short wavelength photocurrent response indicating better electronic properties of the interfaces comparing to reference samples. All the structures suffered from photocurrent suppression at voltages near U_{OC} values suggesting high transversal leakage current. Simple "optimization" of H2 structure resulted in substantial increase of photocurrent to the value J_{SC} = 17.7 mA/cm^2 that was higher than the values in the reference samples. Explanations of characterization data have speculative character and should be considered only as some guidelines and motivation for further study. Although some hybrid structures demonstrated substantial improvement of performance, while other hybrid structures showed poor performance, further R&D efforts seem to be promising and should be focused on deeper study of organic materials and related interface properties.

Acknowledgements

Ismael Cosme Bolaños acknowledges support from "CONACyT Cathedra" project No.2734. Hiram Enrique Martinez and Antonio Olivares acknowledge support from CONACyT through the scholarships number 362152 and 363344, respectively.

Nomenclatures

PECVD	Plasma enhanced chemical vapor deposition
CVD	Chemical vapor deposition
GD	Glow discharge
ESR	Electron-spin resonance
DOS	Density of states
LT	Low temperature
HIT	Heterojunction intrinsic thin configuration
P3HT:PCBM	Poly(3-hexythiophene):methano-fullerenephenyl-C61-butyric-acid-methyl-ester
PEDOT:PSS	Poly(3,4ethylenedioxythiophene):poly(4-styrenesulfonate)
IPA	Isopropyl alcohol
DC	Direct current
VRH	Variable range hopping
1D	One dimension
TCO	Transparent conductive oxide
PLD	Pulsed laser deposition
AZO	Aluminum zinc oxide
PET	Polyethylene terephthalate
PEN	Polyethylene naphthalate
PTFE	Polytetrafluoroethylene
CIOD	Concept for inorganic-organic device
PV	Photovoltaic
AM1.5	Air mass spectrum 1.5
SEM	Scanning electron microscopy

Author details

Andrey Kosarev[1]*, Ismael Cosme[2], Svetlana Mansurova[1], Antonio J. Olivares[1] and Hiram E. Martinez[1]

*Address all correspondence to: akosarev@inaoep.mx

1 National Institute for Astrophysics, Optics and Electronics, Optics and Electronics (INAOE), Puebla, Mexico

2 CONACyT-INAOE, Mexico

References

[1] J.M. Ziman, editor. Principles of the theory of solids. 2nd ed. Cambridge: Cambridge University Press; 1965. 452 p.

[2] H.F. Sterling, R.C.G. Swann. Chemical vapour deposition promoted by R.F. discharge. Solid-State Electronics. 1965;**8**:653-654.

[3] R.C. Chittick, J.H. Alexander, H.F. Sterling. The preparation and properties of amorphous silicon. Journals of Electrochemical Society. 1969;**166**:77-88.

[4] W.E. Spears, P.G. Le Comber. Substitutional doping of amorphous silicon. Solid State Communications. 1975;**17**:1193-1196.

[5] W.E. Spear, P.G. Le Comber, S. Kinmond, M.H. Brodsky. Amorphous silicon p-n junction. Applied Physics Letters. 1976;**28**:105.

[6] R. Martins, H. Aguas, V. Silva, I. Ferreira, A. Cabrita, E. Fortunato. Nanostructured silicon films produced by PECVD. Materials Research Society Symposium Proceedings. 2001;**664**:A9.6.1-A9.6.6.

[7] Y. Nasuno, M. Kondo, A. Matsuda. Passivation of oxygen-related donors in microcrystalline silicon by low temperature deposition. Applied Physics Letters. 2001;**78**:2330-2332.

[8] A. Matsuda. Formation kinetics and control of micro-crystallite in μc- Si:H from glow discharge plasma. Journal of Non-crystalline Solids. 1983;**59-60**:767-774.

[9] R. Cariou, M. Labrune, R.I. Cabarrocas. Thin crystalline silicon solar cells based on epitaxial films grown at 165°C by RF-PECVD. Solar Energy Materials and Solar Cells. 2011;**95**:2260-2263.

[10] A. Kosarev, A. Torres, C. Zuñiga, M. Adamo, L. Sanchez. Nano-structured GeySi1-y:H films deposited by low frequency plasma for photovoltaic application. Materials Research Society Symposium Proceedings. 2009;**1127**:1127-T04-03.

[11] J. Robertson. Deposition mechanism of hydrogenated amorphous silicon. Journal of Applied Physics. 2000;**87**:2608-2617.

[12] W.S. Wong, A. Salleo, editors. Flexible Electronics: Materials and Applications. 1st ed. USA: Springer Science + Business Media; 2009. 473 p.

[13] A. Terakawa. Review of thin-film silicon deposition techniques for high-efficiency solar cells developed at Panasonic/Sanyo. Solar Energy Materials & Solar Cells. 2013;**119**:204-208.

[14] Z. Mrázková, K. Postava, A. Torres-Rios, M. Foldyna, P. Roca i Cabarrocas, J. Pištora. Optical modeling of microcrystalline silicon deposited by plasma-enhanced chemical vapor deposition on low-cost iron-nickel substrates for photovoltaic applications. Procedia Materials Science. 2016;**12**:130-135.

[15] Y. Nasuno, M. Kondo, A. Matsuda. Microcrystalline silicon thin-film solar cells prepared at low temperature using PECVD. Solar Energy Materials & Solar Cells.2002;**74**:497-503.

[16] P. Alpuim, V. Chu, J.P. Conde. Amorphous and microcrystalline silicon films grown at low temperatures by radiofrequency and hot-wire chemical vapor deposition. Journal of Applied Physics. 1999;**86**(7):1.

[17] W.E. Spear, P.G. Le Comber. Electronic properties of substitutionally doped amorphous Si and Ge. Philosophical Magazine. 1976;**33**:935-949.

[18] D. Hauschildt, M. Stutzmann, J. Stuke, H. Dersch. Electronic properties of doped glow-discharge amorphous germanium. Solar Energy Materials. 1982;**8**(1-3):319-330.

[19] J. Xu, S. Miyazaki, M. Hirose. High-quality hydrogenated amorphous silicon-germanium alloys for narrow bandgap thin film solar cells. Journal of Non-Crystalline Solids. 1996;**208**(3):277-281.

[20] W.B. Jordan, B. Wagner. Effects of deposition temperature and film thickness on the structural, electrical, and optical properties of germanium thin films. MRS Proceedings. 2002;**715**:A18.2.

[21] F. Origo, P. Hammer, D. Comedi, I. Chambouleyron. The effect of ion-bombardment on the formation of voids during deposition of a-Ge:H. MRS Proceedings. 1998;**507**:477-482.

[22] A. Belfedal, Y. Bouizem, J.D. Sib, L. Chahed. Films thickness effect on structural and optoelectronic properties of hydrogenated amorphous germanium (a-Ge:H). Journal of Non-Crystalline Solids. 2012;**358**(11):1404-1409.

[23] C.-M. Wang, Y.-T. Huang, Y. Kuo-Hsi, H.-J. Hsu, C.-H. Hsu, H.-W. Zan, C.C. Tsai. Influence of hydrogen on the germanium incorporation in a-Si1-xGex:H for thin-film solar cell application. MRS Proceedings. 2010;**1245**:A04.02.1-04.02-6 .

[24] A. Kosarev, A. Torres, Y. Hernandez, R. Ambrosio, C. Zuniga, T.E. Felter, R. Asomoza, Y. Kudriavtsev, R. Silva-Gonzalez, E. Gomez-Barojas, A. Ilinski, A.S. Abramov. Silicon-germanium films deposited by low-frequency plasma-enhanced chemical vapor deposition: Effect of H2 and Ar dilution. Journal of Materials Research. 2006;**21**(1):88-104.

[25] I. Cosme. Study of GeSi:H materials deposited by PECVD at low temperatures (Td<200 °C) for device applications [thesis]. Puebla, MEX: INAOE; 2013. 157 p.

[26] C. Brabec, V. Dyakonov, U. Scherf, editors. Organic Photovoltaics. 2nd ed. Germany: Wiley-VCH Verlag GmbH & Co. KGaA; 2014. 642 p.

[27] F.C. Krebs. Fabrication and processing of polymer solar cells: A review of printing and coating techniques. Solar Energy Materials & Solar Cells. 2009;**93**:394-412.

[28] I. Cosme, A. Kosarev, S. Mansurova, A.J. Olivares, H.E. Martinez, A. Itzmoyotl. Hybrid photovoltaic structures based on amorphous silicon and P3HT:PCBM/PEDOT:PSS polymer semiconductors. Organic Electronics. 2016;**38**:271-277.

[29] K.A. Nagamatsu, S. Avasthi, J. Jhaveri, J.C. Sturm. A 12% efficient silicon/PEDOT:PSS heterojunction solar cell fabricated at <100°C. IEEE Journal of Photovoltaics. 2014;**4**(1):260-264.

[30] E.L. Williams, G.E. Jabbour, Q. Wang, S.E. Shaheen, D.S. Ginley, E.A. Schiff. Conducting polymer and hydrogenated amorphous silicon hybrid solar cells. Appl. Phys. Lett. 2005;**87**:223504-223506 .

[31] A. Olivares, I. Cosme, S. Mansurova, A. Kosarev, H.E. Martinez. Study of Electrical Conductivity of PEDOT:PSS at Temperatures >300 K for Hybrid Photovoltaic Applications. In: 12th International Conference on Electrical Engineering, Computing Science and Automatic Control (CCE); 2015; Mexico City. Mexico: CCE; 2015. p. 1-3.

[32] D. Ginley, H. Hosono, D.C. Paine, editors. Handbook of Transparent Conductors. 1st ed. Online: Springer US; 2011. 534 p.

[33] K.-H. Choi, J.-A. Jeong, J.-W. Kang, D.-G. Kim, J.K. Kim, S.-I. Na, D.-Y. Kim, S.-S. Kim, H.-K. Kim. Characteristics of flexible indium tin oxide electrode grown by continuous roll-to-roll sputtering process for flexible organic solar cells. Solar Energy Materials & Solar Cells Materials. 2009;**93**:1248-1255.

[34] H.J. Park, J.-W. Park, S.-Y. Jeong, C.-S. Ha. Transparent flexible substrates based on polyimides with aluminum doped zinc oxide (AZO) thin films. Proceedings of the IEEE. 2005;**93**(6): 1447-1450.

[35] S. Sanzaro, A.L. Magna, E. Smecca, G. Mannino, G. Pellegrino, E. Fazio, F. Neri, A. Alberti. Controlled Al3+ incorporation in the ZnO lattice at 188 °C by soft reactive co-sputtering for transparent conductive oxides. Energies. 2016;**9**:443.

[36] K. Tvingstedt, O. Inganäs. Electrode grids for ITO free organic photovoltaic devices. Advance Materials. 2007;**19**:2893-2897.

[37] J. Wu, H.A. Becerril, Z. Bao, Z. Liu, Y. Chen, P. Peumans. Organic solar cells with solution-processed graphene transparent electrodes. Applied Physics Letters. 2008;**92**:26.

[38] Y.H. Kim, C. Sachse, M.L. Machala, C. May, L. Müller-Meskamp, K. Leo. Highly conductive PEDOT:PSS electrode with optimized solvent and thermal post-treatment for ITO-free organic solar cells. Advanced Functional Materials. 2010;**94**(8):1076-1081.

[39] S.-I. Na, B.-K. Yu, S.-S. Kim, D. Vak, T.-S. Kim, J.-S. Yeo, D.-Y. Kim. Fully spray-coated ITO-free organic solar cells for low-cost power generation. Solar Energy Materials and Solar Cells. 2010;**94**(8):1333-1337.

[40] D. Alemu, H.-Y. Wei, K.-C. Ho, C.-W. Chu. Highly conductive PEDOT:PSS electrode by simple film treatment with methanol for ITO-free polymer solar cells. Energy & Environmental Sciences. 2012;**5**:9662-9671.

[41] X. Crispin, F.L.E. Jakobsson, A. Crispin, P.C.M. Grim, P. Andersson, A. Volodin, C. van Haesendonck, M. Van der Auweraer, W.R. Salaneck, M. Berggren. The origin of the high conductivity of poly(3,4-ethylenedioxythiophene) – poly (styrenesulfonate) (PEDOT–PSS) plastic electrodes. Chemistry of Materials. 2006;**18**(18):4354-4360.

[42] A. Elschner, S. Kirchmeyer, W. Lovenich, U. Merker, K. Reuter, editors. PEDOT: Principles and Applications of an Intrinsically Conductive Polymer. 1st ed. London: CRC Press; 2011. 355 p.

[43] Alshammari, A. S., Sam, F. L. M., Rozanski, L. J., Mills, C. A., Alenezi, M. R., Beliatis, M. J., Jayawardena, K. D. G. I., Underwood, J. M. and Silva, S. R. P., Controlled growth and spray deposition of silver nanowires for ITO-free, flexible, and high brightness OLEDs. Phys. Status Solidi A, 2016, DOI 10.1002/pssa.201600561

[44] A. Plichta, A. Habeck, S. Knoche, A. Kruse, A. Weber, N. Hildebrand. Flexible glass substrate. In: Crawford GP (ed), Flexible Flat Panel Displays. John Wiley & Sons Ltd, West Sussex, 2005; pp. 35-55.

[45] C.C. Wu, S.D. Theiuss, G. Gu, M.H. Lu, J.C. Sturm, S. Wagner, C.C. Wu, S.D. Theiuss, G. Gu, M.H. Lu, J.C. Sturm, S. Wagner, S.R. Forrest. Integration of organic LEDs and amorphous Si TFTs onto flexible and lightweight metal foil substrates. IEEE Electron Device Letters. 1997;**18**(12):609-612.

[46] H. Izadi, M. Golmakania, A. Penlidis. Enhanced adhesion and friction by electrostatic interactions of double-level Teflon nanopillars. Soft Matter. 2013;**9**:1985-1996.

[47] M. Wright, A. Uddin. Organic—inorganic hybrid solar cells: A comparative review. Solar Energy Materials & Solar Cells. 2012;**107**:87-111.

[48] X. Shen, B. Sun, D. Liu, S.-T. Lee. Hybrid heterojunction solar cell based on organic inorganic silicon nanowire array architecture. Journal of the American Chemical Society. 2011;**133**:19408-19415.

3

Phosphor-LED-Based Wireless Visible Light Communication (VLC) and Its Applications

Chien-Hung Yeh and Chi-Wai Chow

Additional information is available at the end of the chapter

Abstract

In this chapter, we review our recent works on the phosphor white-light light-emitting diode (LED)-based wireless visible light communication (VLC) and its applications. This chapter is divided into two sections for introduction. In the first section, in order to enhance the transmission rate in phosphor-LED VLC system, we propose and demonstrate a novel multiband orthogonal-frequency-division-multiplexed (OFDM) modulation format for capacity enhancement. Based on the proposed scheme, various bands of OFDM signals are employed to different LED chips of the LED luminary; it can prevent the power fading and nonlinearity effects of transmission signal. Therefore, the maximum enhanced percentage of VLC data rate is 41.1%. In the second section, we also demonstrate a 71.3–148.4 Mbps phosphor-LED wireless VLC system at the free space transmission distance between 1.4 and 2.1 m. Finally, to understand and demonstrate the real-time LED VLC transmission, a commercial OFDM-based digital signal processor (DSP) is used in the LED transmitting side and client side, respectively. Therefore, the proposed real-time half-duplex VLC system can complete around 70 Mbps downstream and upstream traffic throughputs, in a free space transmission distance of 2 m long for practical in-home illumination and smart city applications.

Keywords: phosphor LED, visible light communication (VLC), Li-Fi, OFDM modulation, Internet of thing (IoT), smart city

1. Introduction

Recently, because of the benefits of low power consumption, high-efficiency, and long lifetime for light-emitting diodes (LEDs), the traditional incandescent and fluorescent lamps are inherited by the LEDs [1–5]. Furthermore, the LEDs have a broad modulation bandwidth comparing to these conventional lighting components [6–8]. Hence, utilizing the LED devices to combine illumination and wireless visible light communication (VLC) system has interested many advertences for the future lighting and network access markets [9–12]. Due to the LED-based VLC system without electromagnetic interference (EMI) effect, it also can be employed in wireless communication inhibited areas, such as hospital or aircraft [13–14]. Besides, compared with the red-green-blue (RGB) white-light LED, the white-light phosphor-LED is naturally used for the wireless VLC transmission [15, 16]. However, the long relaxation time of yellow phosphor nature will narrow the modulation bandwidth of white-light LED in a few MHz and reduce the VLC capacity for end-user [1, 17].

In order to increase and extend the modulation bandwidth of phosphor-LED, utilizing a blue optical filter in client side to remove the slow response produced by phosphor [18], employing spectral-efficient advanced discrete multitone (DMT) or orthogonal-frequency-division-multiplexed quadrature-amplitude-modulation (OFDM-QAM) modulations [19, 20], applying multi-input multi-output (MIMO) algorithm in VLC [21], and exploiting analogy equalization and filtering circuits in transmitter (Tx) and receiver (Rx) sides [22] have been proposed and demonstrated, respectively. In addition, utilizing advanced OFDM modulation signal for high data rate VLC transmission has also been demonstrated by Prof. Haas's group. They proposed using asymmetrically clipped optical (ACO) OFDM and DC-biased optical (DCO) OFDM for the indoor LED VLC network system to reducing LED clipping distortions [23, 24]. While the RBG-based white-light LED was employed in the VLC transmission, 110 Mbps OFDM traffic rate was also reached [25].

If the key function of the LED luminary is applied for lighting, it will confront the communication characteristics of optical wireless VLC [26]. Hence, how to use the commercially available white-light LED luminary to optimize both the illumination and communication is important and desirable issue for practical in-home illumination application [6, 13].

In this chapter, we will introduce our recent works on the white-light phosphor-LED-based wireless VLC system and its applications. This chapter will be divided into two sections for introduction. In the first section, to enhance the data rate in phosphor-LED VLC system, we propose and demonstrate an advanced multiband OFDM modulation design for VLC capacity enhancement. According to the proposed VLC scheme, various bands of OFDM signals are applied on different LED chips of the LED luminary; the special modulation design can prevent the power fading and nonlinearity effects of transmission signal by utilizing the same OFDM signal to the entire LED chips. As a result, the maximum enhanced percentage of VLC data rate is 41.1%, when we use the proposed novel new multiband OFDM modulation.

In the second section, we will demonstrate a 71.3–148.4 Mbps phosphor-LED wireless VLC system at the free space transmission distances between 1.4 and 2.1 m. Moreover, a commercial phosphor LED luminary with five phosphor-LED chips in series is applied for illumination and communication simultaneously. In the measurement, optical OFDM modulation signal with bit-loading algorithm is applied for proposed VLC transmission. We also investigate an optimal bias-tee circuit with equalization and filtering method to extend the modulation bandwidth from 1 to 27 MHz. Finally, to understand and demonstrate the real-time LED VLC transmission, a commercial OFDM-based digital signal processor (DSP) is used in the LED Tx side and client Rx side, respectively. Therefore, the proposed real-time half-duplex VLC system can complete the around 70 Mbps downstream and upstream traffic throughputs, in a free space transmission distance of 2 m long. We believe that it can be a cost-effective and promising candidate for practical in-home illumination and smart city applications.

2. Multiband OFDM in phosphor-LED VLC system

In the first section, we will present demonstrating a multiband OFDM modulation format in phosphor-LED VLC system for data rate enhancement [27]. First of all, **Figure 1(a)** and **(b)** presents the proposed optical setup of phosphor-LED VLC transmission by utilizing single-band and triple-band OFDM modulation formats. In this experiment, we use a phosphor-LED luminary with six LED light chips for serving as the VLC transmitter (Tx). Besides, the lenses can be applied in front of the phosphor-LED-based Tx and the receiver (Rx) of client side to focus and enhance the VLC light performance simultaneously. Here, the modulated OFDM signal can be directly applied on the LED luminary via a bias-tee (BT) without analogy equalization for VLC transmission. Furthermore, the proposed BT circuit without pre-equalization function has also been demonstrated in Ref. [10]. Conventionally, the entire OFDM signal would be applied on the six LED chip in Tx-side for VLC transmission, as seen in **Figure 1(a)**. However, due to the power fading and nonlinear effect of LED, the relative electrical power spectrum drops quickly in the higher frequency range. These influences will limit the traffic rate of VLC. To overcome this issue, we propose a multiband OFDM modulation format to increase the total VLC rate. As shown in **Figure 1(b)**, the real architecture of three-band OFDM VLC transmission is investigated. Hence, the total OFDM data rate in VLC system can be divided in three data ranges ($Data_1$, $Data_2$, and $Data_3$) within a total frequency bandwidth. Each modulation bandwidth ($Data_1$, $Data_2$, and $Data_3$) has the same range for data transmission. In the measurement, the phosphor-LED luminary has six lighting chips. The triple-band OFDM signals are employed on the LED luminary directly and divided into three groups of phosphor-LED chip via three BTs for direct OFDM modulation. Hence, each group has two phosphor-LED chips. And each LED chip is driven at 3 V.

Due to the power fading and nonlinear effects, the relative electrical power spectrum drops quickly in the higher frequency range, as shown in **Figure 2(a)**, when single-band OFDM modulation is used. It would also produce poor signal-to-noise ratio (SNR) of each OFDM subcarrier for VLC transmission. Hence, the VLC rate could also be restricted under the available frequency bandwidth. To solve the issues, using triple-band OFDM would be choice for phosphor-LED VLC system. The relative electrical power spectra of triple-band OFDM signals ($Band_1$, $Band_2$, and $Band_3$) are depicted conceptually in **Figure 2(b)**. If the OFDM signal has a high peak-to-average power ratio (PAPR) by separating the OFDM into triple-band OFDM, various OFDM bands are utilized to various groups of LED chips. Therefore, much higher driving power can be employed to each LED group for enhancement. The obtained SNR of each OFDM subcarrier would enhance in each OFDM-band obviously. As a result, the total VLC rate could be improved by using proposed triple-band OFDM modulation scheme.

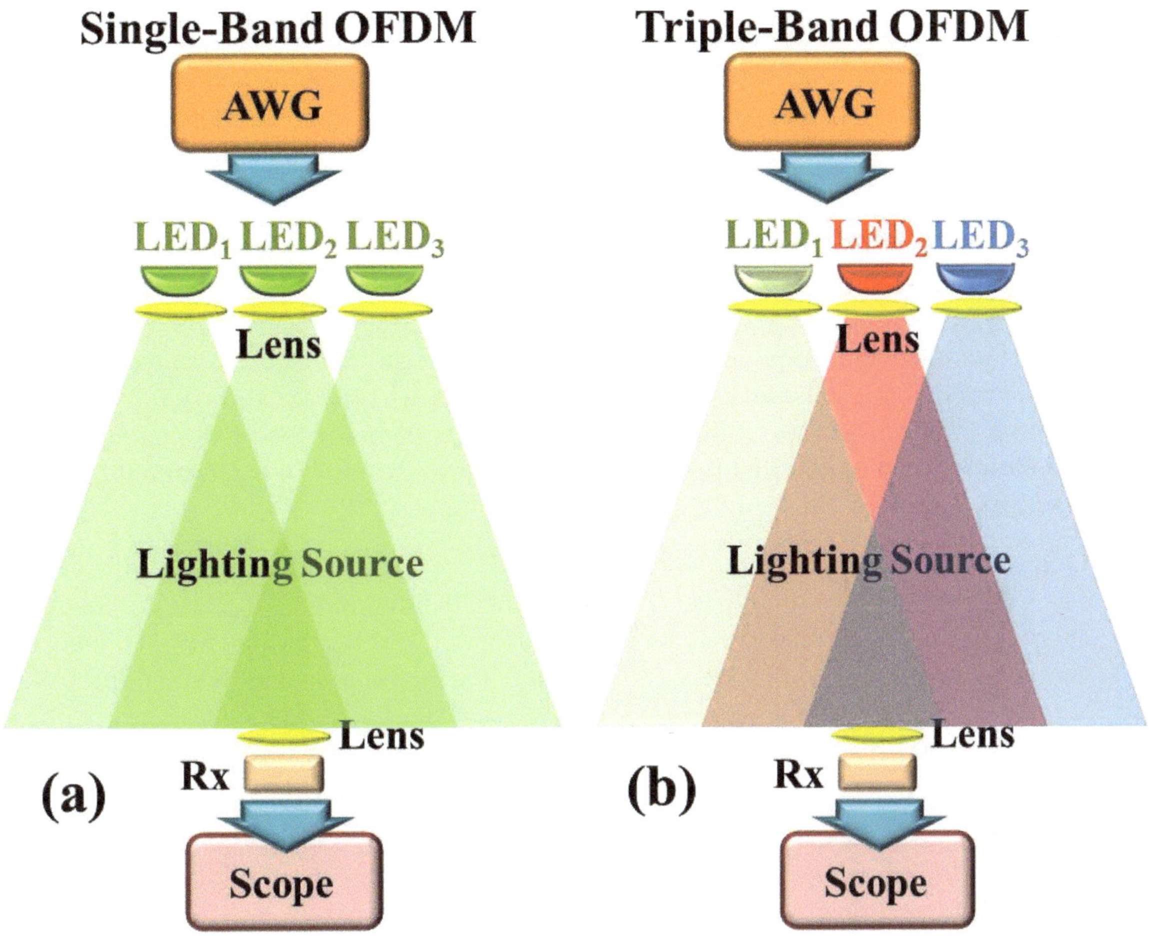

Figure 1. Proposed phosphor-LED VLC architectures using (a) single-band and (b) triple-band OFDM modulations.

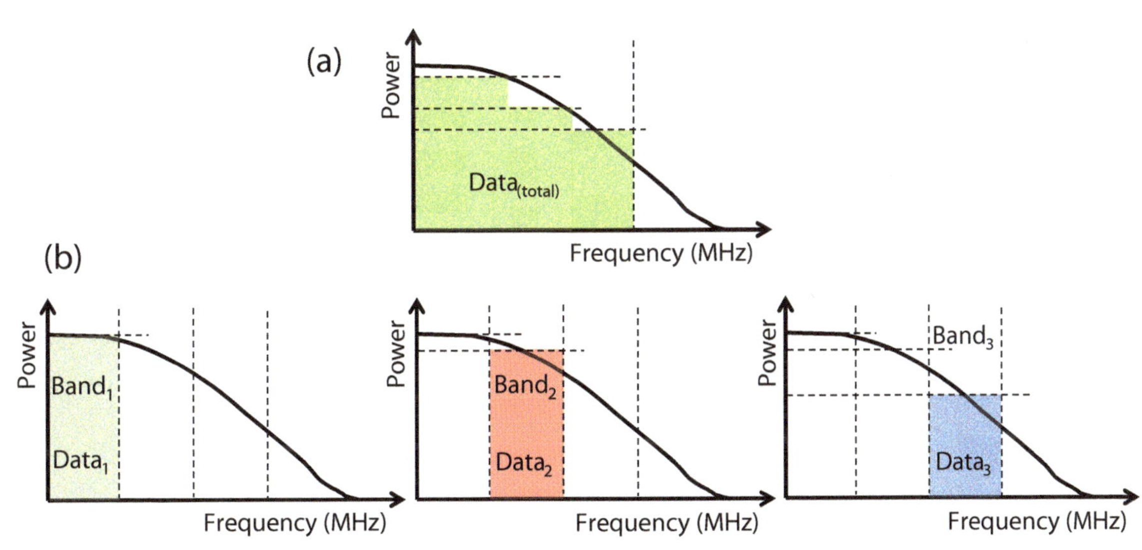

Figure 2. The relative electrical spectra under (a) single-band and (b) triple-band OFDM modulation format.

We utilize six white-light phosphor-LED chips with series connection in the experiment. Thus, the used LED luminary is driven around ~18 V. We utilize an arbitrary waveform generator (AWG) to produce OFDM-QAM signal applying on LED luminary for VLC transmission. As shown in **Figure 1(a)**, the white-light VLC emits from the LED luminary and is detected by a silicon-based PIN Rx. Here, the detection wavelength ranges and active area of detected PIN Rx are between 350 and 1100 nm with responsivity of 0.63 A/W 150 mm^2, respectively. Besides, the operated bandwidth and the root mean square (rms) noise of Rx are 50 MHz of 530 μV, respectively. Then, the VLC wireless signal will be directly detected by a PIN Rx. Next, the detected VLC signal is amplified by a broadband electrical amplifier, and is linked by a real-time oscilloscope for signal demodulation. Finally, the corresponding bit error rate (BER) will be calculated and obtained based on the observed SNR of each OFDM subcarrier. In addition, in order to prevent the power fading and nonlinearity problems of the LED, the multi-band OFDM modulation method is first proposed in VLC system to improve the traffic rate, as seen in **Figure 2(a)**.

Normally, the phosphor-LED only reaches a few MHz modulation bandwidth without any related improvements [1, 12]. To extend the frequency bandwidth of LED, the proposed BT circuit only needs to adjust the resistance (R), inductance (L), and capacitance (C) suitably for matching the impedance of LED. Hence, we measure the modulation bandwidth of proposed VLC system, while the BT circuit is utilized. **Figure 3** displays the frequency response of phosphor-LED, while the detected illumination at Rx side is at 1000 lx. The available modulation bandwidth of 25 MHz can be reached as shown in **Figure 3**. According to the proposed BT in LED Tx side, the frequency bandwidth can be widened from a few MHz to 25 MHz. As a result, the modulation capacity of VLC system should be increased via proposed multiband OFDM signal.

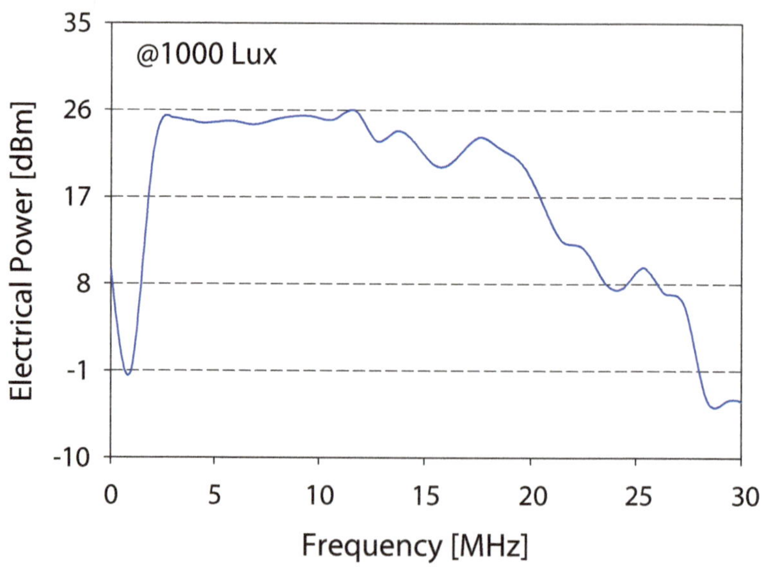

Figure 3. Measured frequency response spectrum of phosphor-LED, as the received illumination of Rx-side is 1000 lx.

Then, to understand the connection of VLC illumination and SNR, we utilize the 16-QAM OFDM modulation signal with 30 subcarriers in a frequency bandwidth of 1.95–30.27 MHz for VLC transmission. Based on the observed results of **Figure 3**, using bit-loading 16-QAM OFDM signal to apply on LED luminary for wireless VLC transmission is around the 30 MHz modulation bandwidth. In the measurement, **Figure 4(a)–(c)** presents the measured SNR of each OFDM subcarrier by using single-, dual-, and triple-band OFDM modulations, since the unlike illuminations of 200, 500, and 1000 lx are detected at the client Rx side, respectively. Here, ~20, 26, and 30 MHz available bandwidths can be employed by using multiband OFDM signal, respectively, while the SNR can be greater than 10 dB, under the different illuminations of 200, 500, and 1000 lx, as seen in **Figure 4(a)–(c)**. Compared with the single- and dual-band OFDM formats, the observed SNR characteristics of triple-band OFDM signal is superior in proposed VLC system. If the measured VLC illumination rises increasingly in the client Rx, the observed SNR would also increase under each OFDM modulation format. Moreover, when proposed triple-OFDM band modulation is applied on LED luminary, the measured SNR can improve by 4–21 dB comparing with the single-OFDM band, as depicted in **Figure 4(a)–(c)**. Therefore, when the proposed triple-band OFDM modulation in phosphor-LED VLC system, the all of received SNR also can be increased. In the experiment, the proposed multiband OFDM modulation can be actually enhanced the VLC traffic rate based on the measured results of **Figure 4**. In addition, the related VLC transmission lengths of 2.1, 2.0, and 1.4 m are also executed and measured, as the detected illuminations of 200, 500, and 1000 lx are observed in the Rx, respectively.

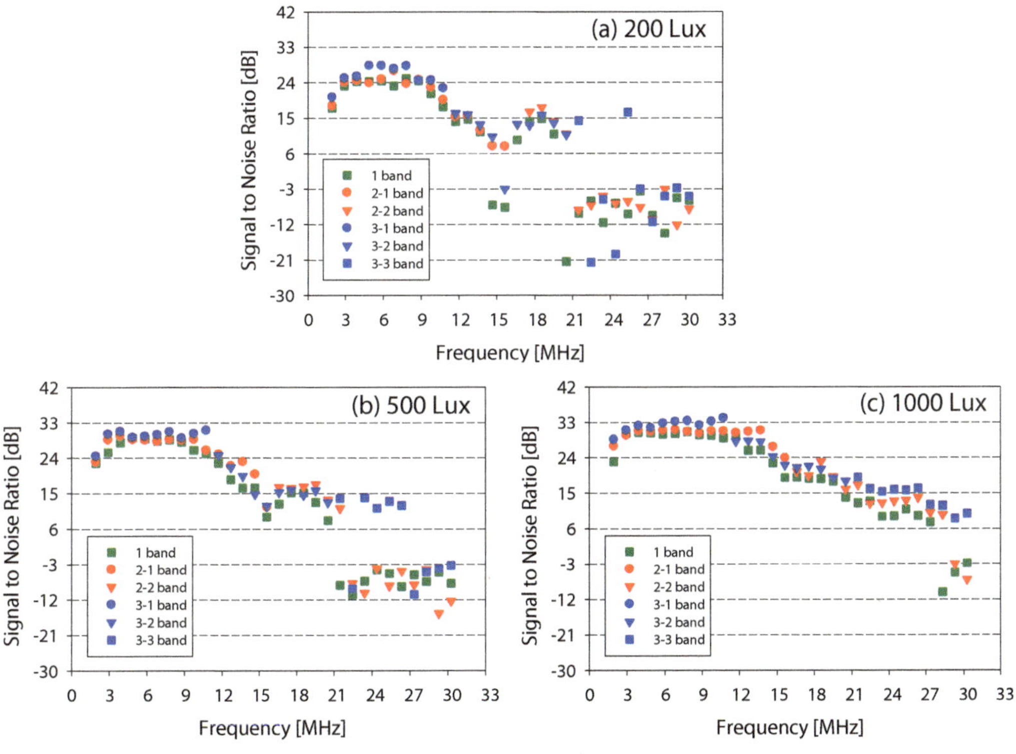

Figure 4. Experimental setup of the proposed fiber laser scheme in linear cavity scheme.

Next, we perform the evaluation of BER experimentally under different observed illuminations. **Figure 5(a)–(c)** shows the measured VLC traffic rate and homologous BER at the observed illuminations of 200, 500, and 1000 lx in the Rx, respectively. Here, the single-, dual- and triple-band OFDM signals are used at the available bandwidth of 1.95–30.27 MHz. As shown in **Figure 5**, if we use the single-band OFDM modulation, the obtained VLC data rates and NERs are 71.3, 109.4, and 148.8 Mbps, and 3.13×10^{-4}, 1.89×10^{-3} and 1.76×10^{-3}, respectively, under the various illuminations of 200, 500, and 1000 lx. Besides, **Figure 5(a)–(c)** also displays the measured VLC data rate and BER of each sub-band, while the proposed multiband OFDM formats are employed. When we use the dual-band OFDM signal, the VLC rates are 79.1 and 15.6 Mbps (total = 94.7 Mbps), 101.6 and 20.5 Mbps (total = 122.1 Mbps), and 113.3 and 46.9 Mbps (total = 160.2 Mbps), respectively, can be achieved under the different illuminations. The measured BER of each sub-band would be less than that of forward error correction (FEC) threshold (BER = 3.8×10^{-3}), and the average BERs are nearly 2.60×10^{-3}, 1.30×10^{-3} and 5.11×10^{-4}, respectively. Finally, while we utilize the proposed triple-band OFDM signal in VLC transmission, the VLC data rates of each band and the average BERs are 69.4, 27.3, and 3.9 Mbps, 76.2, 43.0, and 9.7 Mbps, and 77.1, 63.5, and 34.2 Mbps and 1.71×10^{-3}, 1.20×10^{-3}, and 1.50×10^{-3}, respectively, as shown in **Figure 5**.

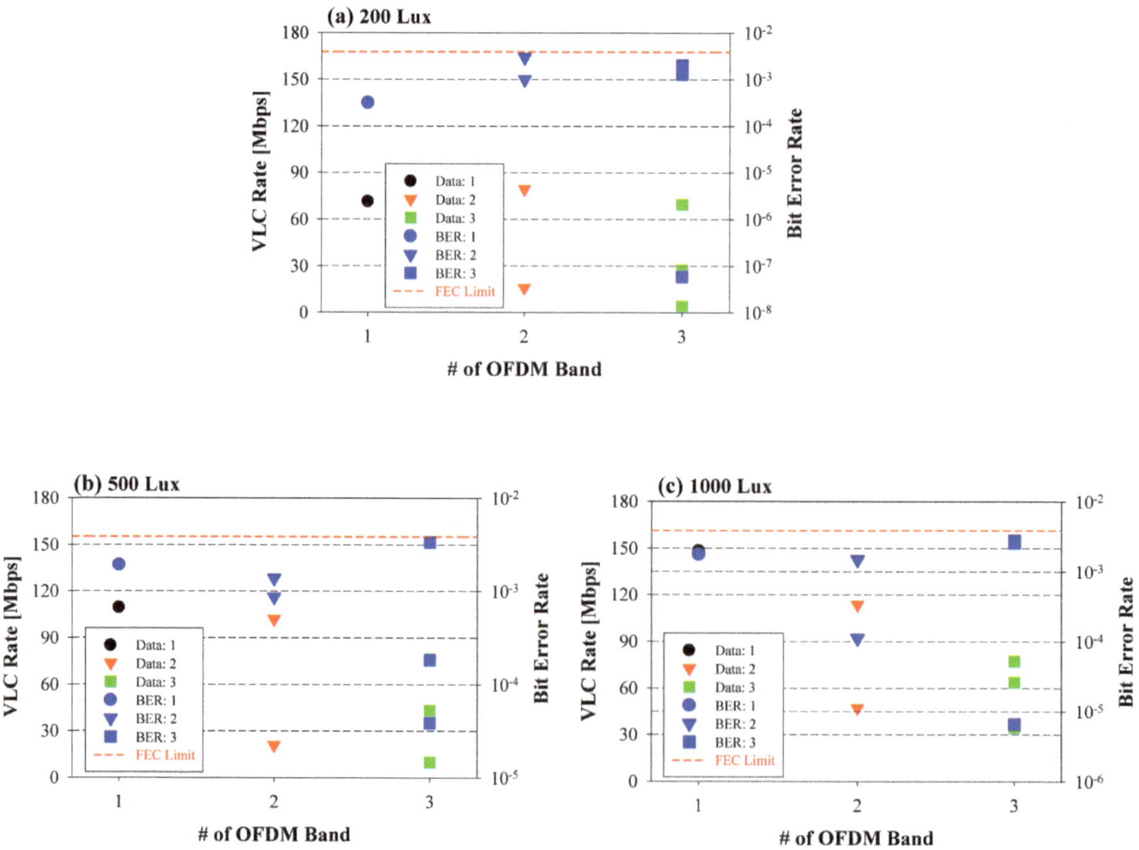

Figure 5. Measured VLC rate and corresponding BER at the detected illumination of (a) 200, (b) 500, and (c) 1000 lx in the Rx, respectively, while the single-, dual-, and triple-band OFDM modulations are applied.

Finally, we will analyze and discuss the VLC performance of obtained triple-band OFDM signal. **Figure 6** shows the detected total VLC rates and increase percentage of rate at the single-, dual-, and triple-band OFDM modulations, respectively, under the detected illuminations of 200, 500, and 1000 lx. In the experiment, when the proposed dual-band and triple-band OFDM signals are used, the VLC traffic data can reach 94.7, 122.1, and 160.2, and 100.6, 128.9, and 174.8 Mbps, respectively, as seen in **Figure 6**. Hence, the greatest increase percentages of VLC rates of 32.9%, 11.6%, and 7.9%, and 41.1%, 17.8%, and 17.8% are observed, while the various illuminations are 200, 500, and 1000 lx, respectively, and the designed dual-band and triple-band OFDM formats are applied on the VLC transmission. Furthermore, based on the results of **Figure 6**, as the number of OFDM-band increases slowly, the retrieved total VLC traffic rate will also enhance. However, if more sub-band OFDM signal is used in the proposed VLC system, the corresponding cost will be higher. Therefore, the proposed multiband OFDM method can improve and enhance the total VLC rate in phosphor-LED VLC wireless transmission.

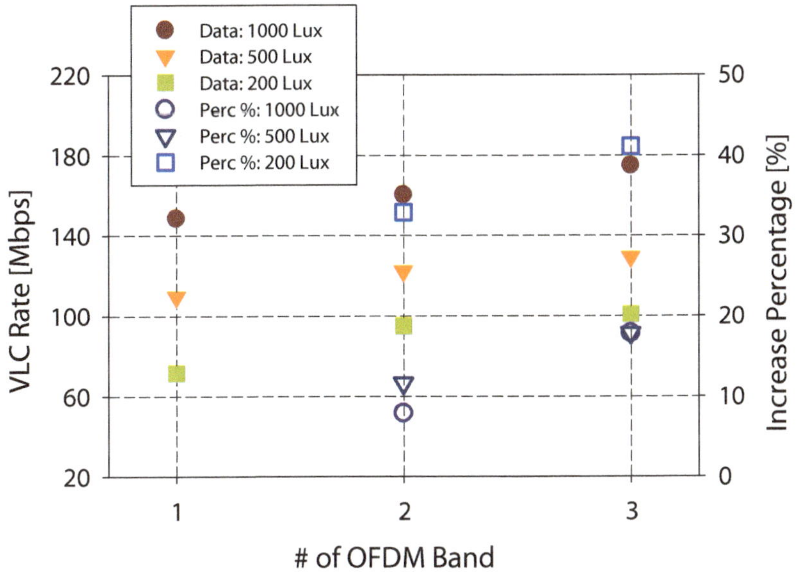

Figure 6. Observed total VLC rate and data increase percentage under the single-, dual-, and triple-band OFDM modulation at the observed illuminations.

In summary, the triple-band OFDM format in phosphor-LED VLC system to improve the VLC data rate have been proposed and demonstrated. In this proposed VLC architecture, various signal bands of OFDM formats are applied on unalike LED chips in the same LED luminary. The proposed VLC technology could avoid the power fading and nonlinearity issue by using the same OFDM signal to all the LED chips. In this measurement, we did not use a blue optical filter to enhance the VLC data rate of phosphor-LED. Furthermore, as we use single-OFDM-band modulation in the LED luminary, which has six white-light phosphor-LED chips, the VLC system could reach 148.4 and 71.3 Mbps traffic rates, respectively, when the detected illuminations were 1000 and 200 lx in the client end-user. If we used the proposed triple-band OFDM method at the same detected illuminations, the total VLC rates could be considerably improved to 174.8 and 100.6 Mbps, respectively. Therefore, employing proposed multiband

OFDM modulation could enhance the 41.1% and 17.8% VLC rates in VLC wireless transmission, while the observed illuminations were 200 and 1000 lx, respectively. In addition, the analysis and verification by experiments have been also executed in this section.

3. Rate improvement of VLC and real-time half-duplex VLC

In this section, we introduce and investigate the high-speed phosphor-LED VLC system together with real-time half-duplex transmission [13]. **Figure 7(a)** presents the optical setup of proposed phosphor-LED VLC architecture. In the measurement, a LED luminary with five phosphor-LED chips for acting as the Tx for VLC wireless transmission. **Figure 7(b)** indicates the related illuminations of LED, which is measured by using a light meter (LUTRON lx-1102) at the different transmission lengths in the field of view (FOV) in proposed VLC transmission system. As the measured illuminances are 200, 500, and 1000 lx, respectively, the corresponding free space transmission lengths of 210, 200, and 100 cm, are also obtained as shown in **Figure 7(b)**.

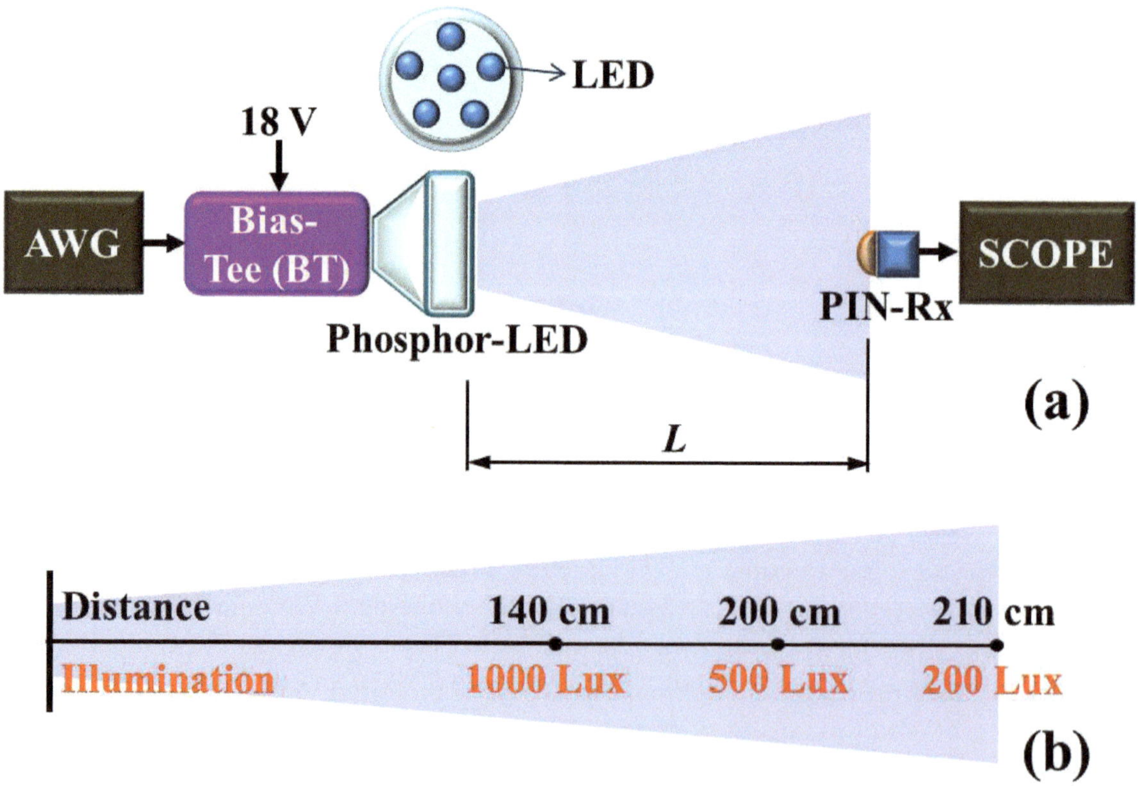

Figure 7. (a) Proposed phosphor-LED VLC wireless system. (b) Corresponding illuminations under different transmission lengths in the field of view of the proposed VLC.

In the VLC-Tx side, the LED luminary has five phosphor-LED chips for lighting and communication. Here, each LED chip is covered by a lens with divergent angle of 45°. Besides, an imaging lens in front of the client Rx is used for focusing and enhancing the detected VLC power. To reach a higher VLC data capacity in phosphor-LED VLC system, we can utilize OFDM-QAM modulation format with bit-loading, which is directly applied on the LED luminary by using proposed BT circuit, as shown in **Figure 8**. The resistance (R) of BT circuit is nearly 3 Ω . The proposed BT circuit without complicate pre-equalization design has been also demonstrated in Ref. [28]. Hence, we adapt the resistance (R), inductance (L), and capacitance (C) suitably for matching the impedance of five LED chips in the LED-Tx side. The AC modulated signal can be generated by the AWG and is connected to the "SMA" port at the lower left corner of **Figure 8**, whereas the LED is linked to the "LED_P" port at the upper right corner of **Figure 8**.

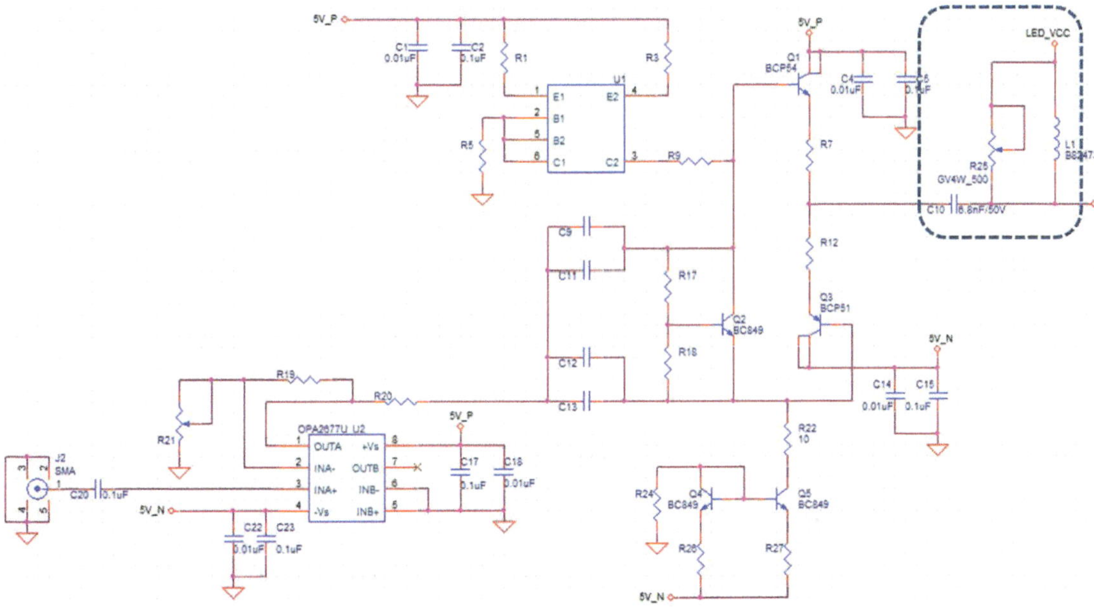

Figure 8. The proposed BT circuit topology.

In the experiment, we employ the five white-light phosphor-LED chips with series connection scheme for indoor illumination. Therefore, the OFDM-QAM signal can be applied on the LED luminary for direct modulation by employing an arbitrary waveform generator (AWG) for VLC wireless communication. In **Figure 7(a)**, the white-light VLC signal emits from the LED-Tx and detected by using the silicon-based PIN Rx. Then, the detected VLC signal will be amplified through a wideband RF amplifier (Picosecond 5867), and connected to a real-time oscilloscope for VLC signal demodulation.

In the measurement, the OFDM-QAM, which uses bit-loading algorithm, and the OFDM symbol can be encoded offline using the MATLAB program. The flow diagram for proposed OFDM-QAM algorithm is shown in **Figure 9**. The serial binary stream can be converted into several parallel streams, and each parallel binary flow is plotted into 16-QAM symbols. The inverse fast Fourier transform (IFFT) process with 256 IFFT size is executed on the QAM symbols to create the digital OFDM symbols. Cyclic prefix (CP) of 1/32 will be added in each OFDM symbol to reduce the dispersion-induced characteristic deprivation. The encoded digital OFDM symbol can be delivered into an AWG and altered into analogue electrical signal via the digital-to-analogue converter (DAC). Here, we use the 250 Msample/s sampling rate and 8-bit DAC resolution from the AWG for measurement. Then, in the Rx-side, the VLC wireless signal can be directly detected via a PIN-PD, which is linked by a real-time oscilloscope with sampling rate of 250 Msample/s for VLC signal decode. Finally, the VLC OFDM signal can be demodulated offline by retreat process of the encoder.

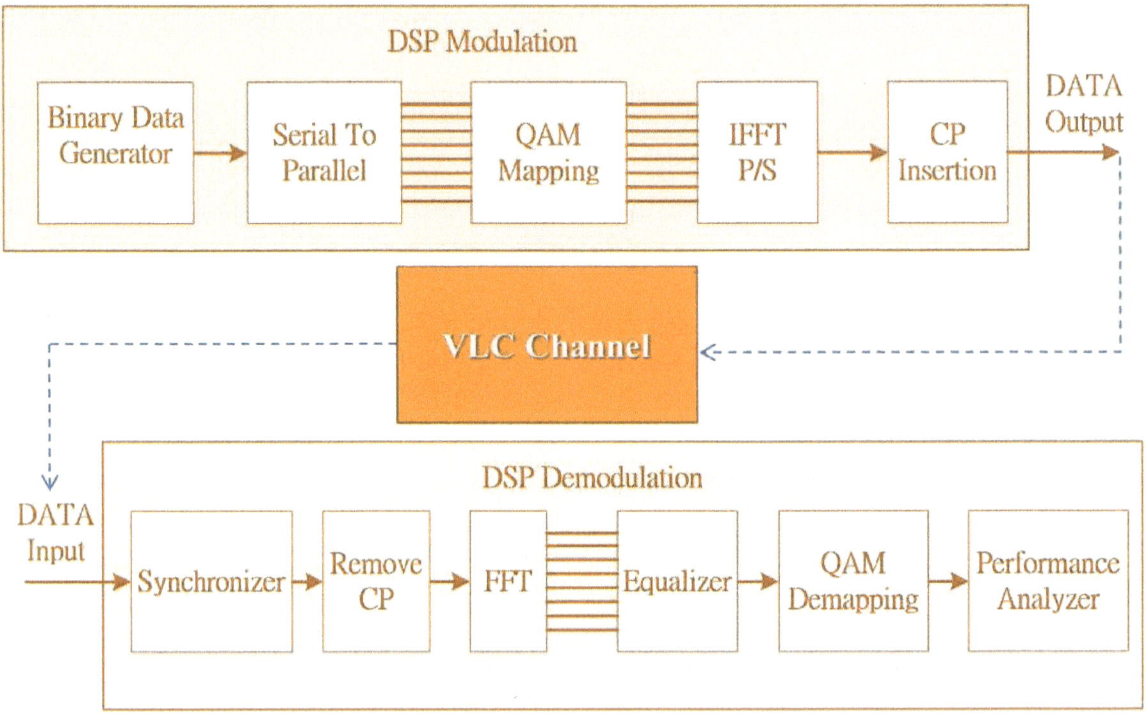

Figure 9. The flow chart of OFDM-QAM algorithm.

Figure 10 presents the observed electrical power spectra of phosphor-LED at the various detected illuminances of 200, 300, 400, 500, and 1000 lx, respectively. And the measured response frequency regions are around 20, 21, 21, 21, and 27 MHz, respectively, as depicted in **Figure 10**. When we decrease the illuminance of client side gradually, the relative powers will also reduce, as shown in **Figure 10**. Furthermore, the available electrical power range will fall rapidly in a higher frequency region, as seen in **Figure 10**. As we know, the smaller electrical power would lead to the poor SNR of each OFDM subcarrier. Thus, the corresponding VLC traffic rate would be restrained in the frequency bandwidth. In the measurement, the SNR is observed from the measured constellation of Rx-side. Moreover, in digital modulation applications, the SNR can be stated usually by measuring the modulation-error ratio from constellation information.

Then, **Figure 11(a)** displays the measured SNR of each OFDM subcarriers using 16-QAM modulation format in the different illuminances of 200, 300, 400, 500, and 1000 lx, respectively. Moreover, we utilize 30 OFDM subcarriers in the modulation range from 1.95 to 30.27

MHz for creating VLC OFDM signal. To completely employ the unflatten SNR, the modulated signal of each OFDM subcarrier can be adapted with bit-loading algorithm based on the obtained SNR value. While we use the bit-loading with QAM order from 4 to 256, the measured SNR spectra can be obtained, as shown in **Figure 11(b)**. Hence, each OFDM subcarriers with their corresponding SNRs are among 9.4 and 25.1, 15.6 and 25.7, 8.3 and 29.1, 8.1 and 29.1, and 7.5 and 30.5 dB in the frequency regions of 1.95–19.53, 1.95–20.50, 1.95–20.50, 1.95– 20.50, and 1.95–26.36 MHz, when the obtained illuminances are 200, 300, 300, 400, 500, and 1000 lx, respectively. Here, the measured corresponding constellations of 4, 16, and 256 QAM under the results of **Figure 11(b)** at 500 lx state are illustrated in **Figure 12**. Therefore, the available modulation range can achieve nearly 27 MHz by using the proposed LED-Tx unit. In addition, the frequency bandwidth can be defined as the available bandwidth. The obtained SNR can satisfy the FEC (i.e., BER = 3.8×10^{-3}) requirement when the bit-loading OFDM is utilized. The useful bit range of the algorithm is adjusted between 4- and 128-QAM adaptively, thus the available frequency range can support at least 4-QAM data traffic (i.e., SNR = ~8 dB).

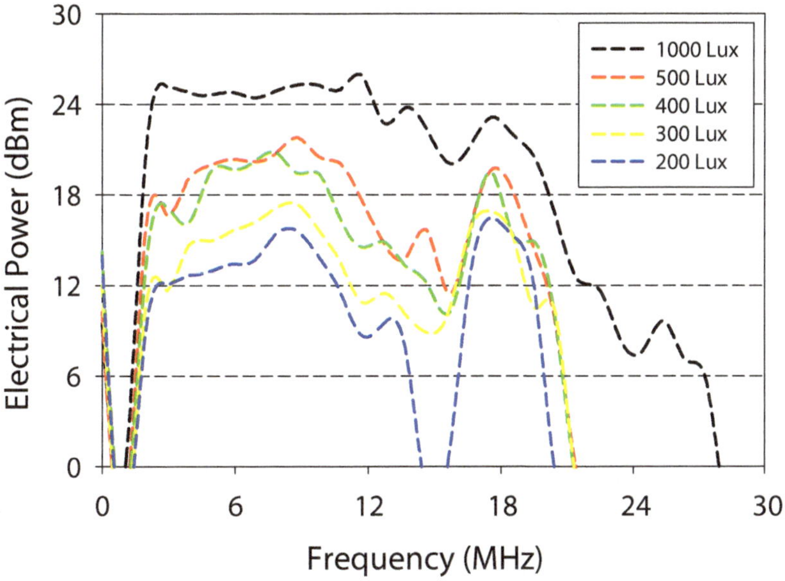

Figure 10. Observed electrical power spectra at the various detected illuminances of 200, 300, 400, 500, and 1000 lx, respectively.

Furthermore, the measured VLC data rates are 71.3, 89.8, 102.5, 109.4, and 148.4 Mbps, respectively, under the different illuminances of 200, 300, 400, 500, and 1000 lx, as also shown in **Figure 13**. In this measurement, all of the measured corresponding BERs are below the FEC threshold under the various detected illuminances, as illustrated in **Figure 13**. In addition, the FEC threshold was also determined as Ref. [29]. EFC would indicate 7% overhead which is used for the coding to measure error-free in the proposed VLC wireless system.

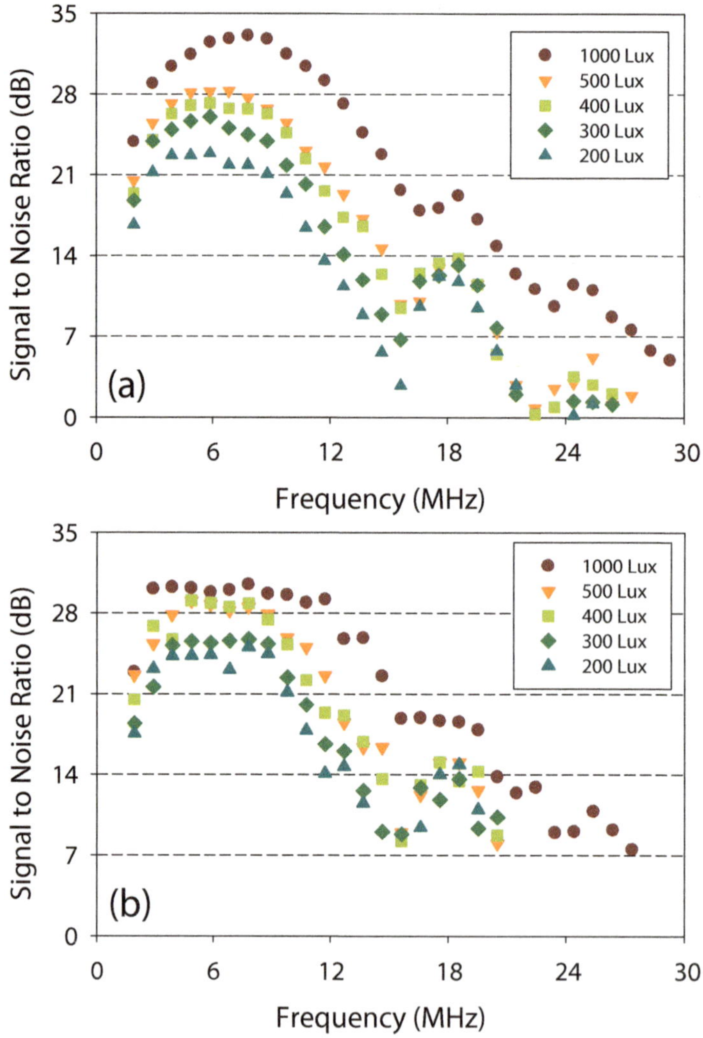

Figure 11. Measured SNR spectra of each OFDM subcarrier (a) without and (b) with bit-loading at the detected illuminances of 200, 300, 400, 500, and 1000 lx, respectively.

Then, a real-time bidirectional phosphor-LED VLC transmission is also proposed and demonstrated. Here, the commercial OFDM-based DSP chip is applied in the Tx and Rx simultaneously sides to replace the AWG and oscilloscope for encoding and decoding VLC channel. However, the commercial DSP chip only has half-duplex transmission function for VLC. In the real-time DSP VLC transmission, we use a commercial DSP chip and Ethernet port for VLC signal processing and connecting, respectively. The modulation bandwidth of DSP chip is from 2 to 100 MHz with adaptively OFDM-QAM formats from QPSK to 16-QAM. The maximum data rate could reach 250 Mbps via the OFDM DSP chip. In the measurement, five phosphor-LED chips in the LED-Tx side carry the same modulated data for VLC transmission. The VLC channel was analyzed and discussed fully in Ref. [2], while the modulation rate of the demonstration was low, the crosstalk influence could be ignored. If some LEDs are utilized in the DC mode or AC mode for VLC transmission, they could generate the background noises affecting the VLC performance within the frequency of 2 MHz, as investigated in Ref. [30]. Hence, the OFDM DSP chip can be used in VLC system to avoid the low-frequency noise interference.

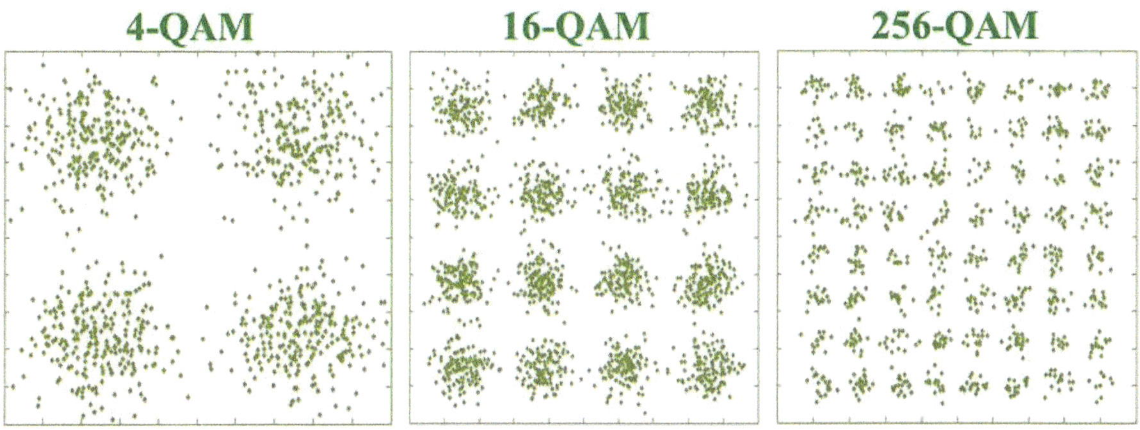

Figure 12. The obtained corresponding constellations at 500 lx of the result in **Figure 11(b)**.

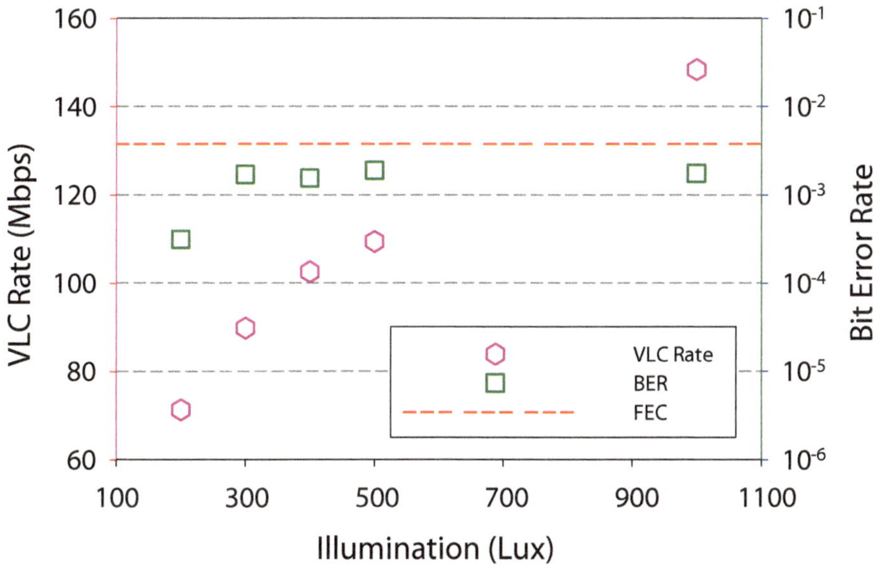

Figure 13. Measured VLC traffic rates and related BERs under different detected illuminances.

Figure 14 presents the new proposed setup of real-time bidirectional phosphor-LED VLC transmission system. In the LED luminary, five phosphor-LED chips with series connection are driven at 15 V for illumination and VLC transmission. In the experiment, the LED Tx-side and client side both contains the analog front-end (AFE) and DSP parts for signal encoding and decoding, as shown in **Figure 14**. Here, we add an infrared (IR) PIN-Rx in the LED luminary, as shown in **Figure 14**, to detect the upstream IR wireless signal emitted from the client end-user. An 850 nm IR-LED with 40 MHz modulation bandwidth is used in client side to act as the upstream emitter for upstream traffic signal. The emission angle of IR-LED is set at 30°. Besides, a PIN-Rx, having 50 MHz bandwidth in the client side, is used to receive the downstream VLC traffic. Here, we also add the lenses in front of each Tx and Rx for maximizing the performance of VLC signal, as also shown in **Figure 14**. In this work, an 850 nm IR-LED is employed serving as an upstream channel for real-time bidirectional VLC transmission.

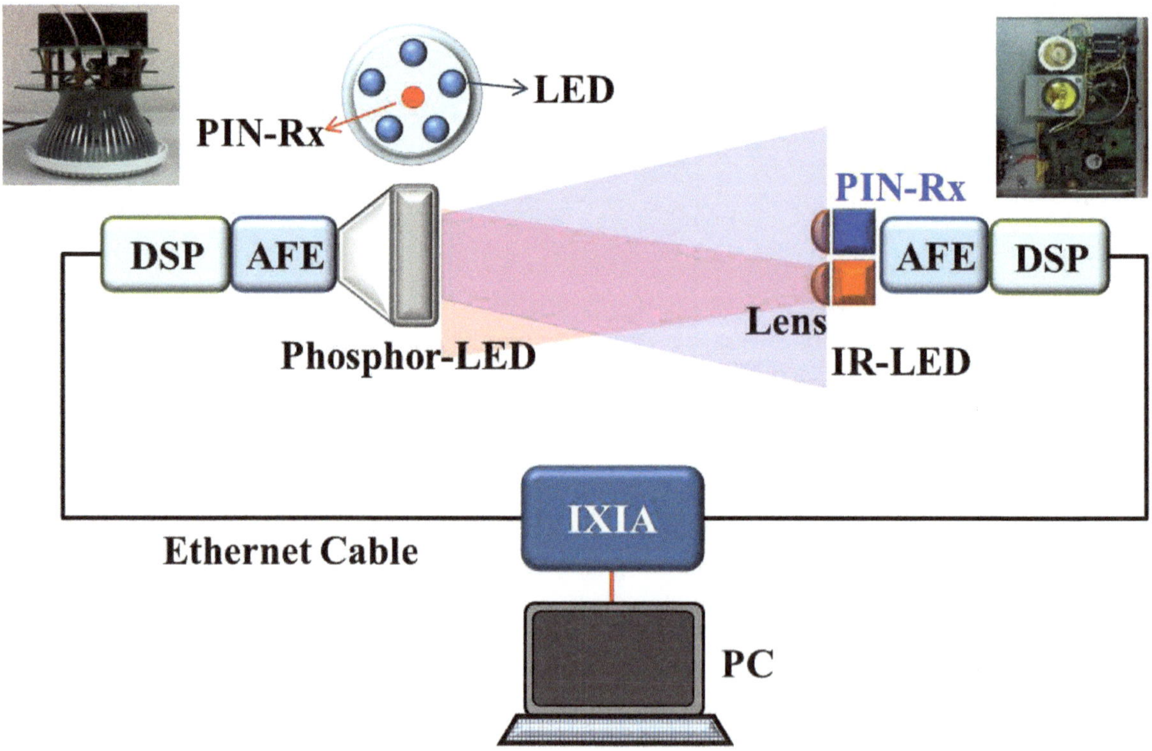

Figure 14. New proposed real-time bidirectional phosphor-LED VLC transmission system.

Furthermore, to understand the connection of the lighting illuminance and the horizontal location of the client side, the VLC vertical transmission distance is set at 200 cm in the central position, as shown in **Figure 15**. And the proposed user client module can be changed at various horizontal positions for detecting VLC signal, as shown in **Figure 15**. In the proposed VLC system, if the horizontal positions are 0, 25, 40, and 60 cm, as depicted in **Figure 15**, respectively, the corresponding illuminances are measured at 500, 400, 300, and 200 lx. In addition, to execute and evaluate the performance of proposed real-time VLC transmission, a network analyzer (IXIA 1600T) is used to test the IP-based throughput for real-time data traffic. Here, a 70 Mbps testing rate of network analyzer can be utilized in the half-duplex VLC transmission for performing VLC performance.

The downstream throughput performance can be performed in relation to the VLC setup of **Figure 15**. Here, the illuminance is 500 lx, when the vertical transmission length is 200 cm.

Figure 16 displays the downstream throughputs under the different data frame sizes of 64, 128, 256, 512, 1024, and 1518, respectively, while the client module is placed to 60 cm away from center location. Hence, we can understand the corresponding illuminance for receiving and decoding the VLC signal under different location. Thus, the illuminances of 500, 400, 300, and 200 lx are measured, when the related horizontal locations are 0, 25, 40, and 60 cm, respectively. Moreover, while the data frame size rises increasingly, the measured traffic throughputs also enhance, as shown in **Figure 15**. The measured throughputs are 15.7, 54.2, 60.2, and 70.0 Mbps under the illuminances of 200, 300, 400, and 500 lx at the 1518 frame size. If the frame size is set at 64, the measured throughputs will be 14.4, 49.5, 55.6, and 63.2 Mbps, respectively.

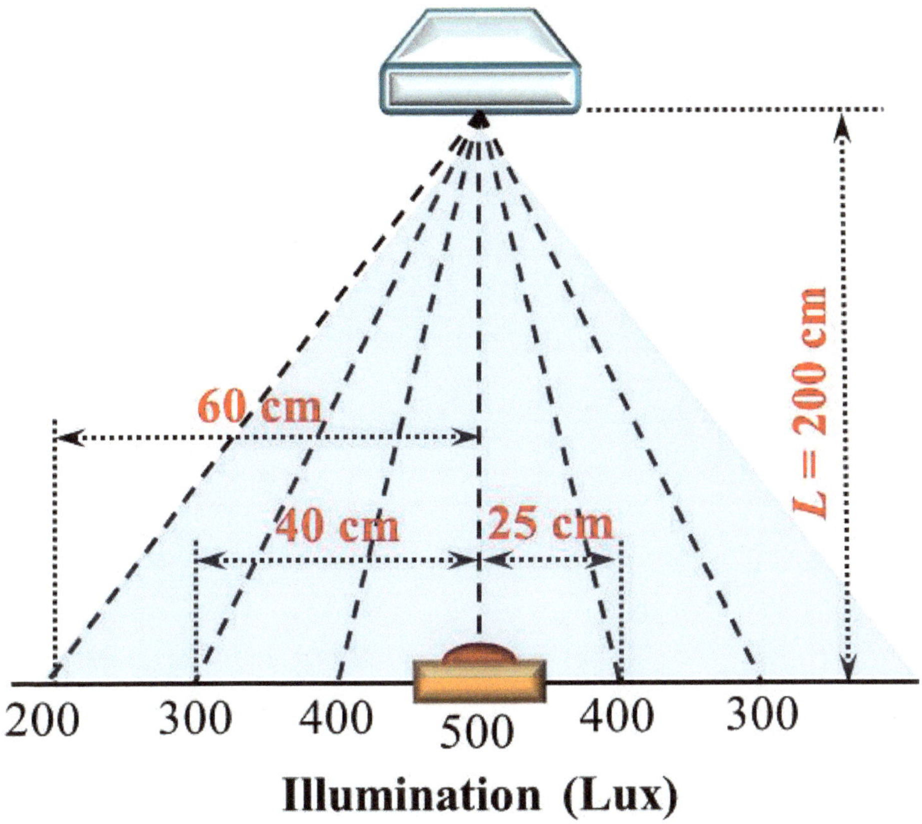

Figure 15. Observed illuminances in the client side at various horizontal positions under a 200 cm vertical transmission distance.

Moreover, to comprehend the practical VLC performance, one to four client modules are used in the same setup of **Figure 15**. Here, the illuminance is set at 500 lx. **Figure 17** present the downstream throughputs under different data frames of 64, 128, 256, 512, 1024, and 1518, respectively, while one to four client modules are employed at the tested LED luminary for measuring VLC traffic. As seen in **Figure 17**, while the frame data size is raised gradually, the retrieved data throughputs also improve. When one client module is utilized in the VLC system, the observed VLC efficiency is the same as **Figure 16**. As the client number increases, the relative throughput would be decreased, as shown in **Figure 16**. While the frame size is 1518, the observed throughput is 70.0, 31.5, 21.7, and 14.8 Mbps, respectively, if one to four client modules are also employed.

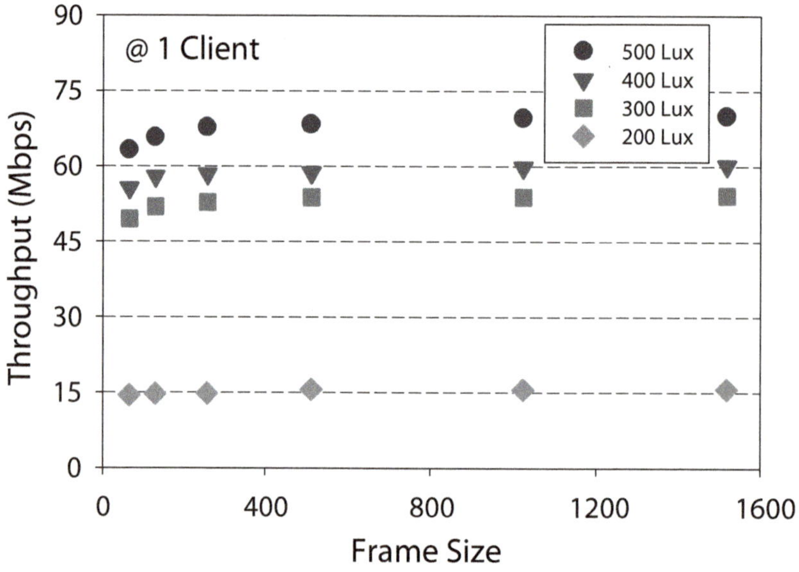

Figure 16. Measured downstream throughputs under the various data frame sizes, when a client module is move outward from central point to 60 cm (200 lx).

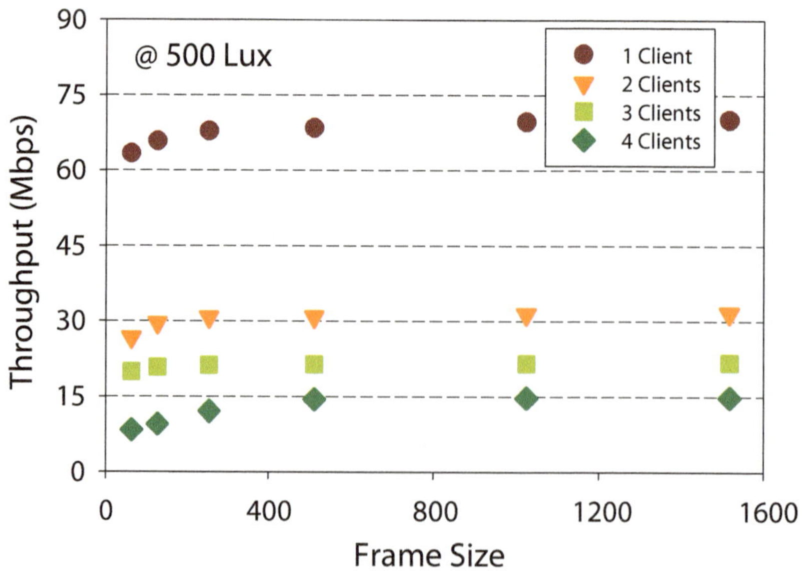

Figure 17. Measured downstream throughput spectra under various data frame sizes of 64, 128, 256, 512, 1024, and 1518, respectively, when one, two, three, and four client modules are used.

Finally, we also execute the 850 nm IR-LED upstream traffic in a 200 cm vertical transmission distance. In this experiment, the real setup is the same as in **Figure 15**. The client module is moved away from the center position to 25, 40, and 60 cm, respectively, for upstream throughput measurement. **Figure 18** shows the maximum upstream throughput with random frame size from 1.7 to 70 Mbps under different illuminances of 200–1000 lx. Therefore, when the commercial half-duplex DSP chip is applied in LED VLC system, the VLC traffic can reach 70 Mbps for real-time bidirectional transmission.

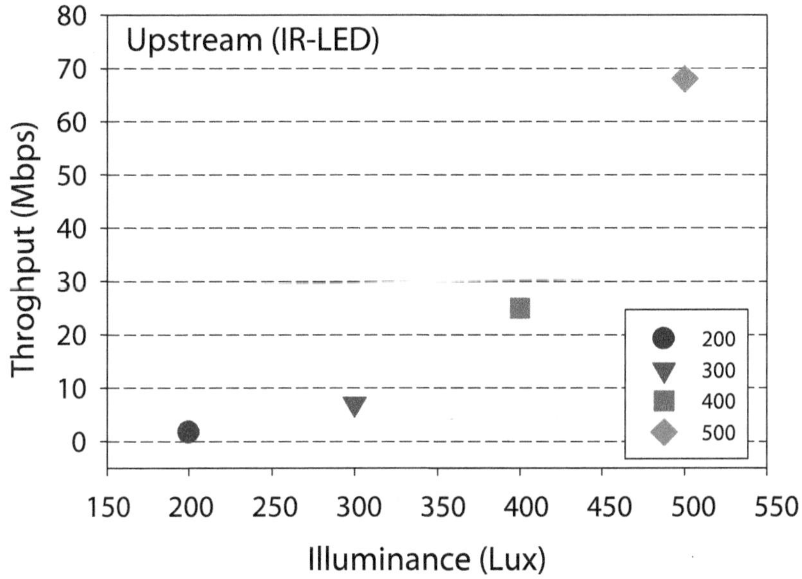

Figure 18. Observed upstream data rate under different detected illuminances in the proposed real-time VLC system.

In summary, we proposed and investigated a 71.3–148.4 Mbps white phosphor-LED VLC system in the free space transmission distance of 140–210 cm in the different illuminances of 200–1000 lx, respectively. Here, a white-light LED luminary, having five cascaded LED chips, was utilized for indoor illumination and wireless access simultaneously. Here, we used the bit-loading OFDM-QAM modulation signal and a novel bias-tee circuit in AFE to improve the modulation bandwidth from 1 to 27 MHz. In addition, we did not utilize a blue optical filter in the client side for improvement. To understand and demonstrate the real-time wireless VLC performance in the proposed phosphor-LED VLC system, a commercial OFDM-based DSP chip was used in the LED lighting side and client sides, simultaneously. Here, the OFDM-DSP chip would result in half-duplex operation. In this investigation, we used a network analyzer to measure the throughput of practical VLC data in the proposed VLC system at different illuminances. As a result, the proposed real-time bidirectional VLC system could accomplish the 70 Mbps data throughputs in a practical transmission distance of 200 cm at the luminance of 500 lx.

4. Conclusion

In conclusion, we reviewed our recent works on the phosphor white-light LED-based wireless VLC and its practical applications. This chapter was divided into two sections. In the first section, to enhance the transmission rate in phosphor-LED VLC system, we proposed and demonstrated a novel multi-band OFDM modulation format for capacity enhancement. According to the proposed architecture, different bands of OFDM signals were used to different LED chips of the LED luminary; it can avoid the power fading and nonlinearity effects of transmission signal by using the same OFDM signal to the whole LED chips. Hence, the maximum enhanced percentage of VLC data rate was 41.1% when the proposed new multi-band OFDM modulation was used.

In the second section, we also demonstrated a 71.3–148.4 Mbps phosphor-LED wireless VLC system at the free space transmission distance between 140 and 210 cm. Moreover, a commercial white-light phosphor-LED luminary with five phosphor-LED chips in series was used for illumination and communication, simultaneously. Here, optical OFDM signal with bit-loading algorithm was applied for VLC transmission. We also investigate an optimal bias-tee circuit with equalization to improve the modulation bandwidth from 1 to 27 MHz. Finally, to understand and demonstrate the real-time LED VLC transmission, a commercial OFDM-based DSP chip was used in the LED transmitting side and client side, simultaneously. Therefore, the proposed real-time half-duplex VLC system could complete around 70 Mbps downstream and upstream traffic throughputs, in a free space transmission distance of 200 cm long for practical in-home illumination and smart city applications.

Acknowledgements

This work was supported by Ministry of Science and Technology, Taiwan, under grants MOST-103-2218-E-035-011-MY3, MOST-104-2628-E-009-011-MY3, and MOST-103-2221-E-009-030-MY3.

Author details

Chien-Hung Yeh[1]* and Chi-Wai Chow[2]

*Address all correspondence to: yehch@fcu.edu.tw

1 Department of Photonics, Feng Cha University, Taichung, Taiwan

2 Department of Photonics, National Chiao Tung University, Hsinchu, Taiwan

References

[1] C. W. Chow, C. H. Yeh, Y. F. Liu, and Y. Liu, "Improved modulation speed of the LED visible light communication system integrated to the main electricity network," Electron. Lett., vol. 47, pp. 867-868, 2011.

[2] T. Komine and M. Nakagawa, "Fundamental analysis for visible-light communication system using LED lights," IEEE Trans. Consum. Electron., vol. 50, pp. 100-107, 2004.

[3] S. Cha, D. Park, Y. Lee, C. Lee, J. Choi, J. Lee, and H. Lee, "AC/DC converter free LED driver for lightings," in Proc. ICCE, pp. 706-708, 2012.

[4] H. H. Tsai, H. C. Chen, C. H. Wang, K. J. Chen, C. W. Hung, C. C. Yeh, C. C. Lin, H. C. Kuo, and T. C. Lu, "Improving the lumen efficiency by air-gap embedded package in white light-emitting diodes," in Proc. CLEO-PR, pp. 138-140, 2011.

[5] N. T. Tran, J. P. You, and F. G. Shi, "Effect of phosphor particle size on luminous efficacy of phosphor-converted white LED," J. Lightwave Technol., vol. 27, pp. 5145-5150, 2009.

[6] Z. Wang, C. Yu, W.-D. Zhong, J. Chen, and W. Chen, "Performance of a novel LED lamp arrangement to reduce SNR fluctuation for multi-user visible light communication systems," Opt. Express, vol. 20, pp. 4564-4573, 2012.

[7] C. W. Chow, Y. Liu, C. H. Yeh, J. Y. Sung, and Y. L. Liu, "A practical in-home illumination consideration to reduce data rate fluctuation in visible light communication", IEEE Wirel. Commun., vol. 22, pp. 17-23, 2015.

[8] W. O. Popoola, "Impact of VLC on light emission quality of white LEDs," J. Lightwave Technol., vol. 34, pp. 2526-2532, 2016.

[9] J. Vucic, C. Kottke, S. Nerreter, K. Habel, A. Buttner, K. D. Langer, and J. W. Walewski, "230 Mbit/s via a wireless visible-light link based on OOK modulation of phosphorescent white LEDs," in Proc. OFC, pp. 1-3, 2010.

[10] C.-H. Yeh, Y.-L. Liu, and C.-W. Chow, "Real-time white-light phosphor-LED visible light communication (VLC) with compact size", Opt. Express, vol. 21, pp. 26192-26197, 2013.

[11] C.-W. Hsu, J.-T. Wu, H.-Y. Wang, C.-W Chow, C.-H. Lee, M.-T. Chu, and C.-H. Yeh, "Visible light positioning and lighting based on identity positioning and RF carrier allocation technique using a solar cell receiver", IEEE Photon. J., vol. 8, #7905507, 2016.

[12] C. H. Yeh, Y. F. Liu, C. W. Chow, Y. Liu, P. Y. Huang, and H. K. Tsang, "Investigation of 4-ASK modulation with digital filtering to increase 20 times of direct modulation speed of white-light LED visible light communication system," Opt. Express, vol. 20, pp. 16218-16223, 2012.

[13] C.-H. Yeh, C.-W. Chow, H.-Y. Chen, Y.-L. Liu, and D.-Z. Hsu, "Investigation of phosphor-LED lamp for real-time half-duplex wireless VLC system", J. Opt., vol. 18, p. 065701, 2016.

[14] Y. F. Liu, C. H. Yeh, C. W. Chow, Y. Liu, Y. L. Liu, and H. K. Tsang, "Demonstration of bi-directional LED visible light communication using TDD traffic with mitigation of reflection interference," Opt. Express, vol. 20, pp. 23019-23024, 2012.

[15] G. Cossu, A. M. Khalid, P. Choudhury, R. Corsini, and E. Ciaramella, "3.4 Gbit/s visible optical wireless transmission based on RGB LED," Opt. Express, vol. 20, pp. B501–B506, 2012.

[16] H. Le-Minh, D. C. O'Brien, G. Faulkner, L. Zeng, K. Lee, D. Jung, and Y. Oh, "High-speed visible light communications using multiple-resonant equalization," IEEE Photon. Technol. Lett., vol. 20, pp. 1243-1245, 2008.

[17] C.-H. Yeh, C.-W. Chow, Y.-L. Liu, H.-Y. Chen, Y. Liu, and J.-C. Hsu, "Investigation of no analogue-equalization and blue filter for 185 Mbps phosphor-LED wireless communication", Opt. Quantum Electron., vol. 47, pp. 1991-1997, 2015.

[18] J.-Y. Sung, C.-W. Chow, and C.-H. Yeh, "Is blue optical filter necessary in high speed phosphor-based white light LED visible light communications?", Opt. Express, vol. 22, pp. 20646-20651, 2014.

[19] E. Costa and S. Pupolin, "M-QAM-OFDM system performance in the presence of a nonlinear amplifier and phase noise," IEEE Trans. Commun., vol. 50, pp. 462-472, 2002.

[20] J.-Y. Sung, C.-W. Chow, and C.-H. Yeh, "Dimming-discrete-multi-tone (DMT) for simultaneous color control and high speed visible light communication," Opt. Express, vol. 22, pp. 7538-7543, 2014.

[21] H.-H. Lu, Y.-P. Lin, P.-Y. Wu, C.-Y. Chen, M.-C. Chen, and T.-W. Jhang, "A multiple-input-multiple-output visible light communication system based on VCSELs and spatial light modulators," Opt. Express, vol. 22, pp. 3468-3474, 2014.

[22] H. Le Minh, D. O'Brien, G. Faulkner, L. Zeng, K. Lee, D. Jung, Y. Oh, and E. T. Won, "100-Mb/s NRZ visible light communications using a postequalized white LED," IEEE Photon. Technol. Lett., vol. 21, pp. 1063-1065, 2009.

[23] H. Elgala, R. Mesleh, H. Haas, and B. Pricope, "OFDM visible light wireless communication based on white LEDs," in Proc. IEEE VCT, pp. 2185-2189, 2007.

[24] R. Mesleh, H. Elgala, and H. Haas, "Performance analysis of indoor OFDM optical wireless communication systems," in Proc. IEEE WCNC, pp. 1005-1010, 2012.

[25] H. Chun, S. Rajbhandari, G. Faulkner, D. Tsonev, H. Haas, and D. O'Brien, "Demonstration of a di-directional visible light communication with an overall sum-rate of 110 Mb/s using LEDs as emitter and detector," in Proc. IEEE IPC, pp. 132-133, 2014.

[26] A. Tsiatmas, C. P. M. J. Baggen, F. M. J. Willems, J. –P. M. G. Linnartz, and J. W. M. Bergmans, "An illumination perspective on visible light communications," IEEE Comm. Mag., vol. 52, no. 7, pp. 64-71, 2014.

[27] C.-H. Yeh, H.-Y. Chen, C.-W. Chow, and Y.-L. Liu, "Utilization of multi-band OFDM modulation to increase traffic rate of phosphor-LED wireless VLC", Opt. Express, vol. 23, pp. 1133-1138, 2015.

[28] C.-H. Yeh, C.-W. Chow, H.-Y. Chen, J. Chen, and Y.-L. Liu, "Adaptive 84.44-190 Mbit/s phosphor-LED wireless communication utilizing no blue filter at practical transmission distance", Opt. Express, vol. 22, pp. 9783-9788, 2014.

[29] Y. Wang, X. Huang, J. Zhang, Y. Wang, and N. Chi, "Enhanced performance of visible light communication employing 512-QAM N-SC-FDE and DD-LMS," Opt. Express, vol. 22, pp. 15328-15334, 2014.

[30] C. W. Chow, C. H. Yeh, Y. F. Liu, and P. Y. Huang, "Background optical noises circumvention in LED optical wireless systems using OFDM," IEEE Photon. J., vol. 5, no. 2, #7900709, 2013.

4

Modeling of Silicon Photonic Devices for Optical Interconnect Transceiver Circuit Design

Binhao Wang

Additional information is available at the end of the chapter

Abstract

Optical interconnect system efficiency is dependent on the ability to optimize the transceiver circuitry for low-power and high-bandwidth operation, motivating co-simulation environments with compact optical device simulation models. This chapter presents compact Verilog-A silicon carrier-injection and carrier-depletion ring modulator models which accurately capture both nonlinear electrical and optical dynamics. Experimental verification of the carrier-injection ring modulator model is performed both at 8 Gb/s with symmetric drive signals to study the impact of pre-emphasis pulse duration, pulse depth, and dc bias, and at 9 Gb/s with a 65-nm CMOS driver capable of asymmetric pre-emphasis pulse duration. Experimental verification of the carrier-depletion ring modulator model is performed at 25 Gb/s with a 65-nm CMOS driver capable of asymmetric equalization.

Keywords: microring modulator, carrier-injection, carrier-depletion, optical interconnects, silicon photonics, co-simulation, Verilog-A

1. Introduction

Due to low channel loss and the potential for wavelength division multiplexing (WDM), optical interconnects are well suited to address the dramatic requirements in high-speed interconnect capacity and transmission distance demanded by mega data centers and supercomputers [1, 2]. For these applications, silicon photonic platforms are attractive due to their large-scale photonic device integration capabilities and potential manufacturing advantages. As shown in **Figure 1**, compact and energy-efficient WDM interconnect architectures are possible with silicon photonic microring resonator modulators and drop filters [3], as these high-Q devices occupy smaller footprints than large-area Mach-Zehnder modulators [4] and offer inherent wavelength multiplexing without extra device structures, such as array waveguide gratings.

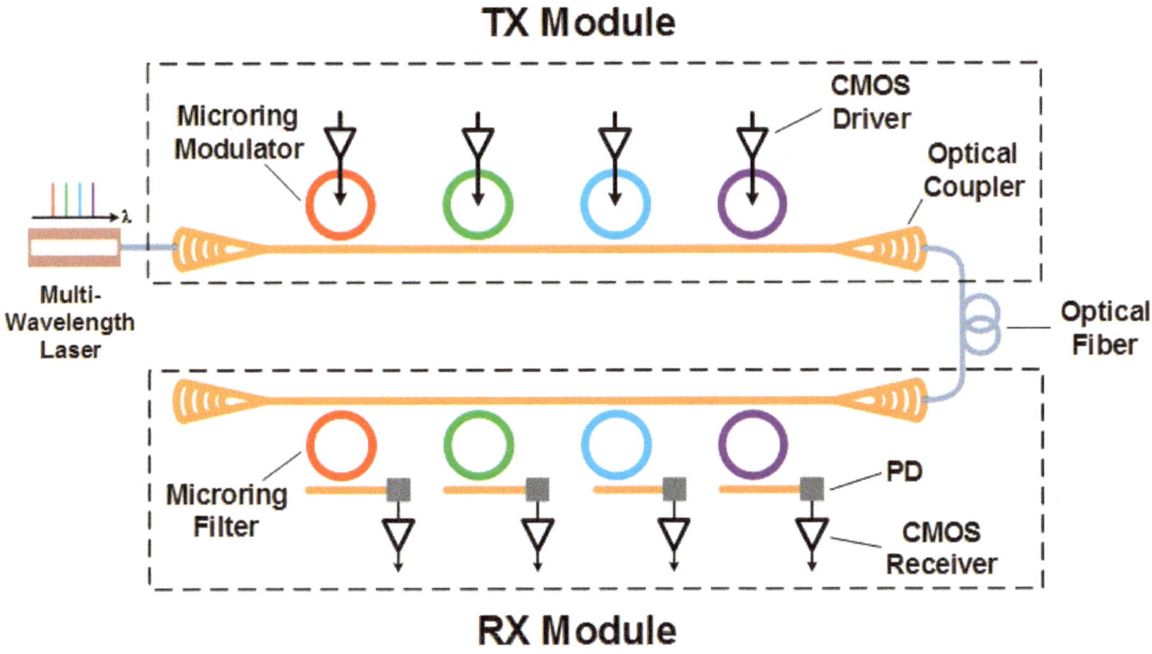

Figure 1. Silicon photonic ring resonator-based wavelength-division-multiplexing (WDM) interconnect system.

The most common high-speed silicon ring modulators operate based on the plasma dispersion effect, with devices based on carrier accumulation, depletion, and injection which display various trade-offs. Accumulation-mode modulators based on metal-oxide-semiconductor (MOS) capacitors can achieve high extinction ratios, but their modulation bandwidth is limited by the relatively high device capacitance [5]. At data rates near 10 Gb/s, injection-mode modulators based on forward-biased p-i-n junctions are an attractive device due to their high modulation depths and rapid bias-based resonance wavelength tuning capabilities [6, 7], but their speed with simple non-return-to-zero (NRZ) modulation is limited by both long minority carrier lifetimes and series resistance effects [8]. Depletion-mode modulators based on reverse-biased p-n junctions can achieve high speed (~40 Gb/s) [9], but require large drive voltages [10]. Injection-mode modulators provide higher energy efficiency and depletion-mode modulators offer higher bandwidth density. Sections 2 and 3 present two compact Verilog-A models for carrier-injection and carrier-depletion ring modulators including both nonlinear electrical and optical dynamics, respectively. Finally, Section 4 concludes the chapter.

2. Silicon carrier-injection ring modulator model

Pre-emphasis signaling, which improves optical transition times, is necessary in order to achieve data rates near 10 Gb/s with carrier-injection ring modulators [7, 8, 11, 12]. As the effective device time constant is different during a rising transition (limited by long minority carrier lifetimes) versus a falling transition (limited by series resistance), nonlinear pre-emphasis waveforms are often used [11, 12]. In addition, the device's optical dynamics must be considered in optimizing the pre-emphasis waveforms [13, 14]. The optical bandwidth is limited by the photon lifetime, which is related to the ring resonator's Q factor, and the ring's phase delay during modulation should be considered to capture the nonlinear optical dynamics. In order to compensate these electrical and optical dynamics, the transmitter circuit must be carefully designed to supply a high-speed pre-emphasis signal with the proper pulse depth, pulse duration, and dc bias. This motivates co-simulation environments with compact optical device simulation models that accurately capture optical and electrical dynamics.

Although previous models have been developed for accumulation-mode [15] and depletion-mode [16] ring modulators, previous models for injection-mode ring devices [8, 17] have lacked accurate modeling of the large-signal p-i-n forward-bias behavior [18] and nonlinear optical dynamics in a format suitable for efficient co-simulation. This section presents a compact Verilog-A model for carrier-injection ring modulators which includes both nonlinear electrical and optical dynamics [19]. The model, which combines an accurate p-i-n electrical [18] and a dynamic ring resonator model [13], will be described and experimentally verified both at 8 Gb/s with symmetric drive signals to study the impact of pre-emphasis pulse duration, pulse depth, and dc bias, and at 9 Gb/s with a 65-nm CMOS driver capable of asymmetric pre-emphasis pulse duration.

2.1. Model description

As shown in **Figure 2**, the modeled carrier-injection ring modulator consists of a ring waveguide coupled to a straight waveguide, with p+ and n+ doping in the inner and outer ring regions, respectively. Accurate high-speed modeling requires inclusion of both electrical and optical dynamics, with **Figure 3** showing a flow chart of the model implementation. When a driving voltage is applied, the dynamic current response is determined by a p-i-n diode SPICE model based on a moment-matching approximation of the ambipolar diffusion equation [18]. After obtaining the current dynamics, the total carriers are calculated by integrating this diode current. However, as some of the carriers recombine and remain inside the waveguide during signal transients, only a portion act as free carriers and impact the effective ring index [20]. Utilizing a subsequent high-pass filter with a time constant equal to the carrier lifetime allows extraction of the free carriers used to calculate the ring index and loss changes due to the plasma dispersion effect [21]. Finally, the optical output power is related to the changes in refractive index and absorption coefficient by a dynamic ring resonator model which accurately considers the ring's cumulative phase shift [13].

The electrical SPICE model is described in **Figure 4**. Electrically, the carrier injection ring modulator is treated as a p-i-n diode (**Figure 4(a)**). As shown in **Figure 4(b)**, the total voltage across the device is distributed across the diode's intrinsic region, $V_{epi}(V(10, 12))$, two junctions, $V_j(V(12, 20))$, and the two terminal contact resistances, V_c. The charge, q_0, required for the total current response, is given by modeling the junction characteristics with the applied voltage, V_j, shown in **Figure 4(c)**. In order to accurately model both the dc and dynamic I-V characteristics, the total current I consists of the current injected into the intrinsic region, I_{epi}, and the current due to the anode recombination effect, I_r. As shown in **Figure 4(d)**, I_r is calculated via q_0 and I_{epi} is modeled by a tenth-order network, modified from Ref. [18] for enhanced accuracy. The tenth-order network is designed according to an approximation of the transfer function (the

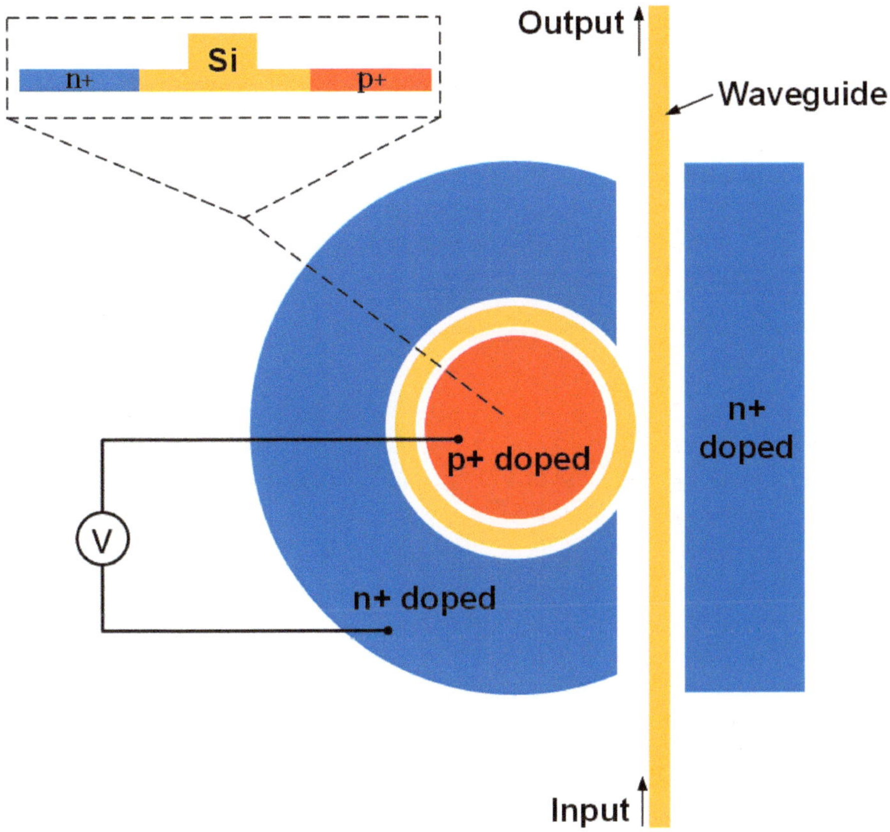

Figure 2. Top and cross-section views of a carrier-injection ring resonator modulator.

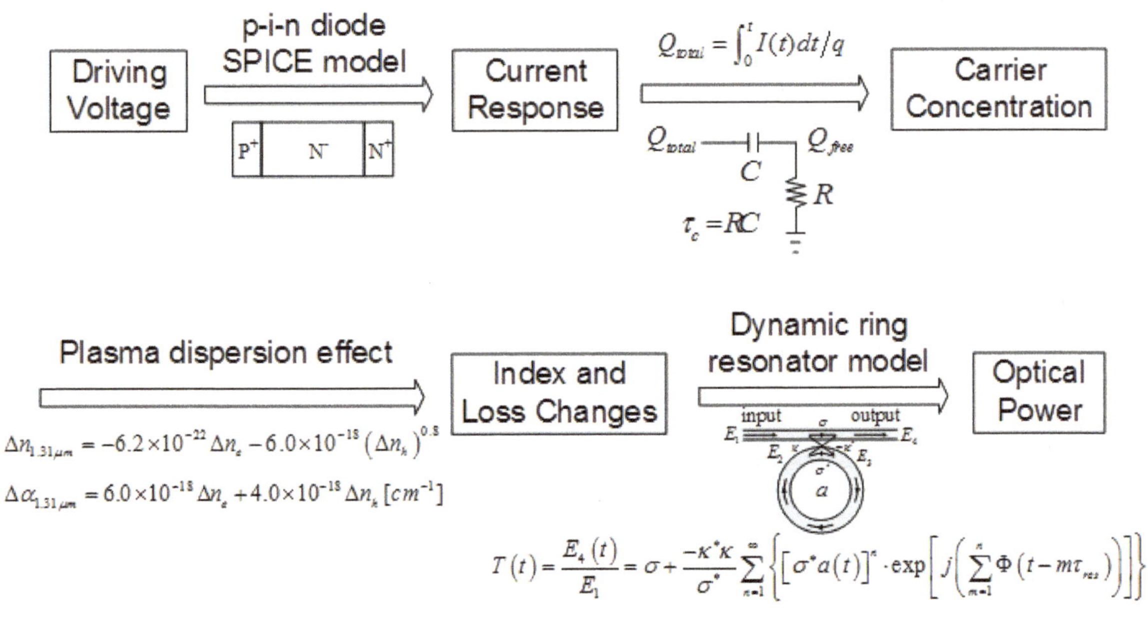

Figure 3. Model flow chart.

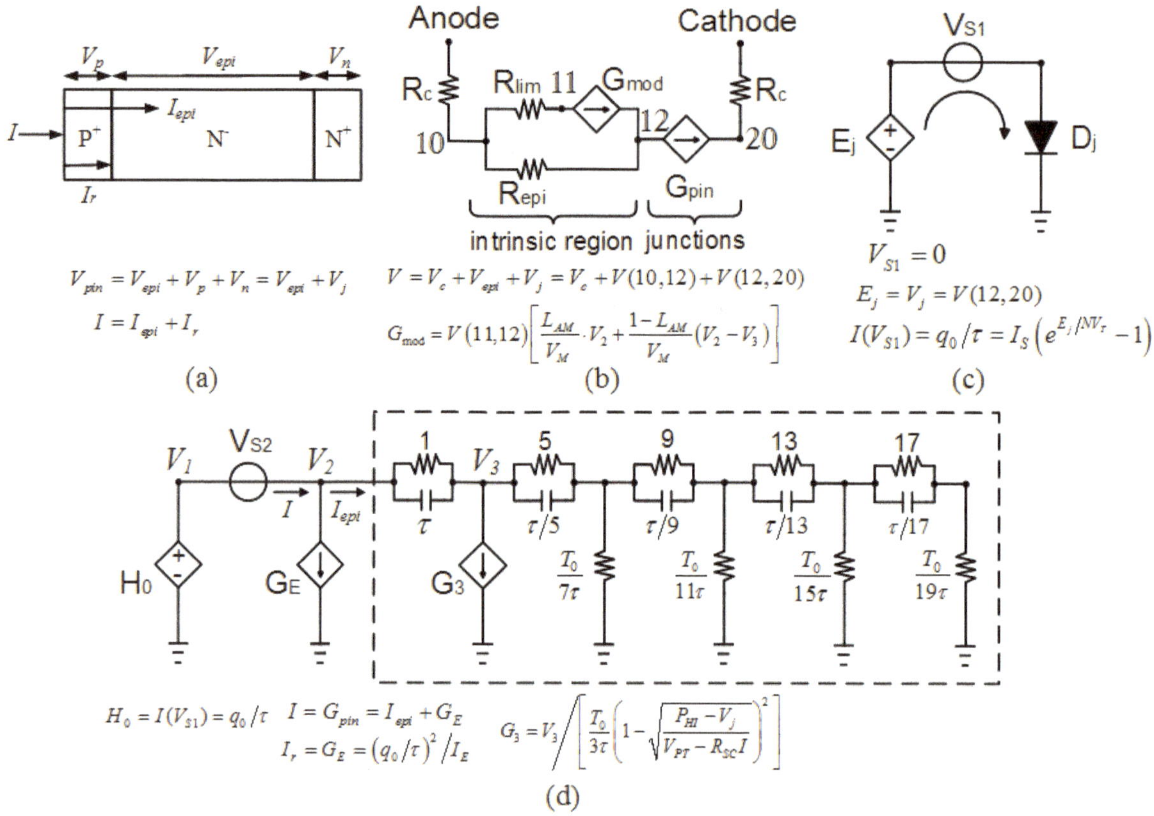

Figure 4. p-i-n Diode and SPICE model schematic: (a) p-i-n diode cross-section, (b) p-i-n voltage distribution model, (c) junction *I-V* characteristic model, and (d) p-i-n current distribution model. The above model is modified from Ref. [18] with the inclusion of contact resistors and by increasing the current distribution model to tenth-order for increased accuracy.

ratio of the intrinsic region current, I_{epi}, and charge, q_0) via asymptotic waveform evaluation (AWE), which is deducted from the ambipolar diffusion equation [22]. The current through the diode D_j, is equal to q_0/τ, is converted to a voltage via the current-controlled voltage source H_0 to drive a network which models the current dynamics in the intrinsic region. Three important nonlinear effects, described by current sources G_{mod}, G_E, and G_3, are included. G_{mod}, which is a function of $V(11,12)$, represents the conductivity modulation in the intrinsic region, G_E expresses the anode recombination effect, and G_3 implements the moving boundary effect during reverse recovery.

Table 1 summarizes the electrical model parameters and shows values for a 5 μm radius device. The extraction procedure for these parameters is described in **Figure 5**. After initializing the parameters with reasonable empirical values, their values are obtained via curve fitting to dc and high-frequency measurements. Eight of the parameters are extracted from the dc characteristic of **Figure 6(a)**. An iterative process is used to curve fit this data, with high sensitivity parameters R_C, I_S, and N first estimated, followed by the low-sensitivity parameters P_{HI}, I_E, V_M, R_{lim}, and R_{epi} related to the previously mentioned nonlinear effects. In the parameter extraction procedure, current levels above 100 μA are given higher weight in the curve fitting since the model is targeted for optical interconnect applications with NRZ modulation.

Parameter	Unit	Description	Empirical range	Value
R_C	Ω	Contact resistance	10–100	50
I_S	A	Saturation current	$1 \times 10^{-14} - 1 \times 10^{-12}$	5.78×10^{-14}
N	–	Emission coefficient	1–2	1.46
P_{HI}	V	Build-in voltage	0.5–1	0.7
T_0	s	Transit time	$1 \times 10^{-10}–1 \times 10^{-9}$	1.046×10^{-10}
I_E	A	Emitter recombination knee current	$1.0 \times 10^{-4}–1.0 \times 10^{-2}$	1.0×10^{-3}
V_M	V	High-injection voltage drop on the base	0–0.5	0.12
R_{lim}	Ω	Carrier-scattering series resistance	$1 \times 10^{-3}–3 \times 10^{-3}$	1.8×10^{-3}
L_{AM}	–	Forward-recovery coefficient	0–0.1	0.03
τ	s	Carrier lifetime in the base	$1.0 \times 10^{-10}–1.0 \times 10^{-8}$	1.0×10^{-9}
R_{epi}	Ω	Base region resistance	$1.0 \times 10^{2}–1.0 \times 10^{3}$	300
V_{PT}	V	Reverse-recovery coefficient	5–20	10
R_{SC}	Ω	Reverse-recovery coefficient	1–100	18

Table 1. Model parameters of the p-i-n diode, with values for a 5 μM device.

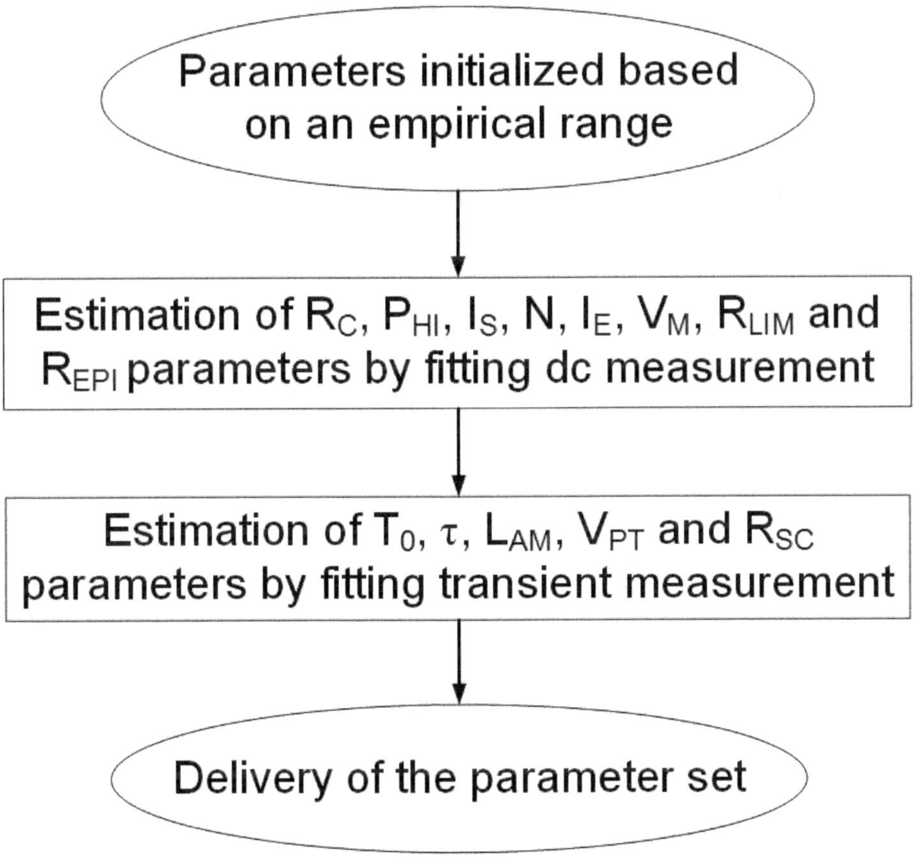

Figure 5. Electrical p-i-n diode parameter extraction procedure.

As shown in **Figure 6(a)**, excellent matching is achieved at these current levels at the cost of some minor error at low current conditions. The remaining five parameters are extracted from **Figure 6(b)** dynamic current response to a 10 Gb/s clock pattern with a voltage swing between −0.5 and 1.5 V. In a similar manner, high sensitivity parameters T_0 and τ are first estimated, followed by the low-sensitivity parameters L_{AM}, V_{PT}, and R_{SC}. Excellent amplitude matching is achieved between the transient simulation and measured results, implying that the current dynamics are captured well. While there is slightly more harmonic content in the measured results, this small error is not deemed critical for NRZ modulation applications. Overall, utilizing the measured dc *I-V* characteristic and transient response for parameter extraction allows for parameters R_C, I_S, N, P_{HI}, T_0, τ, R_{lim}, and R_{epi} to be well defined, while low-sensitivity parameters I_E, V_M, L_{AM}, V_{PT}, and R_{SC} are more softly defined.

After obtaining the dynamic current response, the total carriers are calculated by integrating the diode current with $Q = \int_0^t I(t)dt/q$. The total carriers consist of the following components [20]:

$$Q_{total}(t) = Q_{remain}(t) + Q_{recombine}(t) + Q_{free}(t), \tag{1}$$

which correspond to carriers remaining in the waveguide during signal transients, Q_{remain}, carriers recombining inside the p-i-n diode, $Q_{recombine}$, and the free carriers, Q_{free}, which impact the effective ring index and loss [21]. As shown in **Figure 7**, the remaining and recombining carriers increase with time, while the free carriers can be extracted utilizing a high-pass filter with a time constant equal to the carrier lifetime. These free carriers are then used to calculate the ring index and loss changes due to the plasma dispersion effect. At a wavelength of 1.31 μm, which is near the resonance wavelength of the devices characterized in this work,

$$\Delta n_{1.31\ \mu m} = -6.2\times10^{-22}\Delta n_e - 6.0\times10^{-18}(\Delta n_h)^{0.8} \tag{2}$$

$$\Delta\alpha_{1.31\ \mu m} = 6.0\times10^{-18}\Delta n_e + 4.0\times10^{-18}\Delta n_h\ [\text{cm}^{-1}], \tag{3}$$

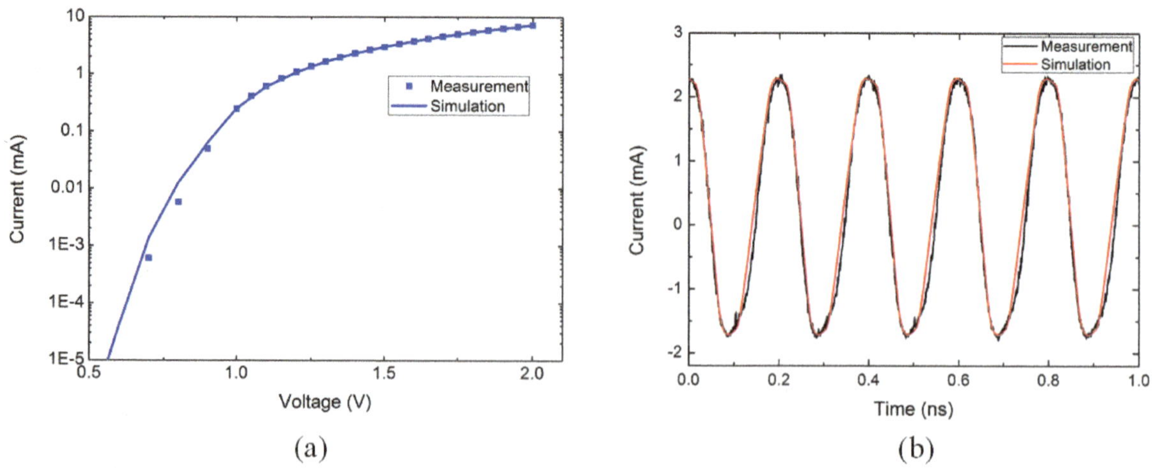

Figure 6. Measured and simulated (a) dc *I-V* characteristic and (b) transient response with a 10 Gb/s clock pattern with voltage swing between −0.5 and 1.5 V.

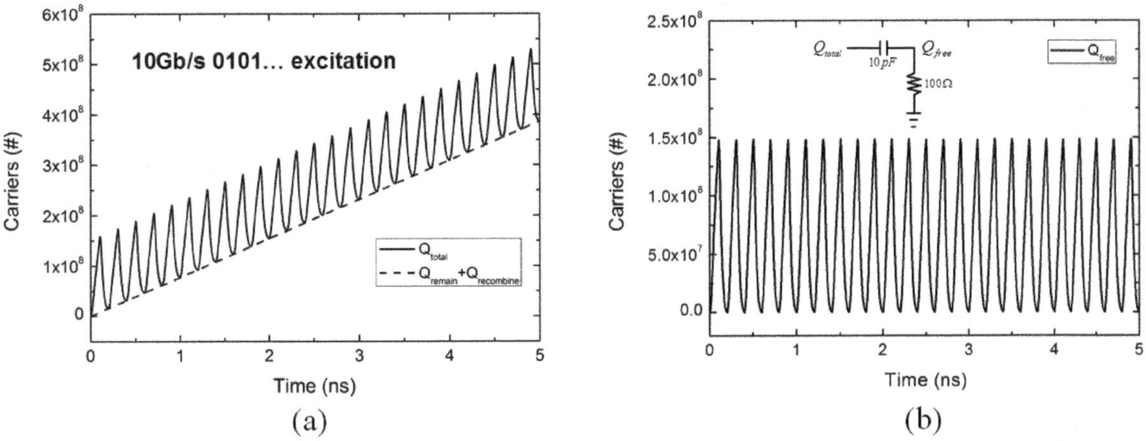

Figure 7. Carriers obtained from the electrical model with a 10 Gb/s clock pattern with voltage swing between −0.5 and 1.5 V: (a) total carriers and (b) free carriers extracted utilizing a high-pass filter with a time constant equal to the carrier lifetime.

where Δn_e and Δn_h are the electron and hole carrier densities [cm^{-3}], respectively. This model assumes $\Delta n_e = \Delta n_h$.

The optical output power is related to the change in refractive index and absorption coefficient by a dynamic ring resonator model which assumes lossless coupling and a single polarization (**Figure 8**). Considering the ring resonator's index dynamics, its time-dependent transmission is described by

$$T(t) = \frac{E_4(t)}{E_1} = \sigma + \frac{-\kappa^*\kappa}{\sigma^*}\sum_{n=1}^{\infty}\left\{[\sigma^* a(t)]^n \cdot \exp\left[j\left(\sum_{m=1}^{n}\Phi(t - m\tau_{res})\right)\right]\right\}, \quad (4)$$

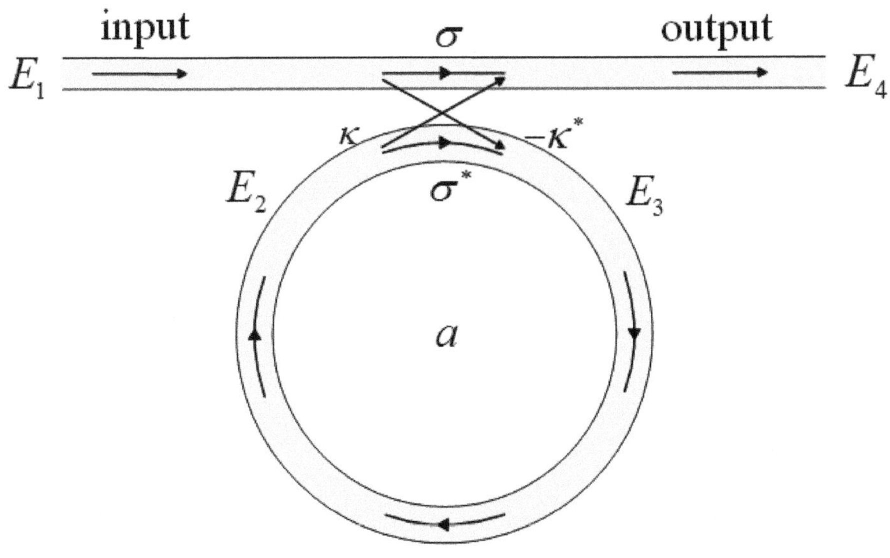

Figure 8. Microring resonator optical model.

where σ and κ are coupling coefficients, $|\kappa^2| + |\sigma^2| = 1$, a is the ring loss coefficient with zero loss corresponding to $a = 1$ and which relates to the absorption coefficient α as $a^2 = \exp(-\alpha L)$, L is the ring circumference, Φ is phase shift, τ_{res} is the resonator round-trip time, and $|T(t)|^2$ is the optical transmission power [13]. The three critical model parameters such as σ, a, and n_{eff} are extracted by curve fitting the steady-state transmission

$$\begin{aligned} T(t) &= \sigma + \frac{-\kappa^*\kappa}{\sigma^*}\sum_{n=1}^{\infty}\{[\sigma^* a(t)]^n \cdot \exp[jn\Phi(t)]\} \\ &= \frac{\sigma - a(t)\cdot \exp[j(\Phi(t))]}{1 - \sigma^* a(t)\cdot \exp[j(\Phi(t))]} \end{aligned} \tag{5}$$

where

$$\Phi(t) = \frac{2\pi}{\lambda} n_{eff}(t) L, \tag{6}$$

λ is the optical wavelength and n_{eff} is the effective index. As shown in **Figure 9**, by fitting the measured through port optical spectrum from a 5-μm ring resonator with applied bias

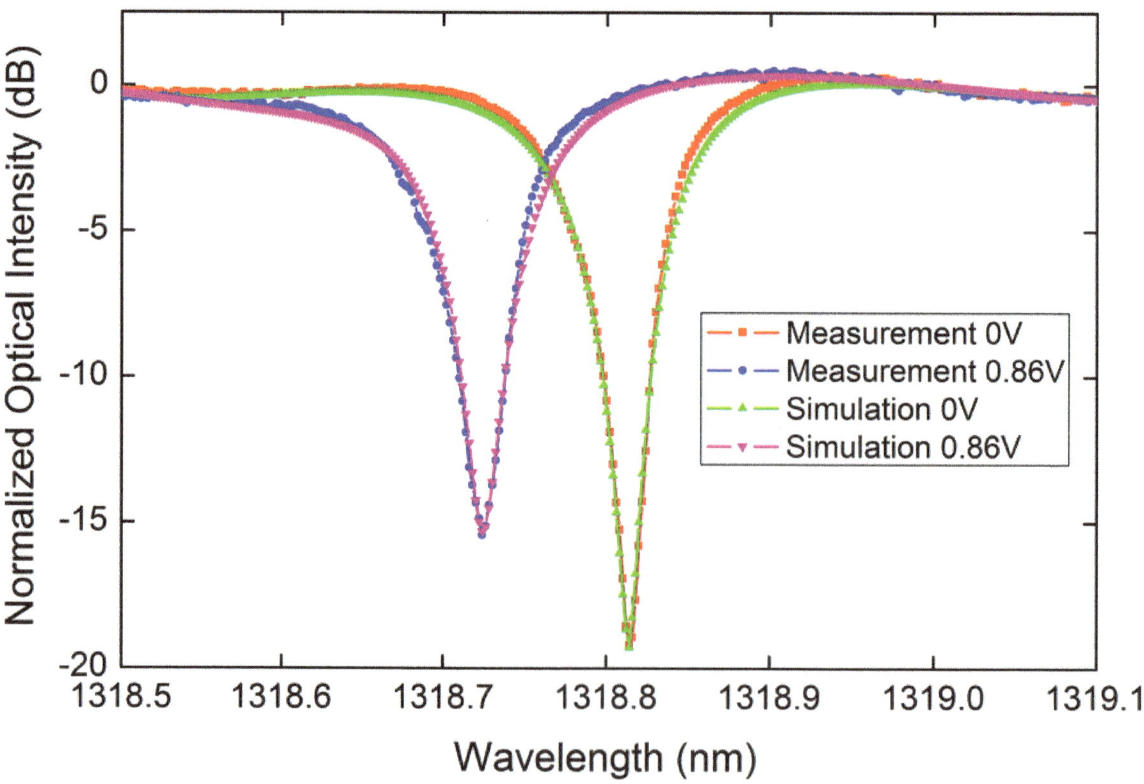

Figure 9. Measured and simulated ring resonator through port optical spectrums. These curves are normalized to the input laser power, accounting for ~10 dB of grating coupler loss.

voltages of 0 and 0.86 V, σ = 0.9944, a = 0.9931, and n_{eff} = 2.5188, are obtained. Utilizing these values in the model described by Eq. (4) allows for excellent matching with measured optical responses with large-signal high-speed modulation.

2.2. Comparison of simulated and measured results

This section presents a comparison of the presented model simulation results with high-speed large-signal measurements. Experimental verification of the model is performed both at 8 Gb/s with symmetric drive signals to study the impact of pre-emphasis pulse duration, pulse depth, and dc bias, and at 9 Gb/s with a 65-nm CMOS driver capable of asymmetric pre-emphasis pulse duration.

2.2.1. Symmetric pre-emphasis modulation with external driver

In order to demonstrate the ring modulator model accuracy, comparisons are made with the measured responses of a 5 μm radius carrier-injection ring modulator operating at 8 Gb/s with pre-emphasis modulation. As shown in the experimental setup of **Figure 10**, differential outputs of a high-speed pattern generator are combined to generate a pre-emphasis NRZ drive signal. The impact of pre-emphasis pulse duration, pulse depth, and dc bias is investigated, with a constant 2 Vpp swing maintained as these parameters are varied. Vertical couplers are used to provide light from a CW laser to the ring modulator input port and direct the modulated light out to a fiber connected to an optical oscilloscope for eye diagram generation.

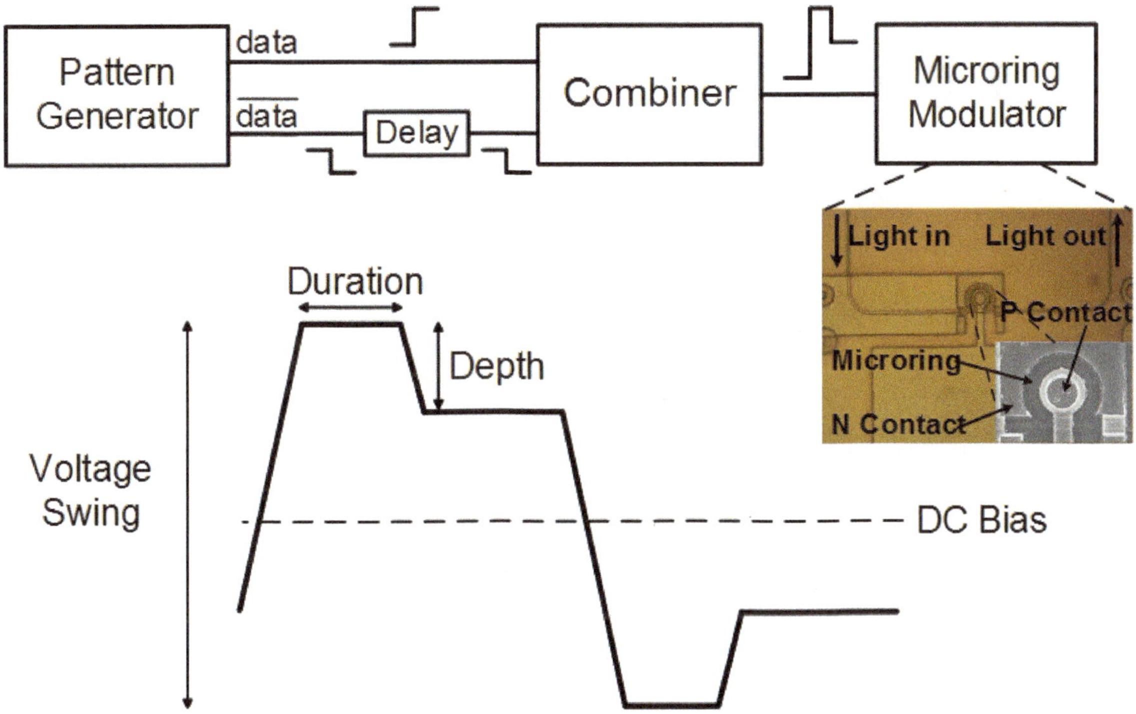

Figure 10. Pre-emphasis NRZ signal generation and waveform.

In all the measured eye diagrams, the CW laser wavelength is tuned to align with the 0 V ring modulator resonance wavelength. Transition times of 40 ps are used in all the external modeling results, which match the equipment used in the measurements.

As predicted by the proposed ring model, utilizing a simple drive signal that is centered at a 0.7 V bias without pre-emphasis results in a very poor eye diagram with a 2^7–1 PRBS data pattern (**Figure 11**). Here the measured eye is completely closed by the system's random jitter, which is not included in the modeling results. Utilizing an optimal 0.8 V pre-emphasis pulse depth, the impact of pulse duration is shown in **Figure 12**. While a 40 ps duration allows the eye to partially open, the height and width are still degraded due to the long rise time caused by the minority carrier lifetime. Increasing the pulse duration to 80 ps provides optimal eye opening, with excellent matching between the simulated and measured eyes observed.

Relative to the optimal eye diagram of **Figure 12(b)**, **Figure 13** shows how the modeling results correlate with measurements as the pre-emphasis pulse depth is varied. An increase in pulse depth to 0.9 V results in excessive overshoot during a rising transition and slow settling to the steady-stage high level due to the relatively low amount of injected carriers after the pre-emphasis pulse. The model's transfer function does introduce some error in these low-carrier recombination dynamics, which results in some offset in the precise positioning of the falling-edge transitions, both **Figure 13** simulated and measured results show similar significant falling-edge deterministic jitter. A decrease in pulse depth to 0.7 V produces excessive charge for the steady-state high level, which results in slow fall times due to the modulator's series resistance limiting carrier extraction (**Figure 13(b)**).

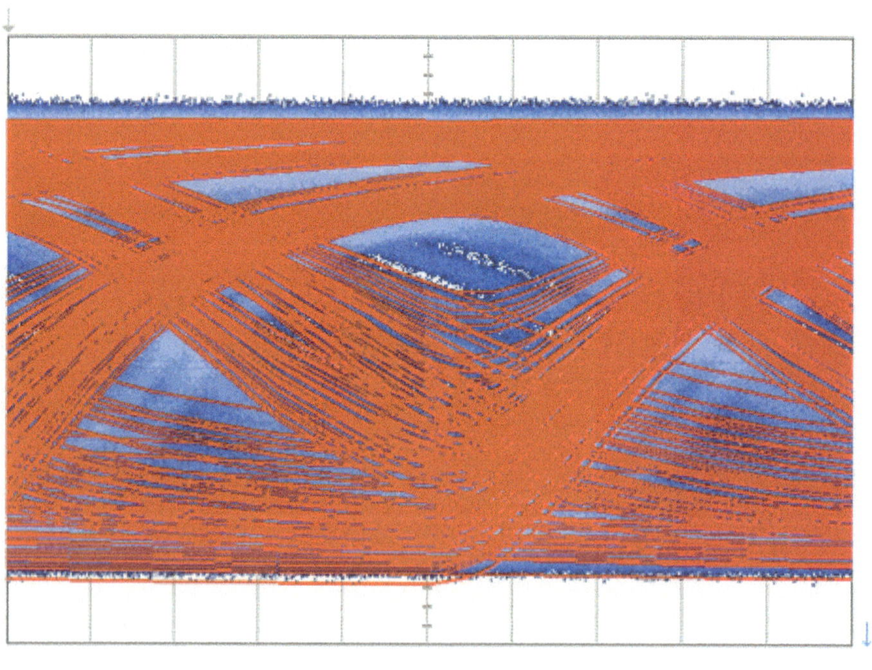

Figure 11. 8 Gb/s measured (blue) and simulated (red) optical eye diagrams with simple NRZ modulation without pre-emphasis [19].

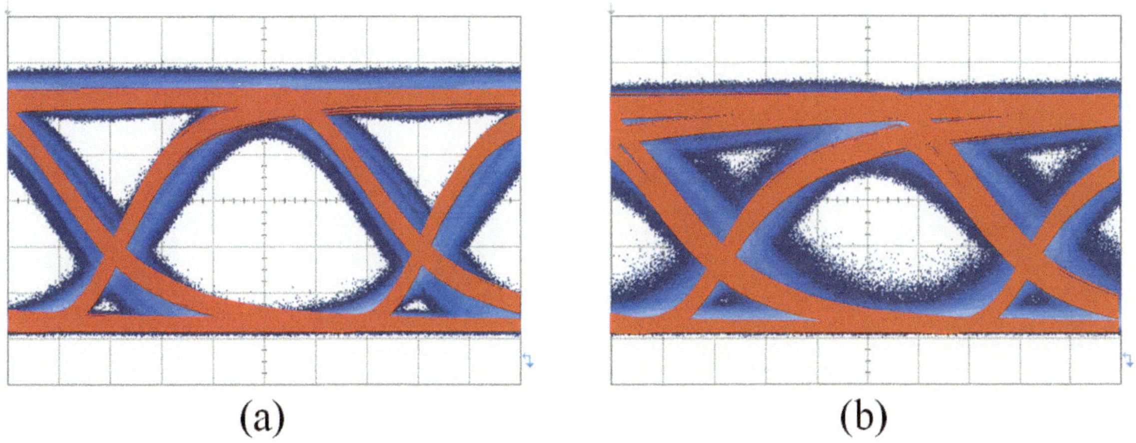

Figure 12. Impact of pre-emphasis pulse duration on 8 Gb/s measured and simulated optical eye diagrams with 0.8 V pulse depth, 0.7 V dc bias, and pulse duration of (a) 40 ps and (b) optimal 80 ps.

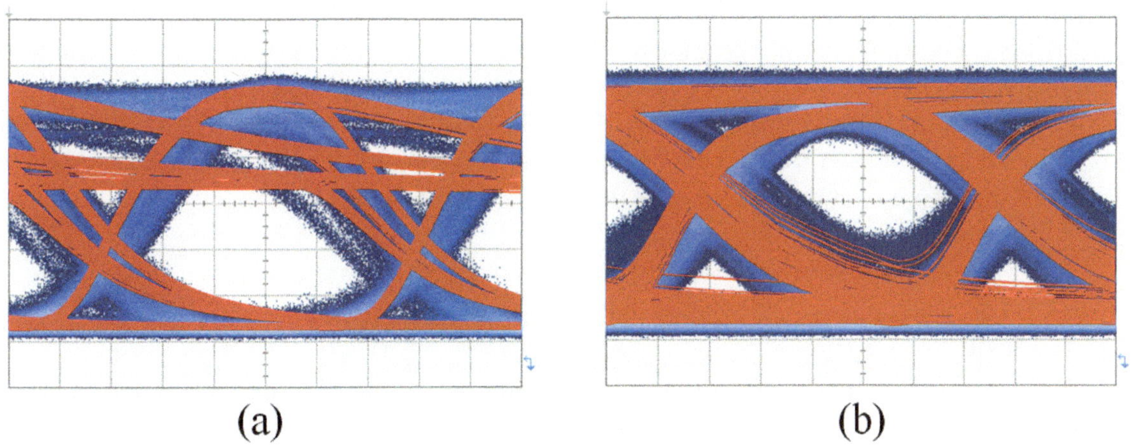

Figure 13. Impact of pre-emphasis pulse depth on 8 Gb/s measured and simulated optical eye diagrams with 80 ps pulse duration, 0.7 V dc bias, and pulse depth of (a) 0.9 V and (b) 0.7 V.

Figure 14 shows the impact of dc bias. As shown in **Figure 14(a)**, an increase in dc bias to 0.75 V produces excessive charge for the steady-state high level which is similar to a decrease in pulse depth to 0.7 V. A decrease in dc bias to 0.65 V results in slower carrier injection and degraded rising transitions (**Figure 14(b)**). Overall, results of **Figures 12–14** show excellent correlation between the proposed ring modulator model and measurements over varying pre-emphasis pulse duration, pulse depth, and dc bias.

2.2.2. *Asymmetric pre-emphasis modulation with CMOS driver*

A key objective of the model is to enable an opto-electronic co-simulation environment which allows for both the optimization of transceiver circuitry and the ability to study the impact of optical device parameters. The co-simulation capabilities are demonstrated by comparing simulated modeling results with the measured responses of the 5 μm radius carrier-injection

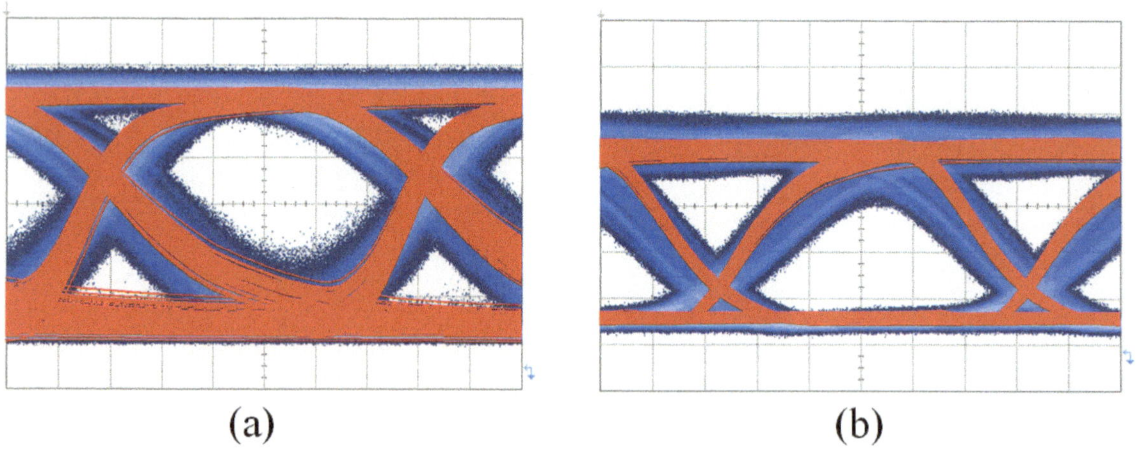

Figure 14. Impact of pre-emphasis dc bias on 8 Gb/s measured and simulated optical eye diagrams with 80 ps pulse duration, 0.8 V pulse depth, and dc bias of (a) 0.75 V and (b) 0.65 V.

ring modulator driven with a custom-designed CMOS driver. As shown in hybrid-integrated prototype in **Figure 15**, the pre-emphasis NRZ driver implemented in a 65-nm CMOS technology [12] is wire-bonded both to the PCB and the silicon ring modulator for testing. While the pre-emphasis pulse depth is fixed in this CMOS driver implementation, the prototype does have the ability to adjust the dc bias and the pre-emphasis pulse duration in an asymmetric manner for independent optimization of the rising and falling responses.

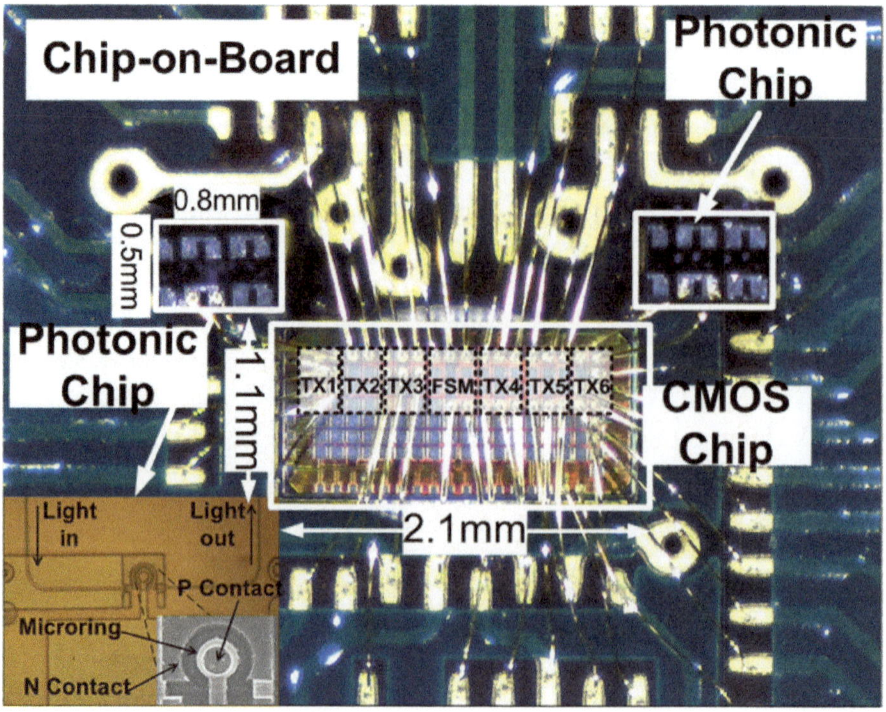

Figure 15. Hybrid-integrated optical transmitter prototype bonded for optical testing.

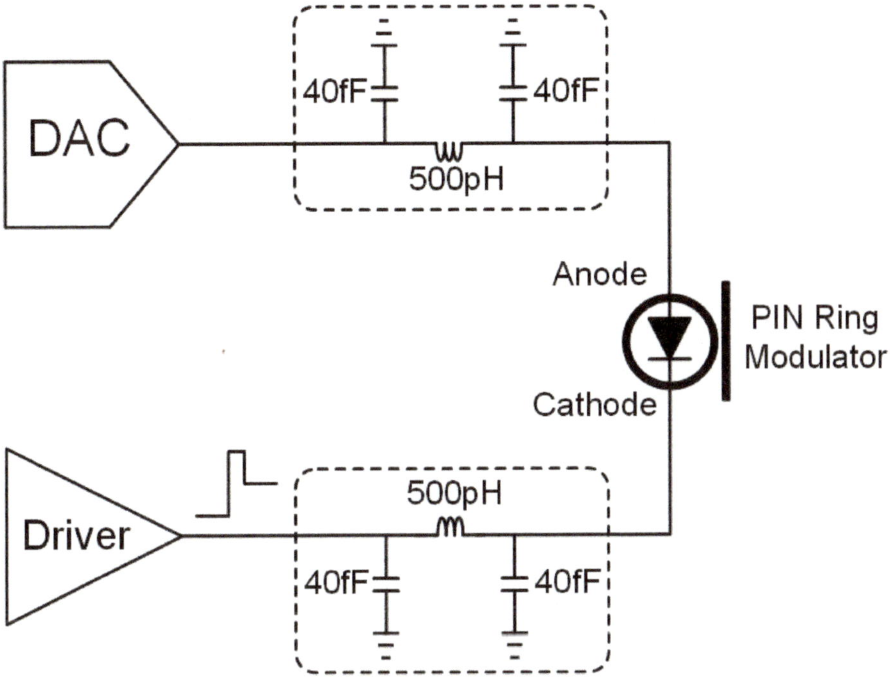

Figure 16. Co-simulation schematic with 65-nm CMOS high-speed CMOS driver, bias DAC, and Verilog-A carrier-injection ring resonator modulator model.

Figure 16 shows the co-simulation schematic in a CADENCE environment, with transistor-level schematics for the high-speed driver and bias digital-to-analog converter (DAC), lumped elements for the wirebond interconnect, and the Verilog-A carrier-injection ring resonator

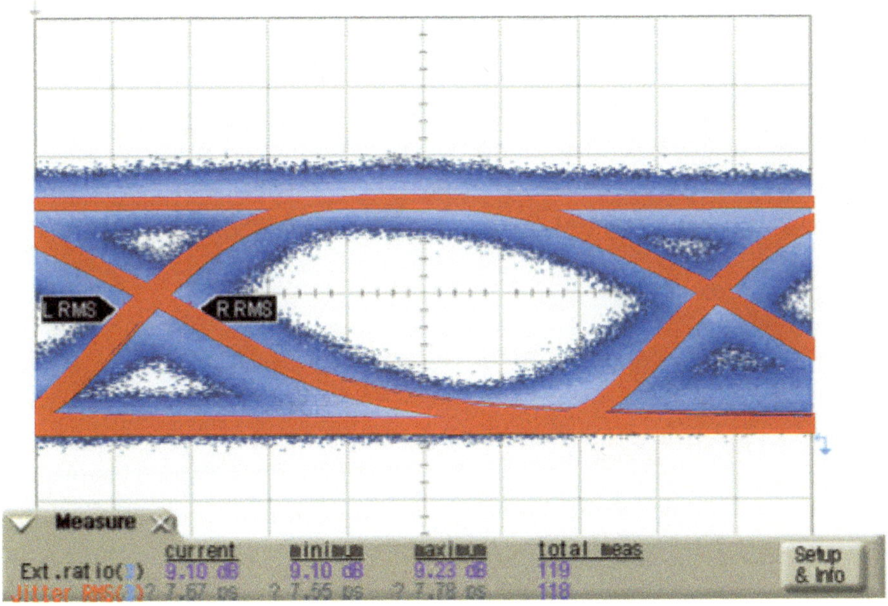

Figure 17. 9 Gb/s measured and co-simulated optical eye diagrams with the ring resonator modulator driven by the 65-nm CMOS driver.

modulator model. The single-ended driver provides a high-speed 2 Vpp output swing with independent dual-edge pre-emphasis duration tuning on the cathode of the ring modulator, while the 9-bit bias tuning DAC is connected to the anode for dc bias adjustment [12]. 500 pH inductors are used to model the ~0.5 mm bondwires that connect the high-speed driver and DAC to the modulator, while 40 fF capacitors model the chips' bondpads. **Figure 17** shows that the optimal 9 Gb/s measured and co-simulated eye diagrams, balancing extinction ratio and eye opening, are achieved when the anode bias is 1.45 V and asymmetric pulse durations for rising and falling transitions are 70 and 50 ps, respectively.

3. Silicon carrier-depletion ring modulator model

For carrier-injection ring modulators, large modulation depth and efficiency are achieved at the cost of relative low modulation bandwidth [12]. It limits application in ultra-high speed data communication. In contrast, carrier-depletion modulators have higher modulation speed ~40 Gb/s. A 320 Gb/s eight-channel WDM transmitter based on carrier-depletion ring modulators was demonstrated in Ref. [9]. The modulation speed of carrier-depletion ring modulators is limited by electrical bandwidth and optical bandwidth. The electrical bandwidth is determined by the RC bandwidth of the ring modulator where the voltage-controlled capacitance results in a nonlinear frequency response with a large voltage swing. The optical bandwidth is limited by photon lifetime related to the Q factor of ring resonators where the time rate of change in ring energy during modulation indicates nonlinear optical dynamics [23]. Therefore, an accurate carrier-depletion ring modulator model is essential to optimize transmitter circuitry while ring modulator models in Refs. [15, 16, 24, 25] did not demonstrate both nonlinear electrical dynamics and optical dynamics.

To design and optimize an optical interconnect transceiver circuitry, an accurate co-simulation environment is required for low-power and high-bandwidth operation. Photonic device models developed in Verilog-A provide the advantage of model compatibility with commercial SPICE circuit simulators. This section presents a Verilog-A carrier-depletion ring modulator model including nonlinear electrical and optical dynamics which provides a co-simulation environment for optical interconnect systems design. The model will be described and verified at 25 Gb/s with a 65-nm CMOS driver capable of asymmetric equalization.

3.1. Model description

The structure of the carrier-depletion ring modulator is shown in **Figure 18**. It consists of a rib waveguide of 500 nm width, 220 nm height, and 90 nm slab height coupled to a ring waveguide with radius of 7.5 μm, p-n junctions formed with outer p+ and inner n+-type doping with doping level near 2×10^{18} cm^{-3} on approximately 75% of the ring waveguide, p++ and n++-type doping utilized for ohmic contact formation, and an integrated heater with 550 Ω resistance formed by doping 15% of the ring with n+-type doping [26]. The ring modulator was fabricated at the IME A*STAR Singapore through OpSIS.

The proposed carrier-depletion ring modulator model is shown in **Figure 19**. The left side is the circuit model in which the electrical bandwidth is dominantly limited by resistances from

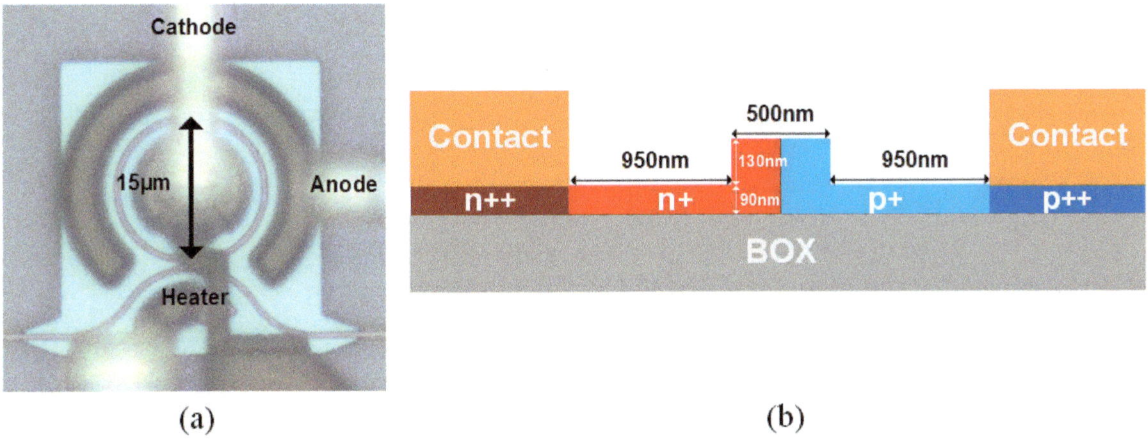

Figure 18. (a) Die photo and (b) cross-section view of carrier-depletion ring modulators.

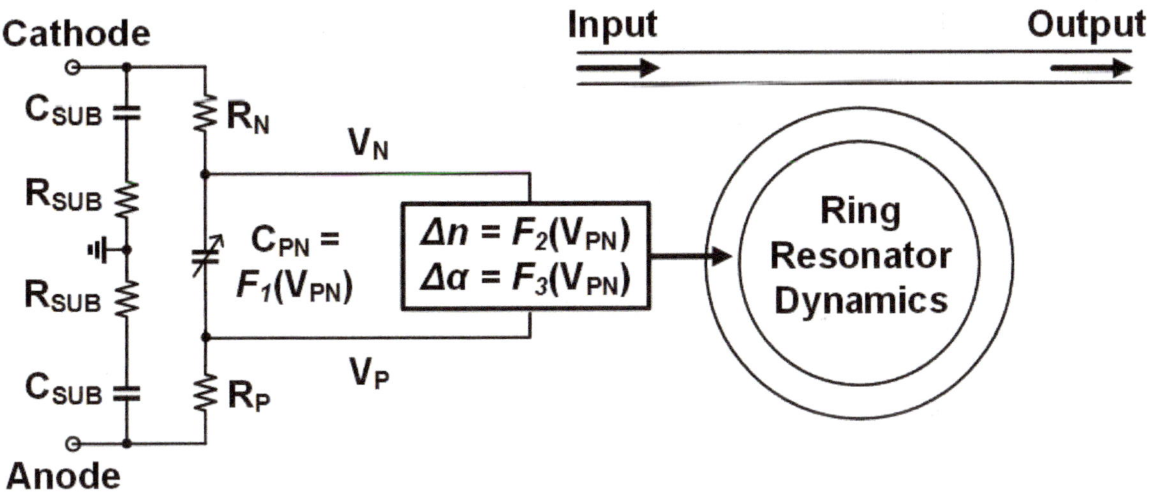

Figure 19. Carrier-depletion ring modulator model.

the electrodes to the junction and capacitance. The right side is the dynamic ring resonator model, and it is related to circuit model by functions of refractive index and absorption coefficient changes versus voltage drop on the junction. By fitting S11 parameter of the device, C_{sub} is 2.5 fF, R_{sub} is 750 Ω-cm, C_{pn} with no bias voltage is 25 fF, and R_p and R_n are 30 Ω [9]. Extracting the devices' carrier densities versus applied voltage with Lumerical allows calculation of the refractive index, n, and absorption coefficient, α, changes by the plasma dispersion effect [21], which for a $\lambda = 1.55$ μm input wavelength are formulated as

$$\Delta n_{1.55\ \mu\text{m}} = -8.8\times10^{-22}\Delta n_e - 8.5\times10^{-18}(\Delta n_h)^{0.8} \tag{7}$$

$$\Delta\alpha_{1.55\ \mu\text{m}} = 8.5\times10^{-18}\Delta n_e + 6.0\times10^{-18}\Delta n_h\ [\text{cm}^{-1}], \tag{8}$$

Figure 20 shows how the single phase shifter ring modulator's effective index, absorption coefficient changes, and junction capacitance change versus applied reverse-bias voltage where $C = \Delta Q/\Delta V$. By curve fitting **Figure 20**, the three parameters Δn, $\Delta\alpha$, and C are then extracted as a polynomial function of voltage.

$$f(V) = a_0 + a_1 V + a_2 V^2 + a_3 V^3 + a_4 V^4 \tag{9}$$

Table 2 gives the a_0-a_4 coefficients, where the valid voltage range is 0–5 V.

The dynamic optical output power is related to the changes in refractive index and absorption coefficient by a ring resonator modeling [23]:

$$\frac{\partial A}{\partial t} = \left(2\pi cj\left(\frac{1}{\lambda} - \frac{1}{\lambda_0}\right) - \frac{1}{\tau}\right)A + j\mu S_i \tag{10}$$

$$S_o = S_i + j\mu A, \tag{11}$$

where $1/\tau = 1/\tau_c + 1/\tau_l$ is an amplitude decay time constant associated with power coupling to bus waveguide τ_c and power lost due to absorption and scattering τ_l, c is light velocity, λ is laser wavelength, λ_0 is the ring resonant wavelength, A is the energy stored in the ring, μ is the mutual coupling between the ring and the bus waveguide, and S_i and S_o are incident and transmitted waves. The coupling factor μ satisfies $\mu^2 = \kappa^2 v_g/2\pi R = 2/\tau_c$, where κ is the coupling ratio, v_g is the ring group velocity, and R is the radius of the ring. The circuit model and ring resonator model are related by $2\pi(n_0 + \Delta n)R = m\lambda_0$ and $\tau_l = 1/\left(v_g exp\left((\alpha_0 + 0.75\Delta\alpha)2\pi R\right)\right)$, where ring effective index with no bias is ~2.57 and ring group index is ~3.89 extracted from Lumerical simulation, and the mode number m = 28 when R = 7.5 μm and λ = 1552.3 nm. By fitting the measured optical spectrum through port applied with reverse bias 0 and 4 V shown in **Figure 21**, three

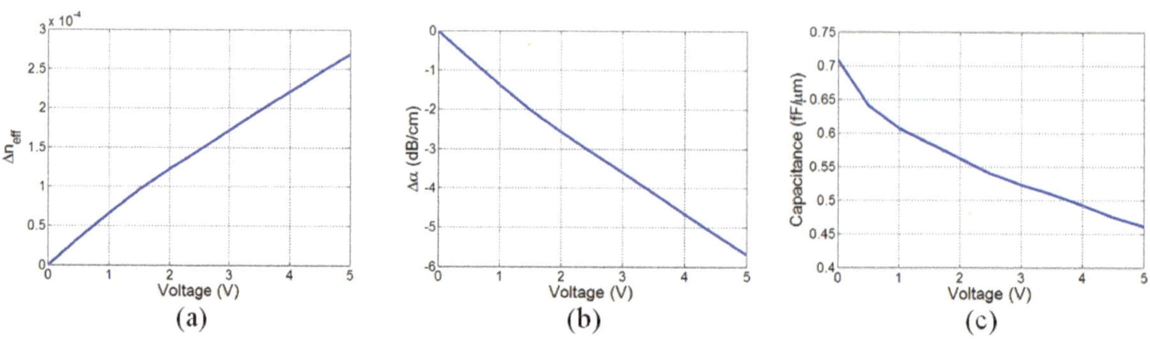

Figure 20. (a) The change of refractive index, (b) the change of absorption coefficient, and (c) the junction capacitance versus applied reverse-bias voltage.

Parameter	Unit	a_0	a_1	a_2	a_3	a_4
Δn_{eff}	–	-4.3×10^{-7}	7.3×10^{-5}	8.0×10^{-6}	1.1×10^{-6}	5.2×10^{-8}
$\Delta\alpha$	dB/cm	0.01	1.5	0.17	-2.3×10^{-2}	1.0×10^{-3}
C	fF/μm	0.71	−0.14	5.5×10^{-2}	-1.2×10^{-2}	1.0×10^{-3}

Table 2. Polynomial coefficients of Δn_{eff}, $\Delta\alpha$, and C.

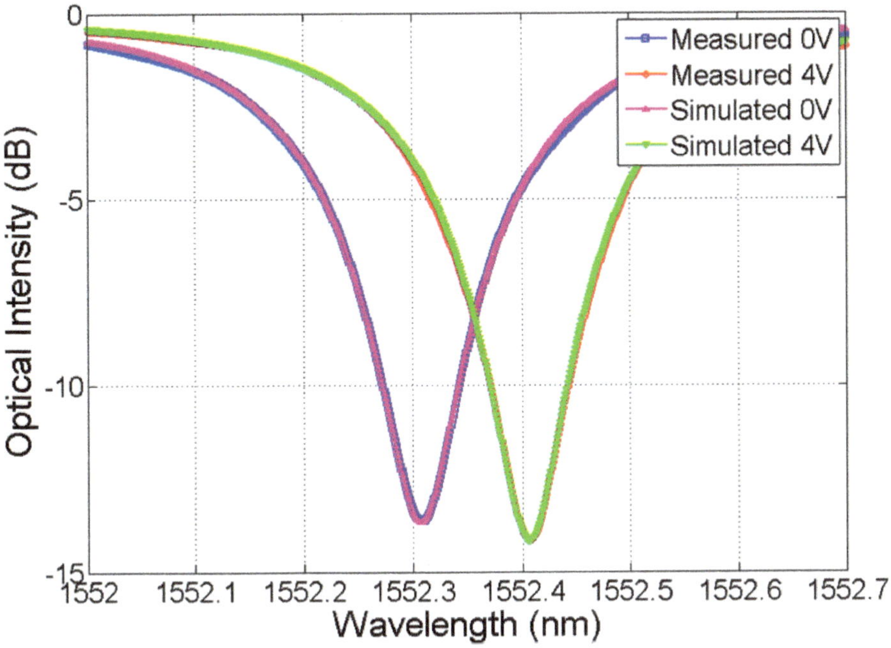

Figure 21. Measured and simulated optical spectrum at through port.

parameters of the model, τ = 9.07 ps, κ^2 = 0.0354, and n_0 = 2.5694, are obtained. The laser wavelength is set to be the resonant wavelength of the ring resonator at 1552.31 nm to maximize the extinction ratio (ER) for NRZ modulation.

3.2. Comparison of simulated and measured results

A key objective of the model is to enable an opto-electronic co-simulation environment which allows for both the optimization of transceiver circuitry and the ability to study the impact of optical device parameters. The co-simulation capabilities are demonstrated by comparing simulated modeling results with the measured responses of the 7.5 μm radius carrier-depletion ring modulator driven with a custom-designed CMOS driver. As shown in hybrid-integrated prototype in **Figure 22**, the AC-coupled differential driver implemented in a 65-nm CMOS technology [27] is wire-bonded both to the PCB and the silicon ring modulator for testing. The prototype has the ability to adjust the equalization to optimize high data rate performance.

Figure 23 shows the co-simulation schematic in a CADENCE environment, with transistor-level schematics for the high-speed differential driver, lumped elements for the wirebond interconnect, and the Verilog-A carrier-depletion ring resonator modulator model. The differential driver provides a high-speed 4.4 Vpp output swing with an asymmetrical feed-forward equalizer (FFE) to compensate the device nonlinearity [27]. 500 pH inductors are used to model the ~0.5 mm bondwires that connect the high-speed differential driver to the modulator, while 40 fF capacitors model the chips' bondpads.

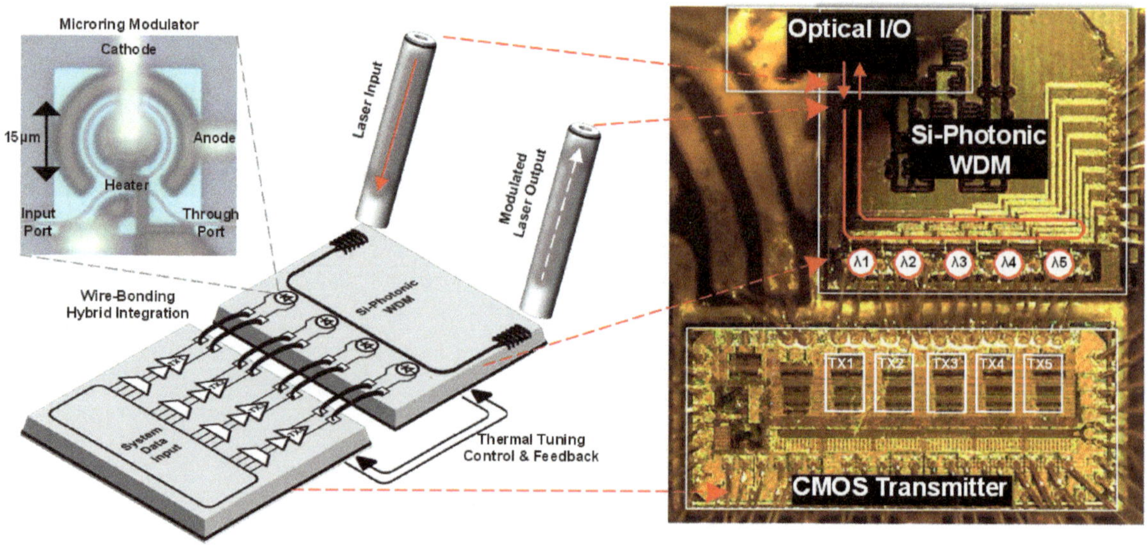

Figure 22. Optical transmitter prototype assembly.

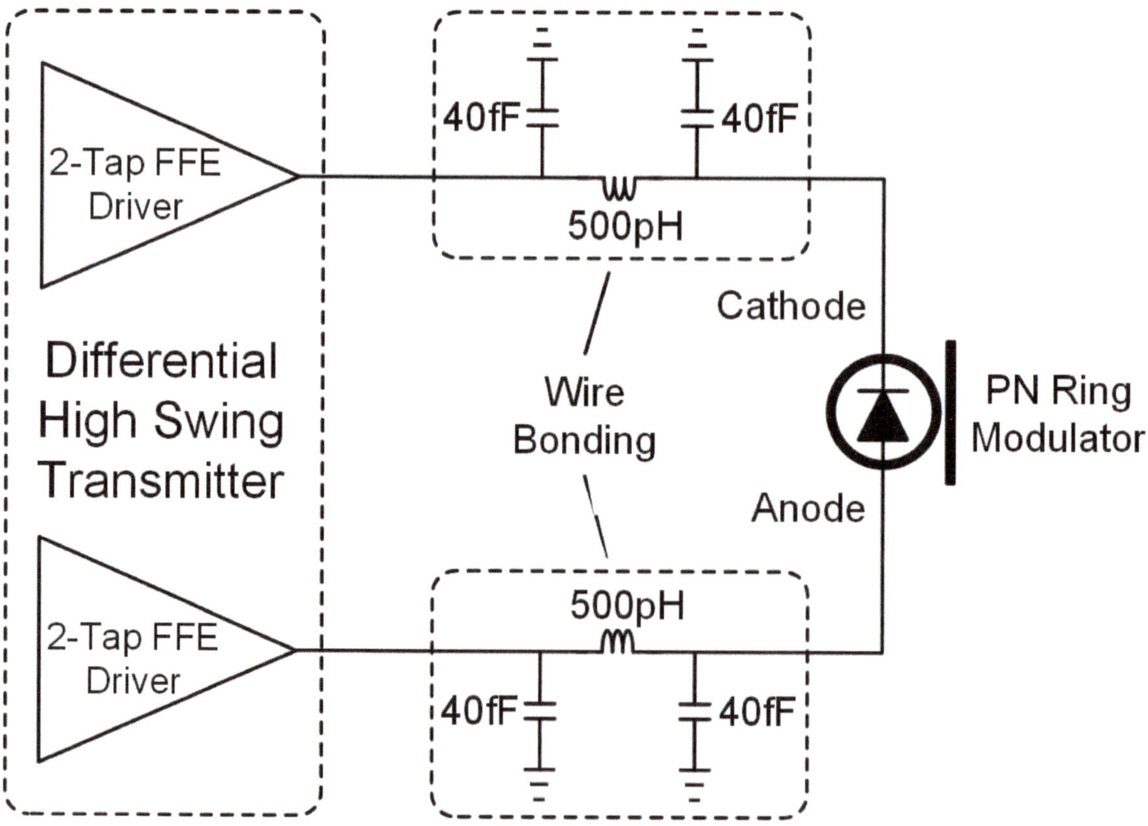

Figure 23. Co-simulation schematic with 65-nm high-speed differential CMOS driver and carrier-depletion ring resonator modulator model.

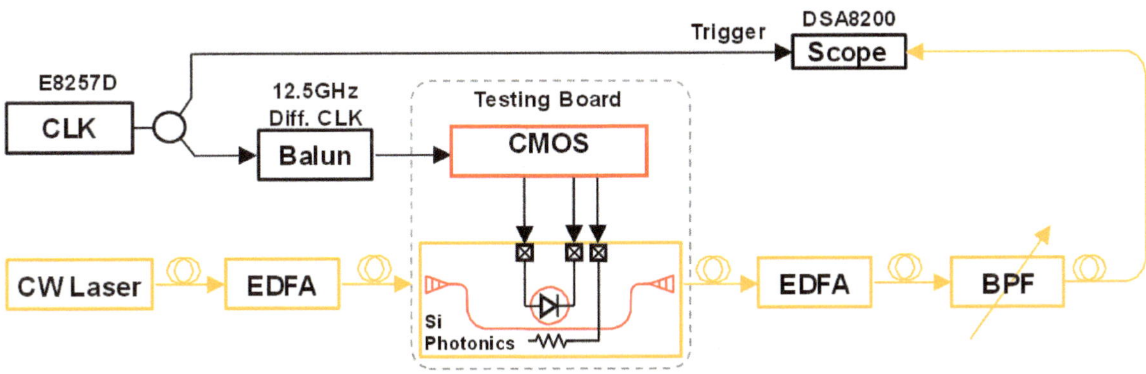

Figure 24. Experimental setup for optical testing.

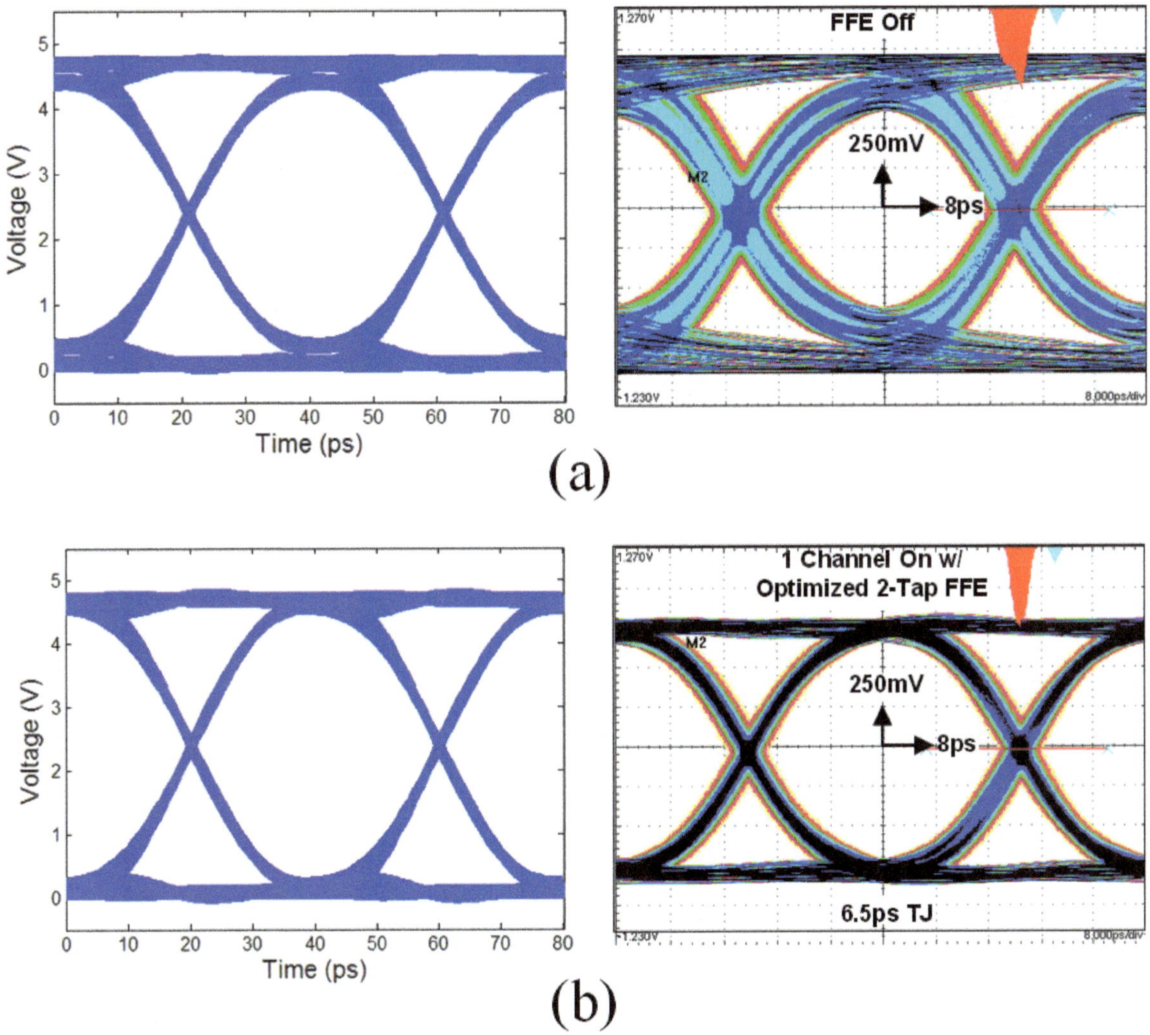

Figure 25. Measured and simulated 25 Gb/s electrical input eye diagrams (a) without equalization and (a) with optimized symmetric equalization.

For model verification, 25 Gb/s measured and simulated eye diagrams are compared. The experimental setup is shown in **Figure 24**. The optical output from the CW laser is amplified by two erbium-doped fiber amplifiers (EDFAs) before and after the photonic chip to compensate input and output insertion losses due to the fiber-to-grating coupler coupling. A bandpass filter is utilized after EDFAs to suppress the amplified noise to increase the signal-to-noise ratio. A clock is utilized for a trigger of an oscilloscope and the input of the CMOS driver. The optical output is received by the oscilloscope.

Figure 25 shows the 25 Gb/s 2^7–1 PRBS CMOS driver signal for model simulation, which matches excellent with the measured eye diagrams including without equalization (**Figure 25(a)**) and with optimized symmetric equalization (**Figure 25(b)**). As shown in the 25 Gb/s eye diagrams of **Figure 26**, excellent matching is achieved between the measured and co-simulated results with and without equalizations. Due to the device bandwidth limitation and nonlinearity, the optical output power is distorted with an unequal amount of inter-symbol-interference (ISI) (**Figure 26 (a)**), which degrades the effective extinction ratio (ER). As shown in **Figure 26(b)**, this asymmetrical ISI is compensated by an optimized nonlinear equalizer.

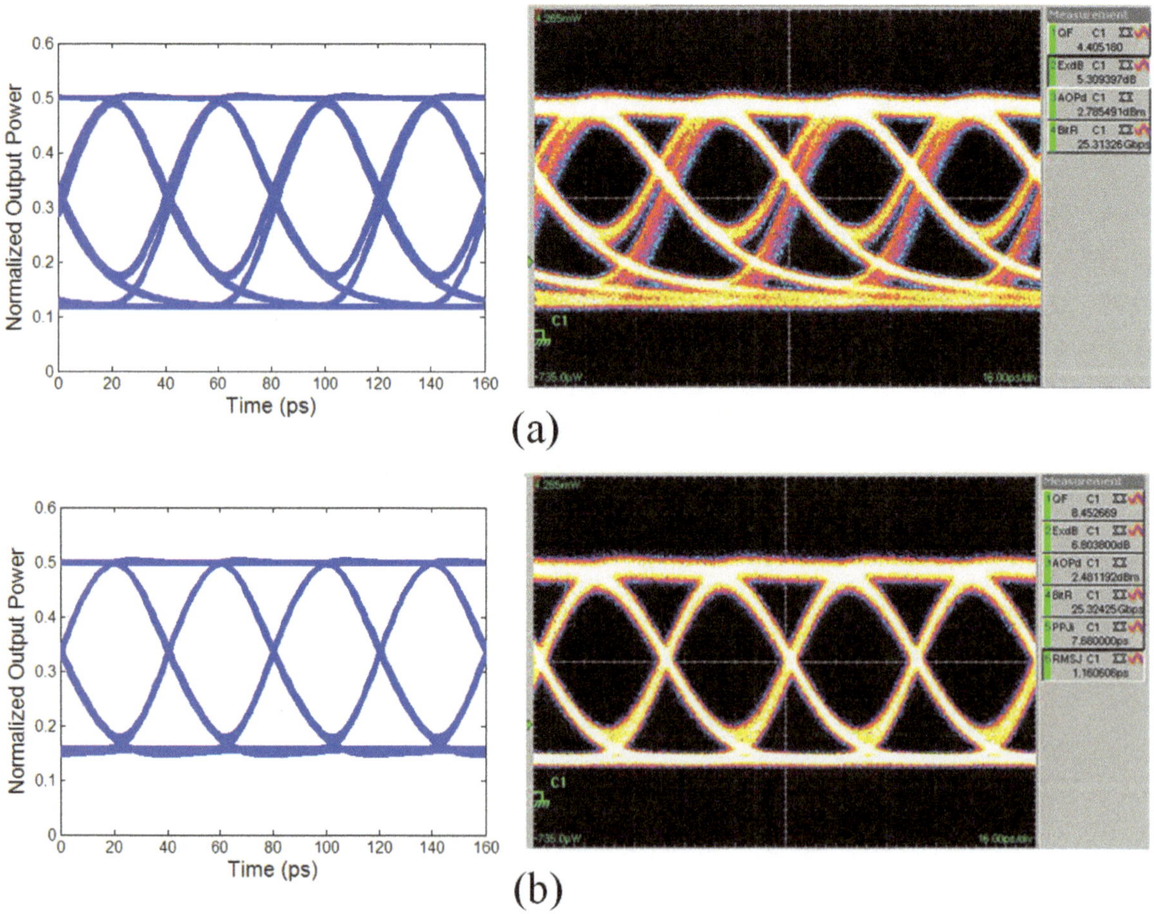

Figure 26. Measured and co-simulated 25 Gb/s optical output eye diagrams (a) without equalization and (b) with the same optimized asymmetric equalization.

4. Conclusions

Optical interconnect system efficiency is dependent on the ability to optimize the transceiver circuitry for low-power and high-bandwidth operation, motivating accurate co-simulation environments. The presented compact Verilog-A models for carrier-injection and carrier-depletion ring modulators include both nonlinear electrical and optical dynamics, allowing for efficient optimization of transmitter signal levels and equalization settings. For the model of carrier-injection microring modulators, excellent matching between simulated and measured optical eye diagrams is achieved both at 8 Gb/s with symmetric drive signals with varying amounts of pre-emphasis pulse duration, pulse depth, and dc bias, and at 9 Gb/s with a 65-nm CMOS driver capable of asymmetric pre-emphasis pulse duration. For the model of carrier-depletion ring modulators, excellent matching between simulated and measured optical eye diagrams is achieved at 25 Gb/s with a 65-nm CMOS driver capable of asymmetric equalization.

Acknowledgements

Parts of this chapter are reproduced from authors' recent journal publication, work, and so on [28].

Author details

Binhao Wang

Address all correspondence to: beehom927@gmail.com

Hewlett Packard Labs, Hewlett Packard Enterprise, USA

References

[1] Barwicz T, Byun H, Gan F, Holzwarth CW, Popovic MA, Rakich PT, Watts MR, Ippen EP, Kartner FX, Smith HI, Orcutt JS, Ram RJ, Stojanovic V, Olubuyide OO, Hoyt JL, Spector S, Geis M, Grein M, Lyszczarz T, Yoon JU. Silicon photonics for compact, energy-efficient interconnects [invited]. Journal of Optical Networking. 2007;**6**(1):63-73

[2] Miller DAB. Device requirements for optical interconnects to silicon chips. Proceedings of IEEE. 2009;**97**(7):1166-1185

[3] Xu QF, Schmidt B, Pradhan S, Lipson M. Micrometre-scale silicon electro-optic modulator. Nature. 2005;**435**(7040):325-327

[4] Liu AS, Jones R, Liao L, Samara-Rubio D, Rubin D, Cohen O, Nicolaescu R, Paniccia M. A high-speed silicon optical modulator based on a metal-oxide-semiconductor capacitor. Nature. 2004;**427**(6975):615-618

[5] Van Campenhout J, Pantouvaki M, Verheyen P, Selvaraja S, Lepage G, Yu H, Lee W, Wouters J, Goossens D, Moelants M, Bogaerts W, Absil P. Low-voltage, low-loss, multi-Gb/s silicon micro-ring modulator based on a MOS capacitor. In: Optical Fiber Communication Conference; 4-8 March; Los Angeles, CA; 2012. p. OM2E.4

[6] Titriku A, Li C, Shafik A, Palermo S. Efficiency modeling of tuning techniques for silicon carrier injection ring resonators. In: Optical Interconnects Conference; 4-7 May; San Diego, CA; 2014. pp. 13-14

[7] Chen CH, Li C, Shafik A, Fiorentino M, Chiang P, Palermo S, Beausoleil R. A WDM silicon photonic transmitter based on carrier-injection microring modulators. In: Optical Interconnects Conference; 4-7 May; San Diego, CA; 2014. pp. 121-122

[8] Xu Q, Manipatruni S, Schmidt B, Shakya J, Lipson M. 12.5 Gbit/s carrier-injection-based silicon micro-ring silicon modulators. Optics Express. 2007;**15**(2):430-436

[9] Liu Y, Ding R, Li Q, Zhe X, Li Y, Yang Y, Lim AE, Lo PGQ, Bergman K, Baehr-Jones T, Hochberg M. Ultra-compact 320 Gb/s and 160 Gb/s WDM transmitters based on silicon microrings. In: Optical Fiber Communication Conference; 9-13 March; San Francisco, CA; 2014. p. Th4G.6

[10] Li H, Li C, Xuan Z, Titriku A, Yu K, Wang B, Qi N, Shafik A, Fiorentino M, Hochberg M, Palermo S, Chiang P. A 5 × 25 Gb/s, 4.4 V swing, AC-coupled, Si-photonic microring transmitter with 2-tap asymmetric FFE and dynamic thermal tuning in 65 nm CMOS. In: IEEE International Solid-State Circuits Conference; 22-26 February; San Francisco, CA; 2015

[11] Moss B, Sun C, Georgas M, Shainline J, Orcutt J, Leu J, Wade M, Chen Y.-H, Nammari K, Wang X, Li H, Ram R, Popovic M, Stojanovic V. A 1.23 pJ/b 2.5 Gb/s monolithically integrated optical carrier-injection ring modulator and all-digital driver circuit in commercial 45 nm SOI. In: IEEE International Solid-State Circuits Conference; 22-26 February; San Francisco, CA; 2015

[12] Li C, Bai R, Shafik A, Tabasy EZ, Wang BH, Tang G, Ma C, Chen CH, Peng Z, Fiorentino M, Beausoleil RG, Chiang P, Palermo S. Silicon photonic transceiver circuits with microring resonator bias-based wavelength stabilization in 65 nm CMOS. IEEE Journal of Solid-State Circuits. 2014;**49**(6):1419-1436

[13] Ioannidis ZK, Radmore PM, Giles IP. Dynamic-response of an all-fiber ring resonator. Optics Letters. 1988;**13**(5):422-424

[14] Sacher WD, Poon JKS. Dynamics of microring resonator modulators. Optics Express. 2008;**16**(20):15741-15753

[15] Zhang L, Li Y, Yang JY, Song M, Beausoleil RG, Willner AE. Silicon-based microring resonator modulators for intensity modulation. IEEE Journal of Selected Topics in Quantum Electronics. 2010;**16**(1):149-158

[16] Buckwalter J, Zheng X, Li G, Raj K, Krishnamoorthy A. A monolithic 25-Gb/s transceiver with photonic ring modulators and Ge detectors in a 130-nm CMOS SOI process. IEEE Journal of Solid-State Circuits. 2012;**47**(6):1309-1322

[17] Wu R, Chen CH, Fedeli JM, Fournier M, Cheng KT, Beausoleil R. Compact models for carrier-injection silicon microring modulators. Optics Express. 2015;**23**(12):15545-15554

[18] Strollo AGM. A new SPICE model of power P-I-N diode based on asymptotic waveform evaluation. IEEE Transactions on Power Electronics. 1997;**12**(1):12-20

[19] Wang B, Li C, Chen CH, Yu K, Fiorentino M, Beausoleil R, Palermo S. Compact Verilog-A modeling of silicon carrier-injection ring modulators. In: Optical Interconnects Conference; 20-22 April; San Diego, CA; 2015. pp. 128-129

[20] Gan F. High-speed silicon electro-optic modulator for electronic photonic integrated circuits [dissertation]. Cambridge, MA: Electrical Engineering and Computer Science, MIT; 2007

[21] Soref RA, Bennett BR. Electrooptical effects in silicon. IEEE Journal of Quantum Electronics. 1987;**23**(1):123-129

[22] Pillage LT, Rohrer RA. Asymptotic waveform evaluation for timing analysis. IEEE Transactions on CAD. 1990;**9**:352-366

[23] Little BE, Chu ST, Haus HA, Foresi J, Laine JP. Microring resonator channel dropping filters. Journal of Lightwave Technology. 1997;**15**(6):998-1005

[24] Smy T, Gunupudi P, McGarry S, Ye WN. Circuit-level transient simulation of configurable ring resonators using physical models. Journal of the Optical Society of America B-Optical Physics. 2011;**28**(6):1534-1543

[25] Kononov E. Modeling photonic links in Verilog-A [thesis]. Cambridge, MA: Eletrical Engineering and Computer Science, MIT; 2012

[26] Ding R, Liu Y, Li Q, Xuan Z, Ma YJ, Yang YS, Lim AEJ, Lo GQ, Bergman K, Baehr-Jones T, Hochberg M. A compact low-power 320-gb/s wdm transmitter based on silicon microrings. IEEE Photonics Journal. 2014;**6**(3):6600608

[27] Li H, Xuan Z, Titriku A, Li C, Yu K, Wang B, Shafik A, Qi N, Liu Y, Ding R, Baehr-Jones T, Fiorentino M, Hochberg M, Palermo S, Chiang PY. A 25 Gb/s, 4.4 V-swing, AC-coupled ring modulator-based WDM transmitter with wavelength stabilization in 65 nm CMOS. IEEE Journal of Solid-State Circuits. 2015;**50**(12):3145-3159

[28] Wang B, Li C, Chen CH, Yu K, Fiorentino M, Beausoleil R, Palermo S. A compact Verilog-A model of silicon carrier-injection ring modulators for interconnect transceiver circuit design. Journal of Lightwave Technology. 2016;**34**(12):2996-3005

5

Properties of GaP Studied over 50 Years

Sergei L. Pyshkin and John Ballato

Additional information is available at the end of the chapter

Abstract

A unique set of GaP semiconductor samples studied for over 50 years has exhibited significant improvement in their properties through the formation of the perfect host crystal lattice and the N-impurity crystal superlattice. This chapter reviews this evolution of properties and discusses their novel utility in advanced optoelectronic devices. More specifically, nitrogen-doped gallium phosphide (GaP:N) crystals that were originally prepared in the 1960s were theorized to form an excitonic crystal (1970s), and the best methods of their bulk, film, and nanoparticle crystal growth have subsequently been developed. The excitonic crystals yield novel and useful properties including enhanced stimulated emission and very bright and broadband luminescence at room temperature, which have been observed. These results provide a new approach to the selection and preparation of "perfect" materials for optoelectronics and offer a unique opportunity to realize a new form of solid-state host—the excitonic crystal—as a high-intensity light source with low thresholds for nonlinear optical effects.

Keywords: long-term evolution of crystal properties, excitonic crystal, close to ideal crystals

1. Introduction

Described herein are studies initiated by one of the authors (SLP) in 1961. Further studies on the same samples were conducted in the later 1960s through 1970s, then again in the 1980s, 1990s, and 2000s making these a very unique set of semi-centennial observations [1–34, 40–42].

One of the most important findings of this study was the observation, first recognized in the 1980s, of a significant improvement over time in the properties of GaP crystals under ambient temperature and pressure. In particular, it was found that ordering of host and impurity atoms improved over time as did the GaP mechanical properties, luminescence with tunable spectral characteristics, and nonlinear optical effects, not observable in the freshly prepared (imperfect) crystals.

Over time, as confirmed by 50 years of experiments on the same samples, driving forces, such as diffusion along concentration gradients, strain relaxation associated with clustering, and minimization of the free energy associated with properly directed chemical bonds between host atoms, results in an ordered redistribution of impurities and host atoms in the crystal. In the particular case of GaP, as well as other compounds possessing highly volatile components, attempt to accelerate these processes through annealing at increased temperatures cannot be successful due to the potential for thermal decomposition (in GaP—due to P desorption). Accordingly, successful thermal processing of these compounds can only take place at temperatures below the sublimation temperatures of their volatile constituents, requiring a longer annealing time. For instance, as evaluated in the framework of the Ising model, the characteristic time of the substitution reaction during N diffusion along P sites in GaP:N crystals at room temperature constitutes 15–20 years [2]. Hence, the observations of highly excited luminescence and some other phenomena in the crystals made in the 1960–1970s and in the 1980–1990s were then compared with the results obtained in 2005–2014 under similar experimental conditions.

The long-term ordering of doped GaP and other semiconductors has been observed as an important accompanying process, which only can be studied using the same unique set of samples and decade time scales. More specifically, the optical and mechanical properties of single crystalline GaP and some other semiconductors, also grown in the 1960s, have been analyzed. Comparison of the properties of the same crystals was performed in the 1960s, 1970s, 1980s, and 1990s [1, 3–17] along with those of newly made GaP nanocrystals [18–20] and freshly prepared bulk single crystals [21–24]. Jointly with Refs. [25, 26, 31–34], this review provides a generalization of the results on these long-term observations of luminescence, absorption, Raman light scattering, and microhardness of the bulk single crystals in comparison with the same properties of the high-quality GaP nanocrystals. It is shown that the combination of these characterization techniques elucidates the evolution of these crystals over the course of many decades. It also clarifies the ordered state brought about by prolonged room-temperature thermal annealing, and the useful optical properties that accompany such ordering. It is demonstrated that long-term natural stimuli that improve the perfection of crystals prevail over other processes and can lead to novel heterogeneous device systems and new semiconductor devices with high temporal stability.

In this chapter, we summarize more than 50 years of results of the study of evolution of the properties of the GaP crystals stored under normal conditions and, on this basis, we propose a new approach to resource-saving production of optoelectronic materials and unique devices based on them.

2. Bulk perfect crystals

The method to grow gallium phosphide crystals from GaP solution melt was chosen due to the significant temperature reduction of the process and the presence of large amounts of solvent which dramatically reduce impurities from the container. Note that lamellar crystals grown from this solution melt are the most convenient and economical material in the manufacture of many semiconductor devices.

The influence of the following factors on the quality of the grown single crystals has been investigated: (1) quantity and chemical nature of impurities; (2) geometric shape of the container and conditions of heat rejection; (3) accuracy of the temperature control; and (4) cooling rate and the law of heat-sink cooling.

The solution-melt method of GaP growth is described in detail in Refs. [1, 3, 7]. The focus here is only on the most essential details of the crystal growth.

Synthesis, alloying, and crystallization of Ga-P mixture were combined in a single cycle. Influence of impurities on the growth of GaP crystals was investigated using the highest-purity industrially produced materials at that time (1960) in the former USSR. The growth of the crystals was conducted in a sealed quartz container that was introduced into the programmable furnace with a vibrating mount to agitate the mixture of Ga, P, and any chosen dopant. This agitation was performed to facilitate chemical reaction between components and to avoid a possible explosion of the container in dangerous temperature points of the process of mass crystallization.

In order to study the form and quality of growing crystals, the crystallization process could be interrupted at any temperature below the point of liquidus. This interruption was achieved with the help of a specially designed growth container and its programmable heating furnace, which allow to stop the change of temperature of the solution melt, to identify and examine grown to this time crystals (see details in Refs. [1, 3]).

The solution, carefully heated to slightly over 400°C, was held at this temperature for about 2 h to transform the red phosphorus precursor into its white form. Then, the heating process may be smoothly prolonged until a temperature of about 1200°C is reached, which exceeds the liquidus point for 5% P solution in Ga (1123°C) and creates appropriate conditions for Ga-P reactions. After about 1–3 h, the samples are gradually cooled at a rate of 30°C/h to 600–700°C and then the furnace is turned off and naturally cooled to room temperature.

The following impurities were doped into the GaP individually: Cu, Zn, Cd, In, Si, Ge, Sn, S, Se, Te, Ni, Cr, Fe, Co, Sm, La, Gd, and Sm_2O_3. Typical features of the doped GaP crystals (size, color, polarity, lamellar structure, twinning, the dislocation structure, and others) were compared with similarly grown but undoped crystals.

It was found [1, 3] that the properties of the crystals depend on the chemical nature as well as on the quantity of the selected impurities. Their small admixtures did not significantly change the form of the crystals, while significant change of the crystallization environment composition created changes in forms of growth. It was also found that the following factors most significantly influenced the morphology of GaP crystals grown from the Ga-P solution: (a) the degree of supersaturation, which depends on the temperature of the solution and (b) its first derivative with time; the conditions of the heat crystallization removal and the accuracy of the temperature control in the solution.

Changing crystallization conditions, mainly the rate of cooling of the melt, significantly affects the morphology of crystals. Accordingly, the increase of the cooling rate thus reduced the size of the crystals, but the dislocation density was not changed and was primarily determined by the concentration and chemical nature of the impurities. The number of crystals having isometric forms increased with decreasing of cooling rate resulting in the simultaneous improvement of the quality of the crystal surface. Uneven removal of heat from the walls of the container led to the formation of up to 30-mm crystal needles elongated in the direction of the heat removal.

Significant impact on the crystal quality provides more accurate temperature control. The crystals obtained with the accuracy of temperature control ±0.5°C had the perfect planes (111), low dislocation density, and dimensions that were two to three times larger than the crystals obtained at the same conditions, but with the accuracy of ±5°C. Crystals reached 25 mm in length and had the dislocation density of $\sim 10^3$ cm^{-2}, which is significantly less than that of the crystals obtained with low accuracy of the temperature control [1, 3].

Based on the findings of this investigation, let us now present an overview of the growth of lamellar GaP crystals. The first crystals, having the form of dendritic needles with the length of the order of 5 mm and thickness of 0.15 mm, were grown at the temperature of 1107°C, which is approximately 16°C of undercooling. The crystals represent thin plates in the form of rhombs, triangles, or hexagons. The twinning on the transverse cross sections was observed at the study of microsections and cleaved facets. It turned out that the lamellar GaP crystals contain the plane of twinning, parallel to the planes (111) [1, 22].

The growth process of lamellar crystals can be formally divided into two components: the tangential growth in the plane (111) and layer-by-layer growth in the direction normal to the plane (111). Because the ratio of Ga and P atoms differed significantly from the stoichiometry, it is assumed that diffusion is the process that determines the rate of growth. Accepting the two-stage model of lamellar crystal growth, let us consider the ratio of the contributions of the tangential growth in the plane (111) and layer-by-layer growth in the direction normal to the plane (111). The growth in the tangential direction is limited by diffusion and by the size of the previous layer, that is, is determined by the initial conditions and the shape and size of the plates, formed during the cooling of the solution melt 10–20° below the point of liquidus. Since the formation of the initial crystals occurs in a very short time, the growth in the tangential direction is difficult to control, as it is in the case of the dendritic crystal growth. The growth in the direction normal to the plane (111) is sensitive to the fluctuations of the degree of supersaturation, resulting from the thermal motion of atoms and temperature fluctuations. In general, the better the homogeneity and perfection of the surface of the lamellar GaP crystals, the lesser the ratio of the time necessary for the formation of flat embryos on it, to the time for which the embryo is created on the next plane. The latter, obviously, in an extreme extent depends on fluctuations of the degree of supersaturation in different points of the growing plane and at different points in time. With the deterioration of precision temperature control, the number of fluctuations increases and low-quality crystals grow in the conditions of "entanglement" of the above-noted characteristic times. Small fluctuations of supersaturation also are probably the cause of twinning planes. Increase of the accuracy of temperature regulation promotes some reduction in the number of twins, and this fact also confirms the notable influence of fluctuations on the twinning [1, 3, 34].

Thus, deteriorations of quality or shape of defects of GaP plates during their growth from Ga-P melt solution were observed in the following cases: (1) at a considerable increase of concentration of impurities in the GaP solution and dependently on the chemical nature of the chosen impurity and (2) at the change of the crystallization conditions, namely, at the sufficient deterioration of the cooling velocity control or in the case of nonuniform heat removal from the container for the crystal growth.

Processes for the formation of dendritic needles and their transformation to platelet crystals take place over a small temperature and time interval. As a result, they are hard to control, while crystal growth is easily controlled in the direction normal to the plate (111), due to the sensitive nature of velocity of the solution cooling and accuracy of the temperature control [1, 3].

The above-stated results imply that only high-quality and carefully controlled growth equipment will yield the highest-quality GaP crystals. However, despite these precautions, the subsequent text shows how to considerably increase the quality and utility of freshly prepared GaP crystals.

3. Elaboration of technologies for fabrication of GaP nanoparticles

GaP nanoparticles have been prepared using white P under mild aqueous low-temperature synthesis conditions using two colloidal methods from mixtures of $GaCl_3 \bullet nH_2O$ and dry $NaBH_4$ or Na_3P and $GaCl_3$ diluted in toluene. Details are provided in Refs. [17–20].

Uniform GaP nanoparticles formed following ultrasonic and other treatments were found to exhibit improved quality in their colloidal stability and brightness of luminescence at room temperature, which is found to be broadband with a maximum at 3 eV. Also discussed is their use in GaP/polymers nanocomposites for light-emissive device structures [28, 30, 31].

Photoluminescence, spontaneous Raman scattering, and X-ray diffraction, together with high-resolution electron microscopy of the nanoparticles prepared under different conditions, were compared with those from bulk single crystals.

For the first time to the best of our knowledge, it is shown that well-aged GaP bulk crystals as well as high-quality GaP nanoparticles have no essential difference in their luminescence behavior, brightness, or spectral position of emitted light.

While bulk and thin-film GaP has been successfully commercialized for many years, their application in nanocomposites as a new optical medium has only recently received attention. This section reviews recent efforts to advance the quality of GaP nanoparticles for light-emissive devices based on polymer/GaP nanocomposites.

This activity is the important milestone in the creation of the nanocomposites for advanced light-emissive device structures because GaP nanoparticles exhibit important luminescent and electroluminescent properties and are compatible with selected polymer matrices.

The quality of GaP nanoparticles was improved using mild aqueous synthesis and different colloidal reactions of Ga and P sources in toluene [26–37]. Ultrasonication and ultracentrifugation were applied during the synthesis and selection of nanoparticles to increase their quality and to control the size of the nanoparticles.

In 2005, the authors developed methods to fabricate GaP nanoparticles [18]. So, the technology and properties of the nanoparticles obtained in 2005–2006 and later [27–30] are a good reference point for comparison of the new data provided herein [31].

The first samples of GaP nanoparticles having a distinct luminescence at room temperature were obtained by hydrothermal method from aqueous solutions at a relatively low temperature (120–200°C). It was found that the composition of the nanoparticles corresponds to stoichiometric GaP. The colloidal method provides a good opportunity to control the conditions of the synthesis, to decrease power inputs, and to increase the quality of nanoparticles concerning their purity and uniformity of their dimensions. In actuality, the single parameter, which may be controlled in the other methods, is temperature, while using colloidal methods one can control nucleation of nanoparticles as well as velocity of their growth. The other important advantage of the colloidal method is the ability of the so-called "capping," that is, to isolate nanoparticles from each other, to prevent their agglomeration during storage, simultaneously inhibiting their further growth. Therefore, we have elaborated the methods of GaP nanocrystals colloidal synthesis using $NaBH_4$ and Na_3P compounds.

4. Development of methods of incorporation of the GaP nanoparticles into polymers

Polyglycidyl methacrylate (PGMA), polyglycidyl methacrylate-co-polyoligoethyleneglycol methacrylate (PGMA-co-POEGMA), and biphenyl vinyl ether (BPVE) polymers were used to synthesize GaP nanocomposites suitable for light-emissive luminescent device structures. Film nanocomposites of good quality with high brightness and broadband luminescence were realized. The thickness of the polymer composite film was within 250–300 nm defined from atomic force microscope (AFM) experiment. The following procedures have been used in the fabrication of the nanocomposites:

1. GaP powder was ultrasonicated in methylethylketone (MEK) using Branson 5210 ultrasonic bath. Then, PGMA was added to the MEK solution. GaP to polymer ratio was less than 1:10.
2. GaP powder was dispersed in water-ethanol mixture (1:1 volume ratio) and ultrasonicated using Branson 5210 bath for 120 min. Then, PGMA-co-POEGMA was added in the form of water-ethanol mixture (1:1 volume ratio) solution. GaP to polymer ratio was less than 1:3. Nanocomposite films were deposited on quartz slides via dip coating.
3. GaP powder was dispersed in the biphenyl vinyl ether/dichloromethane (BPVE/DCM) solution; the solution was stirred and filtered from the excess of the powder. A few milliliter drops of the settled solution were casted onto silicon wafer.

More details on the preparation and characterization of our GaP/polymers nanocomposites can be found in Refs. [31, 33–36].

Figure 1a shows transmission electron microscope (TEM) images of GaP nanoparticles obtained from aqueous synthesis. One can see GaP nanoparticles, having characteristic dimensions less than 10 nm. The washed, thoroughly ultrasonicated, and dried nanopowder contains mainly single nanoparticles, while the same powder obtained without ultrasonic treatment consists of the clusters with the dimensions of the order of 100 nm.

Figure 1b shows atomic force microscope (AFM) topographic images of the GaP/PGMA film nanocomposite deposited by dip coating from a suspension in water-ethanol mixture solution on the surface of a silica substrate. The AFM images demonstrated that no significant aggregation was caused by the polymerization. In general, individual particles were observed. The relevant luminescence spectra are presented in **Figures 2** and **3**.

Figure 2 shows the spectra for GaP/PGMA-co-POEGMA nanocomposites. Comparing the results for the nanocomposites prepared from GaP powder or suspension (**Figure 2**, spectra 1 and 2, respectively), it was established that the best quality has the nanocomposites obtained from the nanoparticles stored as a suspension in a suitable liquid (see spectrum 2).

According to our measurements, the matrix polymers PGMA-co-POEGMA or BPVE used in this work provide no contribution to the luminescence spectra of luminescence based on these matrixes GaP nanocomposites presented in **Figures 2** and **3**, so, the nanocomposite spectra coincide with those obtained from the relevant GaP powders or suspensions. It is noted that in the GaP/BPVE nanocomposite, the position of the luminescent maximum can be changed between 2.5 and 3.2 eV and the brightness is 20–30 more than in the PGMA and PGMA-co-POEGMA matrixes. We explain the broadening of the luminescence band and the shift of its maximum to low-photon energies in luminescence of the nanocomposite based on the GaP powder in **Figure 2**, spectrum 1, by the presence of the nanoparticles with the dimensions of 10–100 nm in the powder.

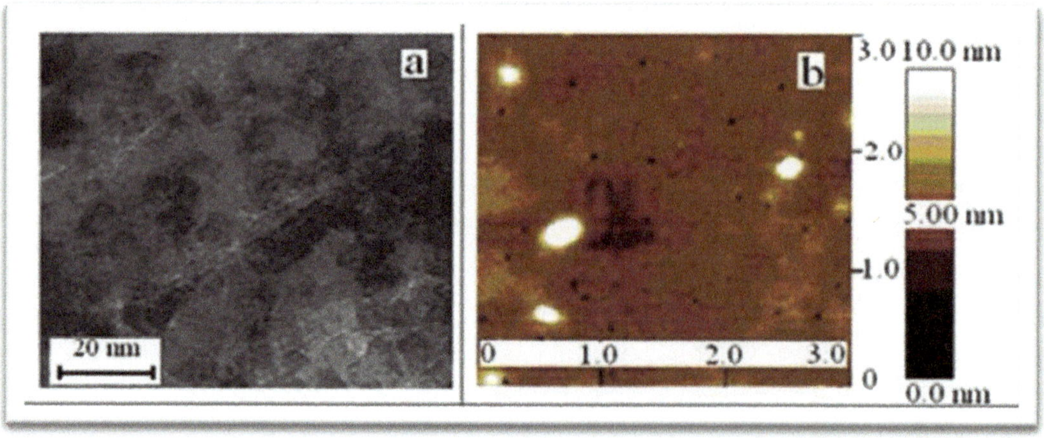

Figure 1. TEM image of GaP thoroughly ultrasonicated and dried nanoparticles obtained by mild aqueous synthesis (a) and AFM topography image of the GaP/PGMA nanocomposite (b) [17].

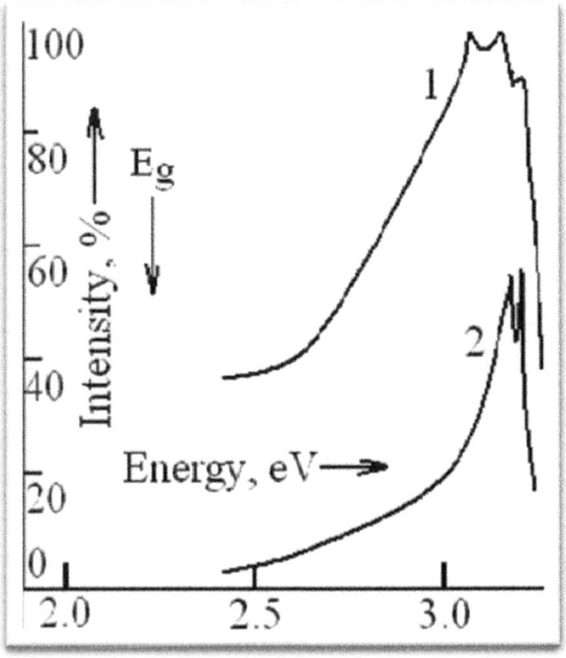

Figure 2. Spectra of luminescence from GaP/PGMA-co-POEGMA nanocomposites. Nanoparticles have been prepared using white P by mild aqueous synthesis and stored as the dry powder (spectrum 1) or suspension in a liquid (spectrum 2) [17].

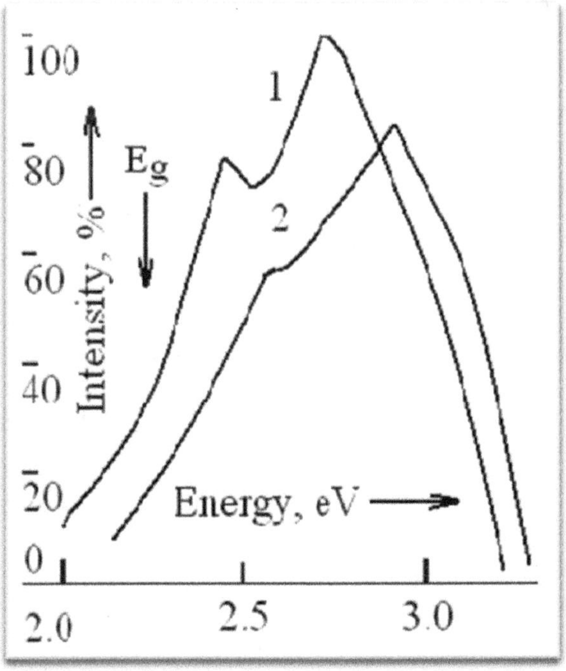

Figure 3. Luminescence spectra of 2 GaP/BPVE nanocomposites produced on the base of two parties of GaP nanoparticles prepared using different conditions [17].

Figures 2 and **3** present a clear image of the quantum confinement effect in the GaP nanoparticles. In accord with our data, the shift is about a few tenths of eV and, obviously, it is impossible to explain only through this effect the dramatic 1 eV enhancement to the region of luminescence at 300°K on the high-energy side of the spectrum.

In order to explain this interesting phenomenon, we postulate that the nanocrystals, much like the ideal long-term-ordered bulk GaP single crystals, exhibit this increase in blue-shifted luminescence due to (a) negligibly small influence of defects and non-radiative recombination of electron-hole pairs and very high efficiency of their radiative annihilation, (b) high perfection of nanocrystal lattice, and (c) high transparency of nanocrystals due to their small dimensions for the light emitted from high points of the GaP Brillouin zones, for instance, in the direct transitions $\Gamma^1{}_c$–$\Gamma^{15}{}_v$ between the conductive and valence bands with the photon energy at 300°K equal to 2.8 eV [39] and (d) high efficiency of this so-called "hot" luminescence. On the basis of these improved technologies for the preparation of GaP nanoparticles and GaP/polymer nanocomposites, we can control within broad limits the main parameters of luminescence and expect therefrom to create a framework for novel light-emissive device structures using this dramatic 1-eV expansion of GaP luminescence to the UV region.

Thus, the film device structures, elaborated by us, demonstrate broadband luminescence in the region from UV until yellow-red with controlled width and position of maximum with the luminous intensity up to 1 cd compared with industrial light-emitting diodes.

5. Optical properties of perfect, long-term-ordered GaP:N in comparison with freshly prepared crystals

Investigating gallium phosphide (GaP) crystals grown by the author in the 1960s, clear improvement of their optical and mechanical properties was noted for the first time only after 10–12 years. It was unusual and interesting in the situation when all inorganic nature around us usually deteriorates in time. Therefore, the decision was taken to investigate, to understand, and to use this phenomenon in the future. In short, the investigation process can be described as follows.

Figure 4 provides a comparison of the evolution in luminescence spectra at low temperatures (80 K and below) from GaP:N over a period of 25 years. It is seen over this period zero-phonon line A of single N-impurity-bound excitons and their phonon replica are narrower in their line widths when compared to the freshly prepared single crystals. Further, as expected, zero-phonon line and replica in samples aged at room temperature for 25 years shift spectral position depending upon the concentration of N impurities (**Figure 4b**, spectra 1–3) according to Refs. [38, 39], while the same freshly prepared crystals exhibited broader luminescence line widths with increasing nitrogen content (**Figure 4b**, spectrum 4). These, along with other half-centennial findings, including modifications in luminescence kinetics, spontaneous Raman scattering, X-ray diffraction, absorption spectra, micro-hardness, and density of dislocations, that are reported elsewhere [1, 3–24], strongly suggest that close-to-ideal GaP:N crystals form over time due to the equally spaced disposition of N impurities instead of their chaotic distribution in the same freshly prepared crystals.

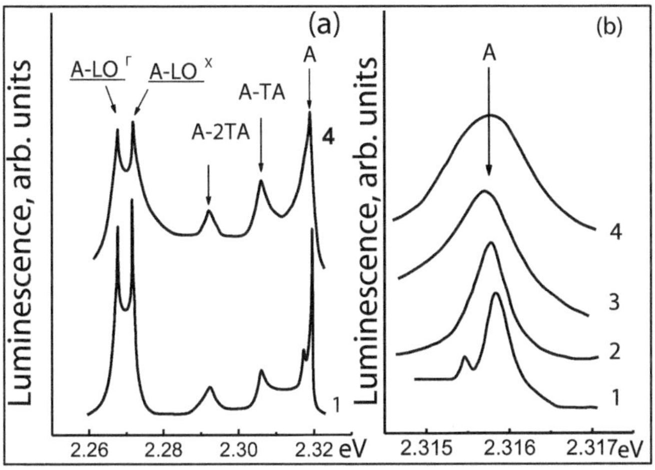

Figure 4. Evolution of the GaP:N luminescence with time and nitrogen concentration at the temperature of 15 K. (a) Zero-phonon line of the bound exciton A and its transversal acoustic (TA) and longitudinal optic (LO) phonon replica in as-prepared (4) and 25-year long-term-ordered (1) crystals. (b) Zero-phonon line A as a function of nitrogen (N) concentration. 1–3: 25-year-old crystals. 4: as-prepared. Curves 1 through 4 represent samples with nitrogen concentrations of 10^{17}, 10^{18}, 10^{19}, and 10^{18} cm^{-3}, respectively [34].

As first noted in Ref. [23], these results suggest a new type of crystal lattice in which the host atoms occupy their proper (equilibrium) positions in the crystal, while the N impurities are periodically substituted into the lattice portion into short chains of equal length. According to the data obtained from Raman light scattering [10, 14], host atoms of this new lattice develop harmonic vibrations, and high degree of lattice perfection leads to an abrupt decrease in the non-radiative recombination and an increase of efficiency and spectral range of luminescence.

Stimulated emission of light in these temporally ordered crystals (**Figure 5b**) also is observed. As was shown in Ref. [23], the GaP:N crystals aged for at least 40 years possess no discrete

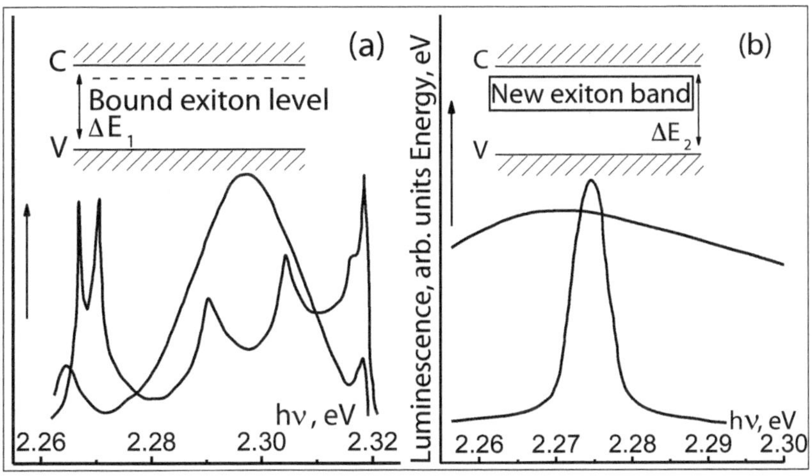

Figure 5. Luminescent spectra and schematic representation of the forbidden gaps (ΔE_1, ΔE_2) in the nitrogen-doped GaP aged for (a) 25 years and (b) 40 years [34].

impurity level for N-bound excitons in the forbidden gap. They also demonstrated a uniform luminescence from a broad excitonic band instead of the narrow zero-phonon line and its phonon replica as observed from the less-aged, 25-year-old crystals.

Thus, long-term-ordered GaP:N crystals demonstrate uniform bright luminescence from a broad excitonic band instead of the narrow zero-phonon line and its phonon replica in disordered and partly ordered (25-year-old) crystals. This is due to the fact that ordered crystals have no discrete impurity level in the forbidden gap. To the best of our knowledge, such transformation of a discrete level within the forbidden gap into an excitonic band (**Figure 5a, b**) is observed for the first time. In this case, the impurity atoms regularly occupy the host lattice sites and affect the band structure of the crystals, which are now a dilute solid solution of GaP-GaN with regular disposition of N atoms instead of freshly prepared GaP doped by occasionally located N atoms. Note that the increase of luminescence excitation in case of partly ordered GaP:N (**Figure 5a**, dotted line) leads to a broad luminescence band as a result of bound exciton interaction [9], while in the case of perfectly ordered crystals (**Figure 5b**), one can see an abrupt narrowing of the luminescence band due to stimulated emission in defect-free crystals. Earlier, in freshly prepared crystals, we observed a clear stimulated emission from a GaP:N resonator at 80 K [5], as well as the so-called superluminescence from GaP single crystals having natural faceting. Presently, our ordered crystals have a bright luminescence at room temperature that implies their perfection and very low light losses. In our studies [14, 23], we demonstrate that the stimulated emission in long-term-ordered GaP is also developed even at room temperature by electron-hole recombination of an electron at the bottom of the conduction band with a hole at the top of the valence band and the LO phonon absorption.

6. Comparison of optical properties of GaP:N nanocrystals and GaP perfect bulk single crystals

It is necessary to note that the very important for optoelectronics long-term ordering and considerable improvement of the semiconductor crystal lattice and accompanying phenomena have been discovered and observed over decade time scales only due to grown in 1960th [1, 3, 7] unique collection of samples and control of evolution of their properties for 50years. And up to now, this collection of long-term-ordered perfect GaP single crystals gives opportunities to find deep fundamental analogies in properties of the perfect single crystals and nanoparticles as well as to predict and to realize in nanoparticles and perfect bulk crystals new interesting properties and applications. Jointly with Refs. [3, 7, 9–17, 21–27, 31–34], this section is a generalization of the results on long-term observation of luminescence, absorption, and Raman light scattering in bulk semiconductors in comparison with some properties of the best to the moment GaP nanocrystals. Combination of these characterization techniques elucidates the evolution of these crystals over the course of many years, as well as the ordered state brought about by prolonged room-temperature thermal annealing, and the interesting optical properties that accompany such ordering. We demonstrate that long-term natural stimuli result in perfection of our crystals, which can lead to novel heterogeneous systems and new semiconductor devices with high temporal stability.

We further improved upon the preparation of GaP nanocrystals using the known methods of hydrothermal and colloidal synthesis [29–31].

The highest-quality GaP nanoparticles have been prepared using hydrothermal or colloidal synthesis from white phosphorus at a decreased temperature (125°C) and intense ultrasonication. It has been established that the maximum shift to ultraviolet and the best quality, in general, are from nanocomposites obtained from the nanoparticles stored as a suspension in a suitable liquid.

As noted earlier, different defects of high concentration in freshly prepared GaP single crystals completely suppress any luminescence at room temperature due to the negligible quantity of free path for non-equilibrium electron-hole pairs between the defects and their non-radiative recombination, while the quantum theory predicts their free movement in the field of an ideal crystal lattice. It was also shown the long-term-ordered and therefore close-to-ideal crystals demonstrate bright luminescence and stimulated emission repeating behavior of the best nanoparticles with pronounced quantum confinement effects.

Figure 6 compares the luminescence spectra of our long-term (up to 50 years)-ordered GaP single crystals (spectrum 1) to that from high-quality GaP nanoparticles and their GaP nanoparticles/polymers nanocomposites [15, 17]. Nanocrystals stored as dry powder demonstrate rather broad luminescent band with maximum at 2.8 eV (**Figure 6**, spectrum 2), while the nanocrystals of about 10-nm sizes, thoroughly separated and distributed in a suspension, which prevents their coagulation, mechanical, and optical interaction, exhibit bright narrow-band

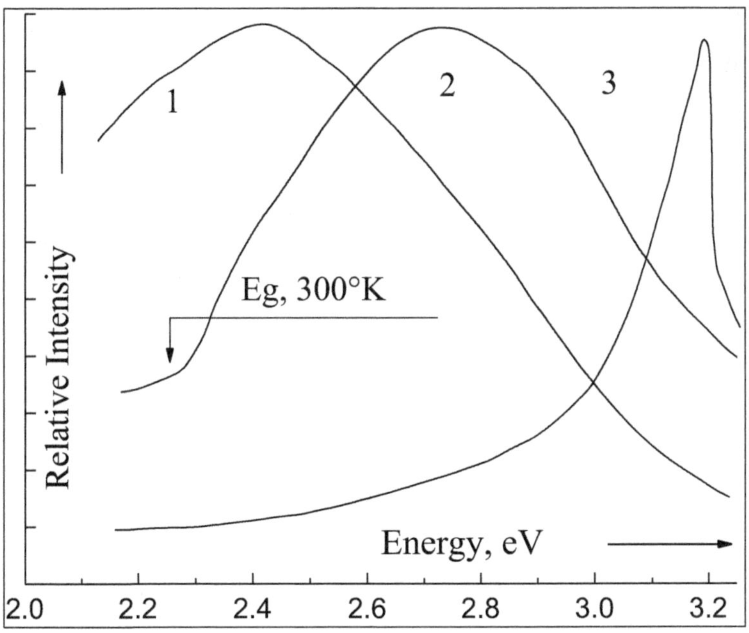

Figure 6. Luminescence of perfect bulk GaP single crystals (1) in comparison with the luminescence of GaP nanoparticles and GaP/polymers nano-composites (2, 3). Nanoparticles were prepared from white P by mild aqueous or colloidal synthesis at decreased temperature, stored as dry powder (spectrum 2) or suspension in a liquid (spectrum 3). Details can be found in Refs. [15, 17].

luminescence with maximum at 3.2 eV, approximately 1 eV above the position of the absorption edge in GaP at 300°K (**Figure 6**, spectrum 3).

The broadening of the luminescence band and the shift of its maximum to low-photon energies in the nanocomposite is explained based on the different dimensions (between 10 and 100 nm) of the GaP powder. Meanwhile, the nanocomposites on the basis of the suspensions containing only approximately 10-nm nanoparticles exhibit bright luminescence with maximum at 3.2 eV due to high transparency of 10-nm nanoparticles for these high-energy-emitted photons and pronounced quantum confinement effect.

Our first attempts to prepare GaP nanoparticles [18, 27] yielded room-temperature luminescence with maximum shifted only to 2.4 eV in comparison with the achieved now new maximum at 3.2 eV achieved today. It confirms significant progress in the technology of GaP nanoparticles and GaP/polymers nanocomposites. On the basis of these improved approaches to prepare GaP nanoparticles and GaP/polymer nanocomposites, we can change the main parameters of luminescence within the broad limits, and to create a framework for novel light-emissive device structures using dramatic 1-eV expansion of GaP luminescence to UV region.

7. Excitonic crystal and its importance in optoelectronics

The role and application of bound excitons in nanoscience and technology are discussed in this section. Bound excitons are well studied in semiconductors, especially in gallium phosphide doped by nitrogen (GaP:N). Doping of GaP with N leads to isoelectronic substitution of the host P atoms by N in its crystal lattice and to the creation of the electron trap with a large capture cross section. Therefore, any non-equilibrium electron in the vicinity of the trap will be captured by N atom, attracting a non-equilibrium hole by Coulomb interaction and creating the bound exciton—short-lived nanoparticle with the dimension of the order of 10 nm (it is the Bohr diameter of bound exciton in GaP:N). Note that none of the nanotechnology methods are used in the creation or selection of dimensions of these nanoparticles—only natural forces of electron-hole interaction and electron capture by the traps are necessary for the creation of these nanoparticles. As a result, we get something like neutral short-lived atom analog—a particle consisting of heavy negatively charged nucleus (N atom with captured electron) and a hole. The so-called "zero vibrations" do not destroy possible solid phase of bound excitons having these heavy nuclei that give an opportunity to reach their crystal state—short-lived excitonic crystal.

Taking into account the abovementioned results, a model for the crystal lattice and its behavior at a high level of optical excitation for well-ordered N-doped GaP (**Figure 7**) can be suggested. At the relevant concentrations of N, the anion sublattice can be represented as a row of anions where N substitutes P atoms with the period equal to the Bohr diameter of the bound exciton in GaP (approximately 10 nm) (**Figure 7a**). At some level of excitation, all the N sites will be filled by excitons, thereby creating an excitonic crystal (**Figure 7b**), which is a new phenomenon in solid-state physics and a very interesting medium for application in optoelectronics and nonlinear optics [4, 15, 17, 25, 26, 32–34].

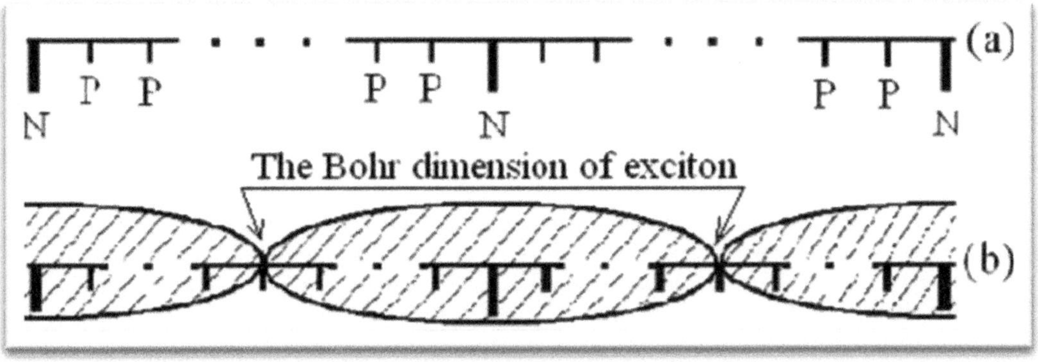

Figure 7. The models of the well-ordered GaP:N [34]. (a) The new type of crystal lattice with periodic substitution of N atoms for the host P atoms. (b) The excitonic crystal on the basis of this lattice. The substitution period is equal to the Bohr diameter of exciton (~100 Å), and optical excitation is enough for complete saturation of the N sublattice with non-equilibrium electron-hole pairs (please see details in Refs. [15, 17]).

Thus, using bound excitons as short-lived analogs of atoms and sticking to some specific rules, including the necessity to build the excitonic superlattice with the identity period equal to the bound exciton Bohr dimension in the GaP:N single crystal, we get a unique opportunity to create a new solid-state media—consisting of short-lived nanoparticles excitonic crystal, obviously, with very useful and interesting properties for application in optoelectronics, nanoscience, and technology.

8. Current approach to selection of materials for electronics and new prospects

The ability of matter to emit light of definite wavelengths and intensity depends on many factors; among them are the details of its band structure and crystal quality. Applied to light-emissive materials and device structures, the quality is characterized by the concentration of the defects damaging light emission due to non-radiative electron-hole recombination on these defects. The ways for minimization of expected losses of the light in non-radiative electron-hole recombination are (1) decrease of concentration of the defects in the material grown by an improved method; (2) considerable quality improvement of GaP crystal structure discovered by us during long-term (tens of years) ordering of the host and impurity atoms, and (3) application of nanocrystals with the dimensions less than the free paths of electrons and holes in the chosen light-emissive material.

Properties that can be used to achieve light emission at a definite spectral region include (1) selection of the material on the basis of its band structure peculiarities and the value of the forbidden gap, which as a rule sizes up the maximum value of the emitting photons; (2) doping of the selected material by impurities or, on the contrary, precise purification of the chosen material for the creation of new channels of light emission; and (3) selection of the temperature and/or excitation level that to change contribution of different channels into light emission due to their disappearance or saturation.

It is impossible or, at least, very difficult to realize all the above-noted possibilities in a single material. Therefore, the present technologies of light-emissive optoelectronics employ diverse range materials using very complicated and expensive technologies for their fabrication with the necessary application parameters. Down-selecting a set of various materials applied currently in optoelectronics at the maintenance or even improvement of their high parameters and quality is one of the most important goals for reduction in price and wide application in the practice of electronic devices. The most impressive demonstration of this approach would be the selection of a suitable material, the development of a peculiar technology of its preparation, its interesting properties and prospects for application in optoelectronics instead of a huge number of various materials and their technologies.

Sophisticated growth technique, knowledge of crystal growth mechanisms, of different processes, improving or deteriorating in time the prepared crystals, and the relevant comprehensive experience are necessary for the preparation of the perfect, free or almost free of contamination and doped GaP single crystals.

However, in the manufacture of artificial crystals, we are faced with the need to fabricate them for a very short time compared with the times for which naturally perfect crystals are grown. For example, a hundred years passes before the growing in the nature crystal acquires the dimensions necessary for its use as feedstock in jewelry making, while the commercial growth of moderately perfect semiconductors takes only a few hours. In such a short time, its own atom or dopant diffusing through the crystal usually cannot absolutely exactly occupy the places intended for them in ideal crystal lattice. This gives rise to lattice defects, which, in turn, hinder the achievement of perfect devices. Manufactured on the basis of such materials, greatly reduce their lifetime, as well as make it impossible manifestation of some very interesting and useful for the application of the effects observed in advanced materials. Significant levels of funds are spent on searching and implementing methods for creating materials, not existing in the nature, that can, even if in a narrow field of application and at short time service, replace the material with a perfect crystal lattice and the necessary concentration and location of dopants. Hence, there is an urgent need to develop methods for the creation of an ideal semiconductor material the required parameters for use in electronics. In this case, it will disappear or substantially decrease the need to develop a plurality of individual costly technology to produce a plurality of different materials with a limited resource and service applications, and as a very important result for any country involved in the production and use of electronic materials—a giant savings and the possibility of redeployment to other needs of funds spent on the production of low-quality materials and products.

Using the long-term-ordered perfect GaP or similar on behavior and properties material in electronic industry instead of current elaboration of very expensive and labor-consuming technologies for diverse materials with their limited application spectral region and other parameters, we get a big commercial, scientific, and technology advantage from their fabrication and application.

We hope that the methods based on the application of this long-term evolution of the important properties of our unique collection of semiconductor single crystals [31–34] as well as the combined methods of laser-assisted and molecular beam epitaxies [40–42] elaborated by us promise a novel approach to the development of a new generation of optoelectronic devices.

The epitaxial methods, elaborated by us [40–42], will be applied to the fabrication of device structures with artificial periodicity; together with classic methods of crystal growth, they can be employed to realize impurity ordering that would yield new types of nanostructures and enhanced optoelectronic device performance.

The long-term, tens of years ordered GaP, its artificial analogs, and nanoparticles have very interesting and properties for different application, which give an opportunity to apply them in optoelectronics with high economic effect instead of a lot of different compounds currently used in optoelectronics.

9. Conclusions

Thus, our half-of-a-century collection of results provides a new approach to the selection and preparation of perfect optoelectronics materials (please see references to this chapter) and a unique opportunity to realize a new form of solid-state host—the excitonic crystal [32–34]. These results confirm expedience of the efforts directed to the formation in GaP of the N-impurity superlattice having a lattice period equal to the bound exciton dimension. As noted in this chapter and the relevant references, high-quality material for industrial electronics can be prepared by storing freshly grown crystals for years in a special storage. Only old crystals with the attained necessary properties will then be annually retrieved for device fabrication, while new portions of fresh crystals will be placed for long-term ordering at room temperature, as is done with wines and fine liquors.

All of the results presented here and included in summary reviews [25, 26, 32–34] may sufficiently change the approach to the selection of materials necessary for electronics, and to make cheaper and simpler technology for the preparation of the selected materials and device structures on which they are based. This study of long-term convergence of bulk- and nanocrystal properties [15, 17] brings a novel perspective to improving the quality of semiconductor crystals. The author's unique collection of pure and doped crystals of semiconductors grown in the 1960s provides an opportunity to observe and understand the physics behind the long-term evolution of properties in these key electronic materials.

Acknowledgements

The authors are glad to note that the broad discussion and dissemination of their joint results stimulate further collaboration with reliable partners from the USA, Russia, Italy, Romania, France, and other countries.

Prof. Sergei L. Pyshkin expresses his cordial gratitude to his teachers, world-known scientists late Profs Nina A. Goryunova, Nobel Prize Laureate Alexander M. Prokhorov, and academicians Rem V. Khokhlov and Sergei I. Radautsan. He is also very grateful to the US Department of State, Institute of International Exchange, Washington, DC, the US Air Force Office for Scientific

Research, the US Office of Naval Research Global, Civilian R&D Foundation, Arlington, VA, the US Science & Technology Center in Ukraine, to his colleagues and coauthors from Clemson University, SC, University of Central Florida, FL, Istituto di elettronica dello stato solido, CNR, Rome, Italy, Universita degli studi, Cagliari, Italy, Lomonosov Moscow State University, Joffe Physico-Technical Institute and State Polytechnical University, St. Petersburg, Ac. Scie. Institute of General Physics, Moscow, Russia, Institute of Applied Physics and Academy of Sciences of Moldova for support and attention to this protracted (1961 to present time) research.

Author details

Sergei L. Pyshkin[1,2*] and John Ballato[2]

*Address all correspondence to: spyshkin@yahoo.com

1 Institute of Applied Physics, Academy of Sciences of Moldova, Kishinev, Moldova, Republic of Moldova

2 Clemson University, South Carolina, USA

References

[1] Goryunova NA, Pyshkin SL, Borshchevskii AS, et al. Influence of impurities and crystallization conditions on growth of platelet GaP crystals. In: Symposium on Crystal Growth at the 7th International Crystallography Congress (Moscow, July 1966). N.N. Sheftal, editor, Journal of Growth of Crystals 1969; 8: 68-72, New York.

[2] Pyshkin SL, Radautsan SI, Zenchenko VP. Processes of long-lasting ordering in crystals with a partly inverse spinel structure. Journal of Soviet Physics – Doklady 1990; 35(4): 301-4.

[3] Pyshkin SL. Preparation and Properties of Gallium Phosphide. Ph.D. thesis 1967, State University of Moldova, Kishinev.

[4] Pyshkin SL, Zifudin L. Excitons in highly optically excited gallium phosphide. Journal of Luminescence 1974; 9: 302-8.

[5] Pyshkin SL (presented by Nobel Prize Laureate A.M. Prokhorov). Stimulated emission in gallium phosphide. Journal of Soviet Physics – Doklady 1975; 19: 845-6.

[6] Pyshkin S. Luminescence of GaP:N:Sm crystals. Journal of Soviet Physics of Semiconductors 1975; 8: 912-3.

[7] Pyshkin SL. Photoconductivity and luminescence of highly optically excited semiconductors. Dr.Sc. thesis 1978, Lomonosov Moscow State University.

[8] Pyshkin S, Anedda A, Congiu F, Mura A. Luminescence of the GaP:N Ordered System. Journal of Pure and Applied Optics 1993; 2: 499-502.

[9] Pyshkin S., Alberto Anedda. Time-Dependent Behaviour of Antistructural Defects and Impurities in Cd-In-S and GaP. Institute of Physics Conference Series 1998, Ternary and Multinary Compounds; 152, Section E: 785-789

[10] Pyshkin S, Ballato J, Chumanov G. Raman light scattering from long-term ordered GaP single crystals. Journal of Optics A: Pure and Applied Optics 2007; 9: 33-6.

[11] Pyshkin SL, Ballato J, Bass M, Turri G. Luminescence of long-term ordered pure and doped gallium phosphide. TMS Annual Meeting, Invited Talk at the Symposium: Advances in Semiconductor, Electro Optic and Radio Frequency Materials, March 9-13, New Orleans, LA. J. Electronic Materials 2008; 37(4): 388-95.

[12] Pyshkin SL, Ballato J, Bass M, Chumanov G. Time-dependent evolution of crystal lattice, defects and impurities in $CdIn_2S_4$ and GaP. Journal of Physica Status Solidi 2009; C(6): 1112-5.

[13] Pyshkin SL, Ballato J, Bass M, Turri G. Evolution of luminescence from doped gallium phosphide over 40 Years. Journal of Electronic Materials 2009; 38(5): 640-6.

[14] Pyshkin S, Ballato J. Evolution of optical and mechanical properties of semiconductors over 40 Years. Journal of Electronic Materials 2010; 39(6): 635-41.

[15] Pyshkin S, Ballato J. Long-term convergence of bulk- and nano-crystal properties. In:Optoelectronics – Materials and Technics, InTech – Open Access Publisher, ISBN 978-953-307-276-0; Chapter 19: 2011; 459-476.

[16] Pyshkin S. Ballato J. Dramatic expansion of luminescence region in GaP/polymer nanocomposites. Supplemental Proceedings of the 2012 TMS Annual Meetings (Orlando, FL, USA, March 11-15); vol. 1: "Materials Processing and Interfaces": 2012; 353-359.

[17] Pyshkin S, John Ballato. Advanced light emissive device structures. In: Optoelectronics – Advanced Materials and Devices, S.L. Pyshkin and J. Ballato, editors, ISBN 978-953-51-0922-8, InTech – Open Access Publisher; Chapter 1: 2013; 1-24.

[18] Pyshkin S, Ballato J, Chumanov G, DiMaio J, Saha AK. Preparation and Characterization of Nanocrystalline GaP. Technical Proceedings of the 2006 NSTI Nanotech Conference; 3: 194-197.

[19] Pyshkin S, Ballato J, Chumanov G, Tsyntsaru N, Rusu E. Preparation and characterization of nanocrystalline GaP for advanced light emissive device structures. In: The 2010 NSTI-Nanotech Conference (Anaheim, CA) Proceedings: www.nsti.org, ISBN 978-1-4398-3401-5; vol. 1: 522-5.

[20] Pyshkin SL, Ballato J, Belevschii S, Rusu E, Racu A, Van DerVeer D. Synthesis and characterization of GaP nanoparticles for light emissive devices. In: The 2011 NSTI-Nanotech Conference, Boston, MA, June 13-16. Proceedings: www.nsti.org, ISBN 978-1 4398-7142-3; vol. 1, 327-30.

[21] Pyshkin SL, Ballato J, Bass M, Chumanov G, Turri G. Time-dependent evolution of crystal lattice, defects and impurities in $CdIn_2S_4$ and GaP. Physica Status Solidi 2009; C(6): 1112-5.

[22] Pyshkin S, Zhitaru R, Ballato J, Chumanov G, Bass M. Structural characterization of long-term ordered semiconductors. Proceedings of the 2009 MS&T Conference, International Symposium "Fundamentals & Characterization": 698-709.

[23] Pyshkin S, Ballato J, Bass M, Chumanov G, Turri G. Properties of the long-term ordered semiconductors. Suppl. Proceedings of the 2009 TMS Annual Meeting and Exhibition; 3: 477-84 (San Francisco, February 15-19, 2009).

[24] Pyshkin S, Ballato J, Chumanov G, Bass M, Turri G, Zhitaru R, et al. Optical and mechanical properties of long-term ordered semiconductors. Moldavian Journal of the Physical Sciences 2009; 8(3-4): 287-95. The 4th Int. Conference on Materials Science and Condensed Matter Physics, Kishinev, Sept 23-26, 2008.

[25] Pyshkin SL. Gallium Phosphide - New Prospect for Optoelectronics. J. Advances in Optoelectronic Materials (AOM), ISSN Online: 2327-7661, ISSN Print: 2327-767X); 2013; 1(4): 59-66.

[26] Pyshkin SL. Excitonic crystal and nanotechnology. Symbiosis Journal of Nanoscience & Technology: Open Access 2014; 1(2), 1-5. Available from http://symbiosisonlinepublishing. com/nanoscience-technology08.pdf

[27] Pyshkin S, John B. Advanced light emissive composite materials for integrated optics. Symposium: The Physics and Materials Challenges for Integrated Optics – A Step in the Future for Photonic Devices, Proc. of the 2005 MS&T Conference, Pittsburgh: 3-13.

[28] Pyshkin S, Ballato J, Luzinov I, Zdyrko B. Fabrication and characterization of the GaP/polymer nanocomposites for advanced light emissive device structures. The 2010 NSTI-Nanotech Conference; 1: 772-5. www.nsti.org, ISBN 978-1-4398-3401-5. Anaheim, CA, June 21-24, 2010.

[29] Pyshkin S, Ballato J, Chumanov G, Tsyntsaru N, Rusu E. Preparation and characterization of nanocrystalline GaP for advanced light emissive device structures. The 2010 NSTI-Nanotech Conference; 1: 522-5. www.nsti.org, ISBN 978-1-4398-3401-5.

[30] Pyshkin SL, Ballato J, Luzinov I, Zdyrko B. Fabrication and characterization of GaP/polymer nanocomposites for advanced light emissive device structures. Journal of Nanoparticle Research 2011; 13: 5565-70.

[31] Pyshkin SL. (Project Manager). Advanced Light Emissive Device Structures. STCU (www.stcu.int) Project 4610, 2009-2012.

[32] Sergei LP. Excitonic crystal, nanotechnology and new prospect for optoelectronics. The Open Optics Journal 2015; 9: 26-37.

[33] Pyshkin S. New prospect for optoelectronics. Nanoscience & Technology 2016; 3(1): 1-5.

[34] Pyshkin SL. Excitonic crystal and perfect semiconductors for optoelectronics. Chapter 1. In: Materials and Devices. SL Pyshkin and J Ballato, editors, Rijeka: InTech; 2015; 1-30.

[35] Shanmin G, Jun L, Nan C, Yan Z, Yi X. Aqueous synthesis of III-V semiconductor GaP and InP exhibiting pronounced quantum confinement. Chemical Communications 2002: 3064-5.

[36] Zhengang L, Yujun B, Deliang C, Qilong W. Preliminary insight into the formation process of InP and GaP nanocrystals. Solid State Sciences 2003; 5: 1037-40.

[37] Zhang Z-C, Wang B-P. Diamond and oxidized disordered graphite on the surface of gallium phosphide nanoparticles. Systems Characterization 2009; 26: 53-7.

[38] Paul W. Band structure of the intermetallic semiconductors from pressure experiments. Journal of Applied Physics 1961; 32: 2082-94.

[39] Zallen R, Paul W. Band structure of gallium phosphide from optical experiments at high pressure. Physical Review 1964; 134: A1628-41.

[40] Pyshkin SL, Fedoseev SA, Lagomarsino S, Giannini C. Preparation and structural properties of some III-V semiconductor films grown on (100) oriented Si substrates, Journal of Applied Surface Science, 1992; 56-58: 39-43.

[41] Budyanu VA, Chechuy SN, Pyshkin SL, et al. Investigation of III-V/Si heterojunction grown by laser deposition. Physica Status Solidi (a) 1985; 91: 737-44.

[42] Budyanu VA, Chechuy SN, Pyshkin SL, et al. Laser vacuum epitaxy of III-V semiconductors on silicon. Revue Roumaine de Physique 1987; 32(1-2): 216-19.

6

Candidate Materials as Gain Media in Organic, Triplet-Based, Room-Temperature masers Targeting the ISM Bands

Erinn van Wynsberghe and Ayse Turak

Additional information is available at the end of the chapter

Abstract

While lasers have enjoyed greater popularity, masers—which emit coherent radiation in the microwave spectrum—are also of critical importance to a variety of applications. Recently, an organic gain medium has been developed, which allows emission at room temperature without the traditional encumbrances of cryogenic cooling or an externally applied magnetic field, at vastly improved power efficiency. This discovery opens up new avenues for applications that were previously impractical. However, further investigation is still required for frequency tuning of the device, through the selection of alternate gain media beyond the original choice of pentacene-doped *p*-terphenyl and some linear acenes similar to the pentacene prototype. This chapter outlines some of the essential criteria necessary to achieve masing with an organic semiconductor gain medium, including zero-field splitting (ZFS), triplet sublevel division, and metastable population inversion. Three tables of possible candidate materials are presented based on this roster of criteria, particularly targeting emission in one of the industrial, scientific, and medical (ISM) bands. A selection of preferred guest molecules is recommended for in-situ testing as room-temperature masers gain media candidates.

Keywords: room-temperature maser, organic semiconductors, triplet sublevels, zero-field splitting, metastable population inversion, candidate

1. Introduction

The maser, the microwave analogue of the laser, has long been a device of considerable interest [1, 2]. With emission frequencies between 0.3 and 300 GHz, masers have had several significant applications, including precision frequency references for atomic clocks [3–5], radio astronomy [6, 7], space and long-distance communication [8–11], radar [12, 13], remote sensing [14], ultrasensitive magnetic resonance spectroscopy [15], and medical imaging [14]. However, widespread use of such devices has been limited by low efficiency and complex technical requirements. Recent advances in room-temperature masers [16–19] have revitalized interest in its potential uses. Particularly, the work of Oxborrow [16, 20] using pentacene *p*-terphenyl as the masing material has opened up a whole new direction of experimental research, through spin manipulation in organic semiconductors materials. Without the traditional encumbrances of cryogenic cooling or an externally applied magnetic field, these new avenues pave the way for a new generation of miniaturized ultra-low-noise high-gain amplifiers, oscillators, and transceivers.

Of particular interest for a variety of applications are materials which have demonstrated emission in the range of 2.4–2.5 GHz, one of the industrial, scientific, and medical (ISM) bands. ISM frequencies are designated as unlicensed and reserved internationally for experimental and short-range applications such as microwave ovens, cordless phones, military radar, and industrial heaters. The ISM bands have seen a dramatic increase in usage over the past decade for wireless connectivity devices [21], including BlueTooth [22], Zigbee and WLAN [23], and Hiperlan, among others [24]. These bands represent an important frequency range for future internet-of-things applications. Although there have been significant advances in recent years, the only viable room temperature masing demonstrated thus far, from pentacene *p*-terphenyl, is limited to a single emission frequency well away from this desirable application range [16]. Additionally, the conversion efficiency (i.e. the energy output in microwaves compared to the energy input in visible light) at the laboratory scale is still only around 3% [20]. To move beyond this prototype material, a comprehensive examination of the necessary features for room temperature masing using organic semiconductors is required.

In this chapter, we examine a few of the essential criteria necessary to achieve masing with an organic semiconductor gain medium, including triplet zero-field splitting (ZFS), metastable population inversion, and triplet lifetime. We enumerate numerous organic components as alternatives candidates, based on advantageous conditions for these three criteria. Specifically, we target materials with ZFS values favorable to emission in the ISM band, as well as highlighting other interesting materials with desirable properties.

1.1. masers

MASER stands for microwave or molecular amplification by stimulated emission of radiation. Like its counterpart, the LASER (light amplification by stimulated emission of radiation), the maser converts an input of electrical or optical energy into a coherent, focused beam of photons. Both devices operate along similar lines and require the stimulation of atoms to elevate electrons to excited states. The electrons migrate to higher orbital levels if the atoms are "pumped," that is, receive energy from an external source. As those electrons lose their energy (after 10^{-8} seconds), they emit photons by emission and retreat to a lower energy level. If this process occurs naturally, it is referred to as spontaneous emission, whereas if it occurs by design, it is called stimulated emission. Stimulated emission occurs when a photon strikes an electron already suspended at an excited state level: the electron releases its energy and the exiting photon will be in phase with the striking photon. That is, the two photons will travel away coherently, with the same wavelength, frequency, and vector.

Key to coherent emission is the maintenance of stable population inversion. The pump must keep a greater percentage of electrons at the excited states, at a rate faster than the natural relaxation rate to the ground state. Excited state electrons must maintain their stimulated position long enough to allow incident photons to strike and cause coherent cascade emission. For masers, due to the small gap between the excited and ground states (1×10^{-6} to 1×10^{-3} eV), it is relatively easy to produce a high ratio of atoms in the excited state. Additionally, as the ratio of the Einstein coefficients (A, spontaneous emission and B, stimulated emission) varies with the cube of the frequency [25–27], spontaneous emission can generally be neglected for the microwave part of the spectrum.

Charles Townes and co-workers showed the first working maser using NH_3 gas as the gain medium [28]. In such a system, the two energy levels used are the two vibrational states of the ammonia molecule given by the oscillation of the nitrogen atom [1]. The difference between the wave function of the two configurations with N above and below the plane of hydrogen atoms yields an output radiation at 1.25 cm wavelengths [28]. By applying an electric field, the electric dipole moment in the ammonia molecules can be used to separate the two molecular configurations, maintaining a stable population inversion. This first maser proved very effective as a low-noise amplifier and was proposed as the first atomic clock standard by the National Institute of Standards and Technology (NIST) [29]. Further developments of gas [30], and then solid state masers [31, 32], focused on providing population inversion through the manipulation of spin states.

1.2. Disadvantages of traditional masers

Although it is relatively easy to produce population inversion with microwave emission, the major bottlenecks for its effective use in most applications are low power efficiency and complex operational requirements.

For the traditional gas or molecular masers, the modes that yield microwaves, either through conformational changes as in ammonia or spin states as in hydrogen [30], are inherently stable and require only physical separation to maintain stable population inversion. However, the population of molecules in the quantum state of interest is relatively low [33]. They also require high vacuum to prevent gas scattering collisions [28, 30, 33]. With a high vacuum, however, the gas molecules are spatially separated to such an extent that effective stimulation is limited, and the power output of gas-based masers is relatively low (pico to nanowatts) [33].

For solid state materials, the limitations are the opposite. If there are non-degenerate spin states, there can be a large population of atoms in the excited state, as thermal energy is generally sufficient to effectively pump the molecules across the small energy gaps. However, the lifetimes of such excited spin states are very short. Spin-lattice relaxation rates increase exponentially with rising temperature [34] to the degree that at room temperature, spin-relaxation times for many solid materials are in the nanosecond range due to rapid phonon scattering [35]. Additionally, the spin population inversion decreases at higher temperatures [16, 36].

Therefore, most solid state masing materials require cryogenic temperatures to maintain stable population inversion and sufficient lifetimes [10, 35, 37].

The requirement of non-degenerate spin states also limits the applicability of masers. If there is no naturally occurring split within the microwave frequency range, magnetic fields must be applied to induce Zeeman splitting of the degenerate energy states. Many gaseous and solid state materials require the application of large magnetic fields in order to emit at desirable frequencies [38–40].

1.3. Organic masers

One method of overcoming the major bottlenecks of room temperature masing is to use a material that has long excited state lifetimes and natural zero-field splitting, such as an organic semiconductor. Despite having low mobility and complex transport properties, organic semiconductors have significant advantages over inorganic semiconductors as a gain medium.

Lifetimes of spin excited states in organics are substantially longer than inorganic materials, microseconds instead of pico or nanoseconds [41]. In some aromatic molecules, lifetimes have been observed as high as milliseconds or even full seconds [42–45]. The long lifetime in organics is due to the weak spin-orbit coupling (SOC), as a result of the low molecular weight of the materials involved, such as carbon and hydrogen in small molecule arrangements (low Z value) or π-conjugated polymers [41]. The strength of spin-orbit interaction is proportional to Z^4 [46].

Oxborrow et al. demonstrated the only room-temperature solid state maser observed thus far based on a pentacene-doped *p*-terphenyl molecular crystal, where the spin lifetime can reach 135 μs at room temperature [16]. This result relies on the excitonic route to forming stable states with suitable separation for microwave emission. The organic maser functions by photoexciting a solid state gain medium composed of an organic guest molecule within a solid polymer crystal (which taken together comprise a Shpolskii matrix [47]) and then emitting photons by exciton decay within the triplet sublevels. Radiative emission is then guided and amplified by a resonance cavity to form a coherent pulsed or continuous maser beam [16].

The gain medium chosen by Oxborrow et al. [16, 20] was the well-studied molecule pentacene, a polycyclic aromatic hydrocarbon with five benzene rings, embedded in a *p*-terphenyl matrix [48–56]. Dispersing a small amount of this guest molecule within the polymer host matrix separates the guest molecules from one another to prevent quenching. Additionally, incorporating pentacene into a matrix frustrates the rotational degrees of freedom and splits the usually degenerate triplet states [48] allowing microwave emission at room temperature.

The key to the long lifetime is the formation of the triplet exciton state, which is quantum mechanically forbidden to decay to the ground state. As shown in **Figure 1**, emission from such a gain medium is based on optical excitation into the dipole-allowed singlet state, followed immediately by an intersystem crossing to the metastable triplet state, and then a return to the ground state (often through phosphorescence). Materials which additionally have non-degenerate triplet states can, as an intermediate step, have transitions between triplet states to produce the required microwaves.

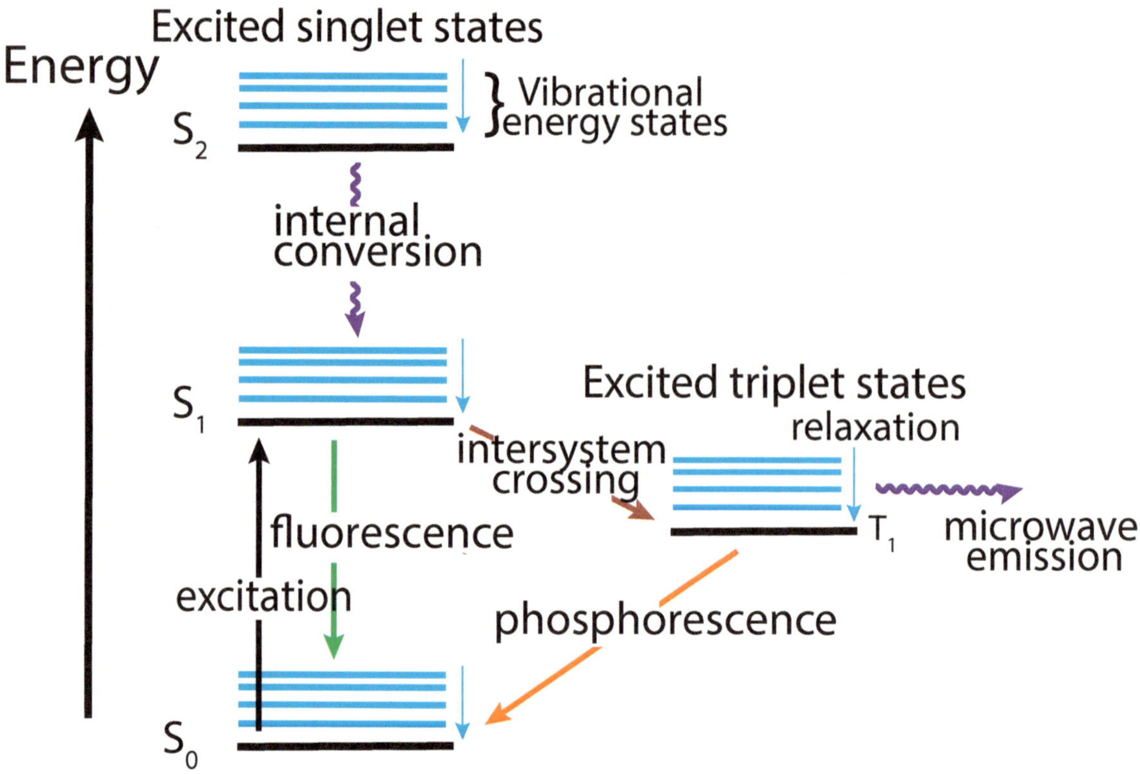

Figure 1. A Jablonski diagram showing the possible transitions that can occur for an organic-based maser.

The mechanism of electron transport in organic semiconductors also makes them advantageous as a gain medium. Semiconduction in organic molecules occurs in a π-conjugated system, where π orbitals are delocalized over some or all of the molecules. Transport of electrons through the π-orbital electrons also further suppresses both SOC and the hyperfine interaction ("HFI", the interaction between electron spin and the adjacent atomic nuclei) [57, 58]. As both singlet and triplet states in organic semiconductors result from room-temperature stable exciton spin pairs [59, 60], emission lifetimes can be substantially longer than those observed in inorganic systems.

Organics are generally cost-effective, easy to fabricate and test, versatile, flexible, plentiful, and lightweight [61–63]. Spin manipulation in organics—for improved optoelectronic devices, for spintronics, for spectroscopy—also has a long history [34, 41, 61, 64–70]. There are many candidate materials to investigate across the vast spectrum of organic molecules, providing numerous choices for applications.

2. Criteria for room-temperature, organic, triplet-based maser

As described above, organic semiconductors provide a new avenue for efficient solid state masers. Though pentacene *p*-terphenyl is the first successful room-temperature organic maser, it still suffers from some limitations. It is limited to a single output frequency of

1.45 GHz, its spin-relaxation lifetime is only 135 μs, its power conversion efficiency is only around 3% [20], it can only operate for pulses of about 300 μs [2], and the molecular crystal is unstable under high-power pumping and at high temperatures. Bogatko et al. [71], building on the success of pentacene, recently identified several possible new candidate materials based on a computational molecular design strategy. Focusing on the linear acenes and some of their derivatives, they were able to suggest new directions to overcome some of these limitations. However, a vast body of organic semiconductors beyond the linear acenes exists that can be explored and optimized as a gain medium, at a variety of emission frequencies, to expand the possibilities of room-temperature maser applications.

In the spirit of Oxborrow's original inspiration [2], we have examined the literature on organic semiconductors to present a list of additional potential gain medium candidates. First, we identify candidates that show solid state microwave emission over a variety of possible emission frequencies without an externally applied magnetic field. Then, we discuss the population inversions and triplet lifetimes that exist in such systems. Finally, we touch on some potential promising candidates that exhibit favorable conditions if a suitable host material can be found to operate effectively at room temperature.

2.1. Emission from triplet state

Fundamentally, to perform masing, photons must first be emitted by stimulation from a gain medium at a desired frequency (or wavelength). The designation of "microwave" applies to electromagnetic waves with wavelengths between 1 m and 1 mm, with frequencies between 300 MHz (100 cm) and 300 GHz (0.1 cm) [72].

In contrast to inorganic semiconductors, the excited states of organic molecules are highly localized. Rather than acting as free carriers, electrons or holes are bound to a molecule, forming a polaron. A polaron consists of the excited state molecule, its internal geometric distortion, and the distortion field with respect to its neighboring molecules, due to the addition or loss of charge. When positive and negative polarons (holes and electrons) interact, they form a room-temperature stable electron-hole pair (exciton) localized on one or a few molecules. This localization and the consequently large exchange splitting generate two distinct states, referred to as singlets (S_1 spin 0) and triplets (T_1 spin 1), depending on the spin interaction of the two carriers. In a singlet state, excitons have opposite spin orientations and the electronic energy levels do not split when the molecule is exposed to a magnetic field. In a triplet state, the electron has the same spin orientation (parallel) as the hole and energy-level splitting becomes possible.

As the name *triplet* implies, there are *three* distinct symmetric spin states, which arise from the interaction of charge carriers with parallel spins (see **Figure 2**). Usually, these levels are degenerate but can be split either through the application of a magnetic field (Zeeman effect) or through the molecular geometry (zero-field splitting).

As the α and β spin states respond oppositely to an applied magnetic field, the Zeeman effect (**Figure 2**) results in a tunable energy separation of the two spin states, which increases proportional to the applied magnetic field according to the expression [73]:

$$E = g\beta M_s H \tag{1}$$

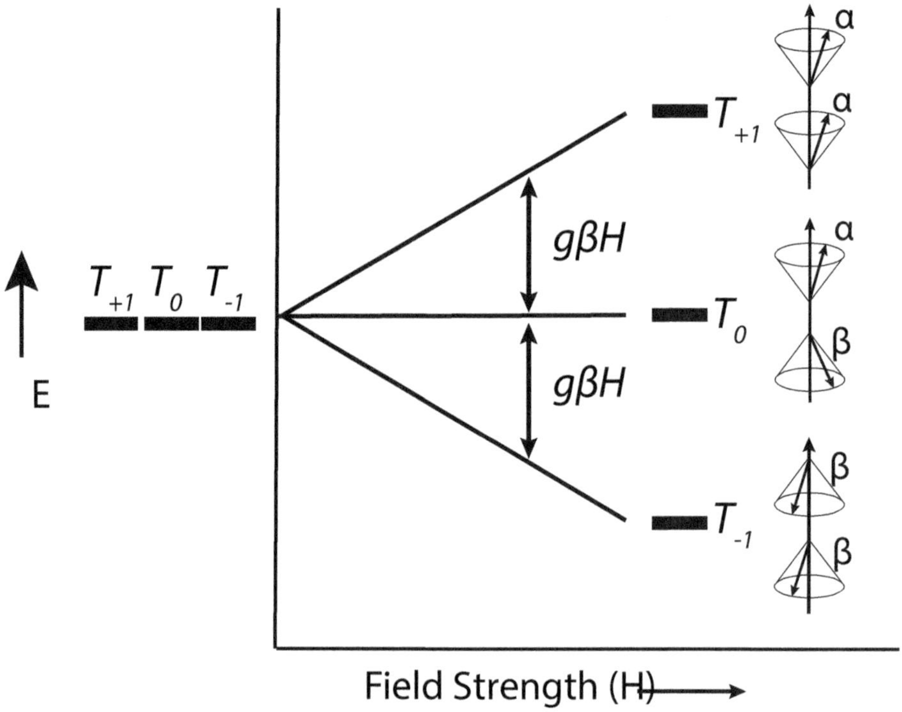

Figure 2. An energy diagram showing the splitting of the degenerate triplet sublevels under an applied magnetic field. Insets show the spin configuration for each of the triplet sublevels.

where g is the electronic g factor, β is the Bohr magneton, M_s is the spin quantum number, and H is the applied field.

However, some materials, particularly organic molecules, exhibit a splitting of the triplet states without an applied magnetic field due to the anisotropic electron distribution from the delocalized orbitals over the molecule [73], as shown in **Figure 3**.

Many organic semiconductors show this type of molecular splitting, with the energy gap described by the zero-field splitting parameters D and E. Generally, shortening along one spatial direction will lead to a decrease in the triplet energy, as seen in **Figure 3(b)** where T_z is lower for a planar symmetric molecule, such as coronene. Conversely, elongating the electrical field distribution along a spatial direction will increase the energy, as for pentacene, which has a long and short axis in the x-y plane. Along the backbone (oriented along X in our example), the triplet energy T_X will be slightly higher than along the transverse direction, T_Y. This arises because the orbitals are distributed over the entire molecule for many organic semiconductors, resulting in an asymmetric electron energy distribution. D is defined as the energy difference between T_Z and degenerate energy states as in (b) or between T_Z and the average of the two other energy states as in (c). It can be positive or negative, depending on whether there is confinement or elongation of the electron distribution over the molecule. The parameter E is half the energy difference of the T_X and T_Y energy states or the gap between X or Y and the degenerate energy level [73].

Of particular interest for a variety of applications are materials which have demonstrated emission in the range of 2.4–2.5 GHz, one of the ISM bands. These bands represent an impor-

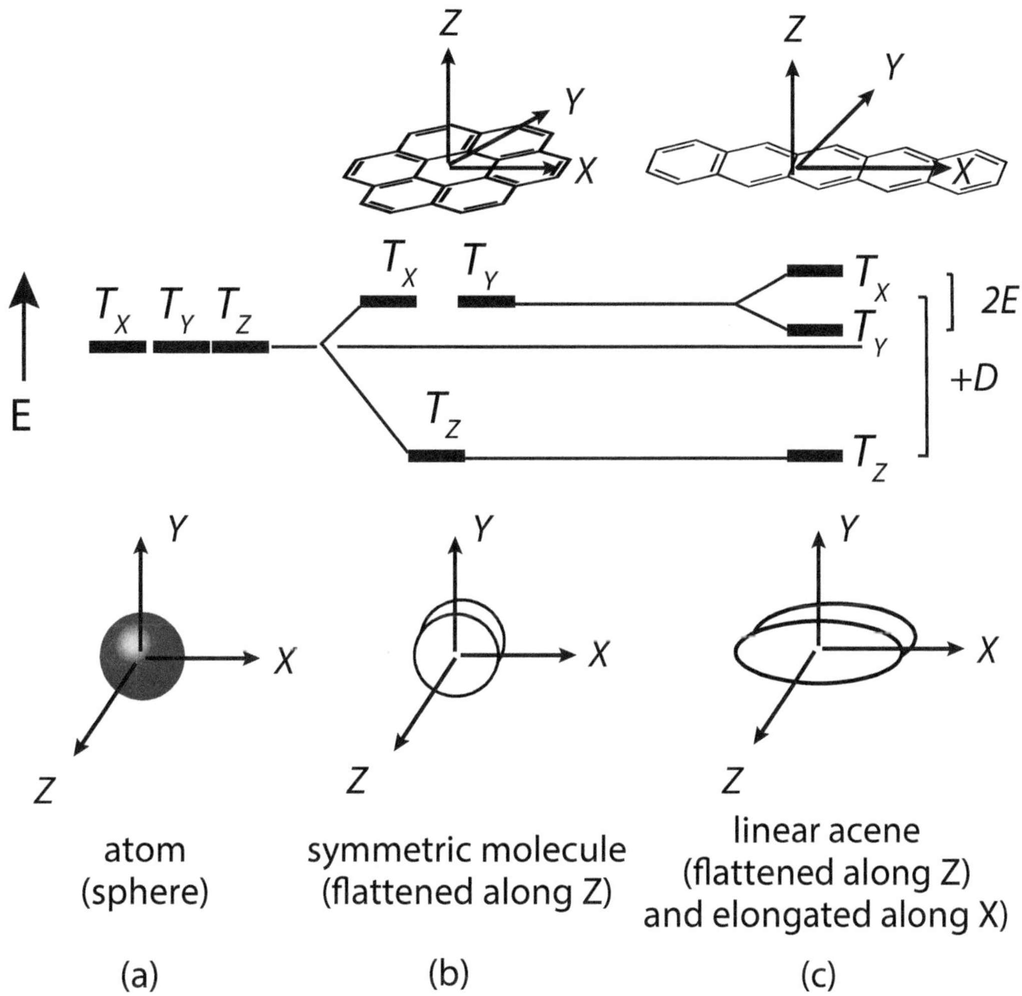

Figure 3. An energy diagram showing the splitting of the degenerate triplet sublevels with changes in molecular conformation. (a) A spherically symmetric electron distribution has degenerate triplet states in three dimensions. (b) A symmetric planar molecule, such as coronene, is symmetric in the *x-y* plane but has one non-degenerate level from shortening along the *Z*-axis. (c) An asymmetric planar molecule, such as pentacene, has three split states. As the molecule is elongated along the *X* axis, T_X has higher energy than T_Y. The zero-field splitting parameters *D* and *E* describe the separation between the triplet states.

tant frequency range for future internet-of-things applications. Candidate guest materials with 2.4–2.5 GHz emission are presented in **Table 1**, along with other candidates with values slightly below that region. Due to the Zeeman effect, discussed above, these materials could also become viable candidates with application of a modest external magnetic field to open up the triplet energy gap.

The organic molecules presented in **Table 1** have calculated triplet emission frequency based on their reported zero-field splitting coefficients *D* and *E* (cm^{-1}). Many of the results are taken from data for randomly oriented molecules in a glassy solution, which only yield the absolute value of the ZFS coefficients [74]. As *D* and *E* can take on positive or negative values, we present two emission frequencies, by either adding to or subtracting *E* from *D*, before converting to frequency. In some of the literature, single emission frequencies were reported without

Guest	Host	ZFS (GHZ)		Ref.
		High	Low	
Azulene	Phenazine	2.191	1.778	[75]
3,4-benzopyrene	EPA or methanol[a]	2.272		[76]
3,4-benzopyrene	Glasses plastics	2.272		[42]
3,4-benzopyrene		2.278		[77]
Fluoroanthene	Glasses plastics	2.278		[42]
Fluoroanthene	Glasses plastics	2.449		[42]
Fluoroanthene	PMMA	2.449		[45]
Fluoroanthene	Ethanol glass	2.458	2.159	[78]
Phenazinium	Sulfuric acid-ethanol[a]	2.317	1.718	[79]
1,2-benzanthracene	Rigid glass	2.368		[77]
Triphenylamine	PMMA	2.401		[45]
Naptho[2,3-a]coronene	Decane[a]	2.413	1.856	[80]
Cata-hexabenzocoronene	PMMA	2.463	2.423	[81]
Aeridine-b9	Biphenyl	2.49	1.959	[80]
Dibenzo[a,g]coronene	Octane[a]	2.491	1.292	[80]
1,3-diazaazulene	Phenazine	2.494		[75]

The high and low ZFS correspond to the absolute difference between positive and negative values of D and E. For entries where either ZSF parameter values were not reported or the molecule was symmetric (hence $E = 0$), the single emission frequency is given under "high."

[a]An experiment performed at or near cryogenic or liquid N_2 temperatures to ensure host is a solid matrix.

Table 1. Candidate materials for organic semiconductor gain medium sorted by zero-field splitting emission frequency.

giving D and E values and are accordingly represented as one value here in the high column. If the molecule was symmetric about the x-y plane, yielding a 0 value for E, the single emission wavelength is listed also in the high column. Though the table is mostly sorted by "high" ZFS emission values, some variation was found in the reported values for a single material, due potentially to variation in temperature, host material, or experimental setup. For clarity, we have grouped molecules from different sources together rather than sorting them by their reported value. Note that we have made our calculations based on the values given in the referenced sources, even if the experiments were originally reported elsewhere.

2.2. Triplet relative population inversion

For masing action, a system must have a top-heavy population imbalance with more excitons in the upper states than lower states. Such a configuration will allow for stimulated emission of the heavily populated upper states, with plenty of openings within the lower states where excitons can decay down. For organic molecules, numerous examples exist of high ratio excitons in the highest triplet sublevel, with a vastly smaller population in the lowest triplet state. For example, pentacene in *p*-terphenyl has a demonstrated a ratio of 0.76:0.16:0.08 relative population in the T_X: T_Y: T_Z sublevels. Other examples of notable population inversion are presented in **Table 2**.

Guest	Host	Relative	Population		High/low ratio	Ref.
		High	Medium	Low		
Naphthalene	Naphthalene-d8	0.82	0.16	0.02	41	[82]
Naphthalene	*n*-pentane[a]	0.64	0.25	0.11	5.82	[83]
Naphthalene	Durene[a]	0.54	0.2	0.26	2.08	[84]
Pyrimidine-h4	Benzene[a]	0.92	0.05	0.03	34	[80]
Anthracene	*n*-heptane[a]	0.65	0.33	0.02	32.5	[82]
Anthracene	Biphenyl	0.39	0.55	0.06	6.5	[82]
Anthracene-d14	*p*-terphenyl-d14	0.38	0.43	0.19	2	[85]
Anthracene-h2d8	Biphenyl	0.34	0.35	0.31	1.09	[80]
Dibenzothiophene trap[b]	–	0.92	0.05	0.03	29.74	[80]
Benzophenone-b10[b]	–	0.57	0.38	0.05	10.74	[80]
Benzophenone-b10	Benzophenone-d10[a]	0.38	0.33	0.3	1.27	[80]
1,4-dibromonaphthalene		0.87	0.04	0.09	10	[80]
Pentacene	*p*-terphenyl	0.76	0.16	0.08	9.5	[16, 54]
p-dichlorobenzene	*p*-xylene[a]	0.5	0.4	0.1	4.87	[80]
Dimer	Biphenyl	0.64	0.22	0.14	4.6	[80]
Dimer	*p*-dichlorobenzene[a]	0.63	0.18	0.18	3.5	[80]
Benzene	Cyclohexane[a]	0.46	0.43	0.11	4.18	[86]
Benzene	Cyclohexane[a]	0.43	0.46	0.11	3.91	[82]

The convention of highest:middle:lowest energy Zeeman triplet line was used, and the entries are sorted in descending order of their respective ratios of high excitons, capable of emission, to low excitons, the receiving state of emitted excitons.
[a]The experiment is performed at or near-cryogenic or liquid N_2 temperatures to ensure host is a solid matrix.
[b]Neat film or crystal.

Table 2. Candidate materials for organic semiconductor gain medium with triplet sub-level relative populations.

In **Table 2**, relative populations are shown for the three triplet sublevels of each material, based on a variety of sources, and sorted in descending order of their respective ratio of high excitons (capable of emission) to low excitons (the receiving state of emitted excitons). The traditional model presents these three stages as *X*, *Y*, and *Z* descending, but here we use the more generic (and more representative) terms high, mid, and low as *X*, *Y*, and *Z* may be split to differing relative energies with respect to one another in any given material (negative *D* and *E* values will determine the placement of *X*, *Y*, and *Z* lines). Again, the entries are grouped by molecular name.

2.3. Long triplet lifetime

An essential requirement for masing is that population inversion must be maintained consistently for a longer period of time than the ordinary excitation time scale. This is one of the most significant advantages of using an organic semiconductor as a gain medium for room temperature masing (see Section 1.3). In an organic maser, long lifetimes are achieved through the creation of the metastable triplet state, through intersystem crossing from the excited singlet. The advantage of operating within the triplet regime is that the excitons are dipole forbidden from decaying to the ground state due to the total spin moment $S = 1$. As Pauli exclusion forbids two electrons with parallel spins from occupying the same orbital, triplet excitons cannot decay to the ground state without spin flipping, phosphorescence, or nonradiative phonon perturbation. It can take significant time for those conditions to be met such that the exciton returns to the ground state (nanoseconds for pentacene and full seconds or tens of seconds for other organic molecules). This impediment to decay creates a natural metastable state, ideal for stimulated emission.

One of the most critical features for microwave emission is the spin-relaxation rate, which is substantially higher in organic materials compared to inorganic ones due to the low spin-orbit coupling in most organic molecules. For microwave emission, the most critical lifetime is the relaxation from the higher exciton sublevel (traditionally T_X and T_Y) down to the lowest triplet sublevel (i.e. T_Z). The triplet lifetime is also defined relative to the final decay from T_Z down to the ground state. This generally is an order of magnitude longer than the desired microwave frequency of emission. **Table 3** lists the reported triplet lifetimes for a variety of materials emitting in the microwave regime.

Guest	Host	Lifetime (s)	Ref.
Coronene	PMMA	56	[87]
Coronene-d12	Octane[a]	34.5	[87]
Coronene	Rigid glass	9.4	[69]
Coronene	Alphanol 79[a]	7.9	[88]
Coronene	PMMA	4.2	[45]
Benzophenone	Carbon tetrachloride crystal	52.1	[89]
Benzene[b]	–	26	[84]
Benzene	Cyclohexane[a]	16	[90]
Benzene	3-methylpentane sol.[a]	7	[91]
Benzene	Rigid glass	7	[69]
Phenanthrene (d)		25	[84]
Phenanthrene	Rigid glass	3.5	[77]
Phenanthrene	Rigid glass	3.3	[69]

Guest	Host	Lifetime (s)	Ref.
Triphenylene	PMMA	15.9	[69]
Triphenylene	Alphanol 79[a]	13.3	[88]
Triphenylene	PMMA	8	[45]
s-Triazine	3-methylpentane sol.[a]	13	[91]
Biphenyl	3-methylpentane sol.[a]	8	[91]
Tryptophan	Ethylene glycol-H_2O[a]	5.5	[92]
1,3,5-triphenylbenzene	Alphanol 79[a]	5.1	[88]
1,3,5-triphenylbenzene	3-methylpentane sol.[a]	4.5	[91]
Pentacene	*p*-Terphenyl	0.000135	[16]
Pentacene	PMMA	0.000045	[93]

[a]An experiment performed at or near cryogenic or liquid N_2 temperatures to ensure host is a solid matrix.
[b]Neat film or crystal.

Table 3. Candidate materials for organic semiconductor gain medium with long triplet lifetime.

2.4. Potential candidate materials

Table 4 summarizes the top candidate materials that should be studied further to exploit their long spin-relaxation lifetimes and high population inversion. Currently, complete information at room temperature is limited for these materials other than pentacene. However, as with pentacene [51], the low temperature behavior suggests that if a suitable Shpolskii matrix [47] can be found for these molecules which preserve the ZFS, then each would be an excellent candidate for a room-temperature maser gain medium.

Guest	Host	Emission (GHz)	Sublevels			High-low ratio	Temp (K)	Lifetime (s)	Refs.
			High	Medium	Low				
Naphthalene	Naphthalene-d8	2.968	0.82	0.16	0.02	41	1.3	3 [88]	[82]
Naphthalene	*n*-pentane[a]		0.64	0.25	0.11	5.8			[82]
Pyrimidine	Benzene[a]	2.187	0.92	0.05	0.03	34	1.2	0.14, 0.017, 0.012	[80]
Anthracene	*n*-heptane[a]	2.159	0.65	0.33	0.02	32.5	1.3	0.05 [77]	[82]
Anthracene	Biphenyl		0.39	0.55	0.06	6.5			[82]
Dibenzothiophene[b]	–	3.29, 1.76	0.92	0.05	0.03	30*	1.3	9, 0.36, 0.29	[80]
Pentacene	*p*-terphenyl	1.42	0.76	0.16	0.08	9.5	RT	0.000135	[16]
Phenazine	ETOH[a]	2.236		0.9	0.1	9*	77	0.011	[79]

Guest	Host	Emission (GHz)	Sublevels			High-low ratio	Temp (K)	Lifetime (s)	Refs.
			High	Medium	Low				
Phenazinium	ETOH[a]	2.018		0.9	0.1	9*	77	0.009	[79]
Phenanthrene	TBB[a]	2.384*	0.1	0.28	0.62	2.8*	1.6	0.26, 0.023	[80]
Triphenylene	Hexane[a]	1.443	0.4	0.43	0.17	2.34	1.2	8.3, 8.3, 27.8	[80]
Tryptophan	Glass	2.95	0.39	0.38	0.23	1.7		5.5	[94]
Tryptophan	Ethylene Glycol-H_2O[a]		0.282	0.347	0.371			4.17, 8.43, 26.5	[92]
Coronene	*n*-hexane	2.827	0.43	0.41	0.16	2.7**	1.6	4.24, 3.94, 59.9	[86]
Coronene	Octane		33.3	33.3	33.3	1.0**	1.35		[80]
Trans-stilbene[a]	–	3.79, 1.51	0.5	0.2	0.3	1.67	1.3	0.014, 0.014, 0.15	[82]

Three distinct triplet lifetimes of *X* (high), *Y* (mid), and *Z* (low) sublevels are reported.

[a]An experiment performed at or near cryogenic or liquid N_2 temperatures to ensure host is a solid matrix.

[b]Neat film or crystal.

*Emission occurs from *Y* down to *X*.

**Uniform shapes (such as coronene) have degenerate *X* and *Y*, so they both emit at the same wavelength, thereby increasing the population ratio.

Table 4. Top candidate materials for an organic semiconductor gain medium.

3. Summary

With significant improvements to power efficiency and operational parameters through the selection of alternative candidate materials, the maser could expand beyond its current niche applications for use in laser-to-radio frequency (RF) or light-to-RF conversion, for coherent emission in the millimeter or centimeter wavelengths, or for coherent transmission in radio frequencies. Possible areas of application include enhanced communication security (narrow casting as opposed to broadcasting), organic radio components such as low-noise or solar-powered amplifiers, direct optical powering of transceivers without an inefficient electrical conversion stage (suitable for sensors and emerging "internet-of-things" applications), radiative diagnoses and therapy, directed energy tools and weapons for military and defense applications, and wireless power transmission over vast distances including energy to and from planetary orbit (space-based solar power).

In this chapter, we have presented a list of possible candidate materials that have zero-field triplet emission over a variety of frequencies, including technologically interesting ISM bands. We have also shown promising materials that exhibit high population inversion and long triplet lifetimes, which are required to achieve room temperature masing.

Funding sources

This research was financially supported by Natural Science and Engineering Research Council, through the grant 436100-2013 RGPIN, and by the Canadian Foundation for Innovation, through the grant 31605 from the John R. Evans Leaders Fund.

Acknowledgements

Special thanks to Lindsay Vasilak, Kaijie Zhang, Fariha Husain, and Haley Glavina for their contributions to data collection.

Author details

Erinn van Wynsberghe and Ayse Turak*

*Address all correspondence to: turaka@mcmaster.ca

McMaster University, Hamilton, Ontario, Canada

References

[1] RP Feynman, RB Leighton, M Sands. Chapter 9: The ammonia maser. In: The Feynman lectures on physics (Vol. III: The new millennium edition: quantum mechanics). New York: Basic Books; 2011.

[2] G Brumfiel. Microwave laser fulfills 60 years of promise. Nature News. doi:10.1038/nature.2012.11199 http://www.nature.com/news/microwave-laser-fulfills-60-years-of-promise-1.11199 (2012).

[3] HTM Wang. Subcompact hydrogen maser atomic clocks. Proceedings of the IEEE 77 (7) (1989) 982-992. doi:10.1109/5.30750.

[4] J Vanier. Atomic clocks based on coherent population trapping: a review. Applied Physics B 81 (4) (2005) 421-442. doi:10.1007/s00340-005-1905-3.

[5] Y Haibo, W Zhengming, D Shaowu. Performance of hydrogen maser and its usage in local atomic time at NTSC. In 2007 IEEE International Frequency Control Symposium Joint with the 21st European Frequency and Time Forum, 2007, pp. 889-892. doi:10.1109/FREQ.2007.4319206.

[6] H Peters, B Owings, T Oakley, L Beno. Hydrogen masers for radio astronomy. In: 41st Annual Symposium on Frequency Control, 1987, pp. 75-81. doi:10.1109/FREQ. 1987. 201004.

[7] JA Giordmaine, LE Alsop, CH Mayer, CH Townes. A maser amplifier for radio astronomy at x-band. Proceedings of the IRE 47 (6) (1959) 1062-1069. doi:10.1109/JRPROC. 1959.287134.

[8] JG Smith. Ka-band (32-GHz) downlink capability for deep space communications. The Telecommunications and Data Acquisition Report Vol. 42- 88 (Oct-Dec) (1987) 96-103.

[9] MS Reid, RC Clauss, DA Bathker, CT Stelzried. Low-noise microwave receiving systems in a worldwide network of large antennas. Proceedings of the IEEE 61 (9) (1973) 1330-1335. doi:10.1109/PROC.1973.9271.

[10] RC Clauss, JS Shell. Ruby masers. In MS Reid (Ed), Low-noise systems in the deep space network (Deep space communication and navigation series). Jet Propulsion Laboratory; Pasadena, California 2008.

[11] JC Walling, FW Smith. Solid state masers and their use in satellite communication systems. Philips Technical Review 25 (11/12) (1963) 289-310.

[12] M Dillard. The amazing maser: the jewel that conquers space. Popular Electronics 12 (4) (1960) 41-45, 122-123.

[13] R Forward, F Goodwin, J Kiefer. Application of a solid-state ruby maser to an X-band radar system. In: WESCON/59 Conference Record, Vol. 3, 1959, pp. 119-125. doi:10.1109/WESCON.1959.1150334.

[14] IV Konoplev, L Fisher, K Ronald, AW Cross, ADR Phelps. Two-dimensional periodic lattice Cherenkov maser: scalability from 37.5GHz to 350GHz. In: 2010 Abstracts IEEE International Conference on Plasma Science, 2010, pp. 1-1. doi:10.1109/PLASMA. 2010.5533924.

[15] JC Mollier, J Hardin, J Uebersfeld. Theoretical and experimental sensitivities of ESR spectrometers using maser techniques. Review of Scientific Instruments 44 (12) (1973) 1763-1771. doi:10.1063/1.1686050.

[16] M Oxborrow, JD Breeze, NM Alford. Room-temperature solid-state maser. Nature 488 (7411) (2012) 353-356. doi:10.1038/nature11339.

[17] L Jin, M Pfender, N Aslam, P Neumann, S Yang, J Wrachtrup, R-B Liu. Proposal for a room-temperature diamond maser. Nature Communications 6 (2015) 8251. doi:10.1038/ncomms9251.

[18] H Kraus, VA Soltamov, D Riedel, S Väth, F Fuchs, A Sperlich, PG Baranov, V Dyakonov, GV Astakhov. Room-temperature quantum microwave emitters based on spin defects in silicon carbide. Nature Physics 10 (2) (2014) 157-162. doi:10.1038/nphys2826.

[19] SM Watts, BJ van Wees. A solid state paramagnetic maser device driven by electron spin injection. Physical Review Letters 97 (11) (2006) 116601. doi:10.1103/PhysRevLett.97. 116601.

[20] J Breeze, K-J Tan, B Richards, J Sathian, M Oxborrow, NM Alford. Enhanced magnetic Purcell effect in room-temperature masers. Nature Communications 6 (2015) 6215. doi:10.1038/ncomms7215.

[21] KR Foster. A world awash with wireless devices: radio-frequency exposure issues. IEEE Microwave Magazine 14 (2) (2013) 73-84. doi:10.1109/MMM.2012.2234641.

[22] JL Sevillano, D Cascado, F Díaz del Río, S Vicente, G Jimenez, A Civit-Balcells. Soft real-time communications over Bluetooth under interferences from ISM devices. International Journal of Communication Systems 19 (10) (2006) 1103-1116. doi:10.1002/dac.796.

[23] RG Garroppo, L Gazzarrini, S Giordano, L Tavanti. Experimental assessment of the coexistence of Wi-Fi, ZigBee, and Bluetooth devices. In: 2011 IEEE International Symposium on a World of Wireless, Mobile and Multimedia Networks, 2011, pp. 1-9. doi:10.1109/WoWMoM.2011.5986182.

[24] D Raychaudhuri, X Jing. A spectrum etiquette protocol for efficient coordination of radio devices in unlicensed bands. In: 14th IEEE Proceedings on Personal, Indoor and Mobile Radio Communications, 2003. PIMRC 2003, Vol. 1, 2003, pp. 172-176. doi:10.1109/PIMRC.2003.1264255.

[25] CC Davis. The relation between the Einstein A and B coefficients. In: Lasers and electro-optics: fundamentals and engineering. Cambridge University Press, Cambridge, UK 2014, pp. 25-30.

[26] RC Hilborn. Einstein coefficients, cross sections, f values, dipole moments, and all that. American Journal of Physics 50 (11) (1982) 982-986. doi:10.1119/1.12937.

[27] M Elitzur. Physical characteristics of astronomical masers. Reviews of Modern Physics 54 (4) (1982) 1225-1260. doi:10.1103/RevModPhys.54.1225.

[28] JP Gordon, HJ Zeiger, CH Townes. The maser-new type of microwave amplifier, frequency standard, and spectrometer. Physical Review 99 (4) (1955) 1264-1274. doi:10.1103/PhysRev.99.1264.

[29] SA Diddams, JC Bergquist, SR Jefferts, CW Oates. Standards of time and frequency at the outset of the 21st century. Science 306 (5700) (2004) 1318-1324. doi:10.1126/science.1102330.

[30] D Kleppner. Hydrogen-maser principles and techniques. Physical Review 138 (4A) (1965) A972–A983. doi:10.1103/PhysRev.138.A972.

[31] N Bloembergen. Proposal for a new type solid state maser. Physical Review 104 (2) (1956) 324-327. doi:10.1103/PhysRev.104.324.

[32] C Kikuchi, J Lambe, G Makhov, RW Terhune. Ruby as a maser material. Journal of Applied Physics 30 (7) (1959) 1061-1067. doi:10.1063/1.1776979.

[33] DC Lainé. Molecular beam masers. Reports on Progress in Physics 33 (3) (1970) 1001. doi:10.1088/0034-4885/33/3/304.

[34] S Reineke, MA Baldo. Room temperature triplet state spectroscopy of organic semiconductors. Scientific Reports 4 (2014) 3797. doi:10.1038/srep03797.

[35] AE Siegman. Microwave solid-state masers. New York: McGraw-Hill; 1964.

[36] A Blank, H Levanon. Toward maser action at room temperature by triplet-radical interaction and its application to microwave technology. RIKEN Review Vol. 44 (2002) 128-130.

[37] HJ Gerritsen. Recent developments in maser devices and materials. Applied Optics 1 (1) (1962) 37-44. doi:10.1364/AO.1.000037.

[38] HED Scovil, G Feher, H Seidel. Operation of a solid state maser. Physical Review 105 (2) (1957) 762-763. doi:10.1103/PhysRev.105.762.

[39] S. Antipov, W. Gai, O. Poluektov, J. G. Power, A. Kanareykin, P. Schoes- sow, L. Sch ächter, Microwave active media studies for paser, in: Proceedings of PAC09, Vol. FR5RFP, TRIUMF, Vancouver, BC, Canada, 2009, pp. 4535-4537.

[40] P. Schoessow, A. Kanareykin, L. Schachter, Y. Bogachev, A. Tyukhtin, E. Bagryanskaya, N. Yevlampieva, Studies of Particle Acceleration by an Active Microwave Medium, AIP Conference Proceedings 877 (1) (2006) 452-460. doi:10.1063/1.2409169.

[41] R Geng, TT Daugherty, K Do, HM Luong, TD Nguyen. A review on organic spintronic materials and devices: I. Magnetic field effect on organic light emitting diodes. Journal of Science: Advanced Materials and Devices 1 (2) (2016) 128-140. doi:10.1016/j.jsamd.2016.05.002.

[42] C Thomson. Electron spin resonance studies of the triplet state. Quarterly Reviews Chemical Society 22 (1) (1968) 45-74. doi:10.1039/QR9682200045.

[43] C Thomson. Zero-field splittings in aromatic triplet states: I. Naphthalene, pyrene and 1,12-benzperylene. Molecular Physics 10 (4) (1966) 309-315. doi:10.1080/00268976600100421.

[44] C Thomson. Zero-field splittings in aromatic hydrocarbons: II. Calculations with n-electron wave functions including overlap and extensive configuration interaction. Molecular Physics 11 (2) (1966) 197-203. doi:10.1080/00268976600101031.

[45] C Thomson. ESR studies of triplet states in plastics: effect of temperature on the spectra. The Journal of Chemical Physics 41 (1) (1964) 1-6. doi:10.1063/1.1725605.

[46] VA Dediu, LE Hueso, I Bergenti, C Taliani. Spin routes in organic semiconductors. Nature Materials 8 (9) (2009) 707-716. doi:10.1038/nmat2510.

[47] ÉV Shpol'skĭ. Line fluorescence spectra of organic compounds and their applications. Soviet Physics Uspekhi 3 (3) (1960) 372. doi:10.1070/PU1960v003n03ABEH003277.

[48] K Takeda, K Takegoshi, T Terao. Zero-field electron spin resonance and theoretical studies of light penetration into single crystal and polycrystalline material doped with molecules photoexcitable to the triplet state via intersystem crossing. The Journal of Chemical Physics 117 (10) (2002) 4940. doi:10.1063/1.1499124.

[49] J Köhler, ACJ Brouwer, EJJ Groenen, J Schmidt. On the intersystem crossing of pentacene in p-terphenyl. Chemical Physics Letters 250 (1) (1996) 137-144. doi:10.1016/0009-2614(95)01393-8.

[50] A Corval, C Kryschi, S Astilean, HP Trommsdorff. Resonant intersystem crossing in pentacene. The Journal of Physical Chemistry 98 (30) (1994) 7376-7381. doi:10.1021/j100081a024.

[51] JL Ong, DJ Sloop, TS Lin. Temperature dependence studies of the paramagnetic properties of the photoexcited triplet states of pentacene in p-terphenyl, benzoic acid, and naphthalene crystals. The Journal of Physical Chemistry 97 (30) (1993) 7833-7838.

[52] I Rebane. Difference values of spontaneous emission rate of a single-impurity molecule of pentacene in biaxial p-terphenyl crystal. Journal of Luminescence 122-123 (2007) 359-361. doi:10.1016/j.jlumin.2006.01.187.

[53] ACJ Brouwer, J Köhler, AM van Oijen, EJJ Groenen, J Schmidt. Single-molecule fluorescence autocorrelation experiments on pentacene: the dependence of intersystem crossing on isotopic composition. The Journal of Chemical Physics 110 (18) (1999) 9151. doi:10.1063/1.478837.

[54] T-C Yang, DJ Sloop, SI Weissman, T-S Lin. Zero-field magnetic resonance of the photoexcited triplet state of pentacene at room temperature. The Journal of Chemical Physics 113 (24) (2000) 11194. doi:10.1063/1.1326069.

[55] DJ Sloop. Electron spin echoes of a photoexcited triplet: pentacene in p-terphenyl crystals. The Journal of Chemical Physics 75 (8) (1981) 3746. doi:10.1063/1.442520.

[56] WP Ambrose, T Basché, WE Moerner. Detection and spectroscopy of single pentacene molecules in a p-terphenyl crystal by means of fluorescence excitation. The Journal of Chemical Physics 95 (10) (1991) 7150. doi:10.1063/1.461392.

[57] WJM. Naber, S Faez, WG van der Wiel. Organic spintronics. Journal of Physics D: Applied Physics 40 (12) (2007) R205. doi:10.1088/0022-3727/40/12/R01.

[58] S Sanvito. Molecular spintronics: the rise of sp interface science. Nature Physics 6 (8) (2010) 562-564. doi:10.1038/nphys1714.

[59] W Barford. Excitons in conjugated polymers. In: Electronic and optical properties of conjugated polymers, Clarendon Press; Oxford University Press, Oxford, UK, Vol. 66, 2002, pp. 90-111.

[60] S Brazovskii, N Kirova. Physical theory of excitons in conducting polymers. Chemical Society Reviews 39 (7) (2010) 2453. doi:10.1039/b917724h.

[61] F Pulizzi. A new face for organics. Nature Materials 8 (9) (2009) 696-697. doi:10.1038/nmat2519.

[62] A Facchetti. π-conjugated polymers for organic electronics and photovoltaic cell applications†. Chemistry of Materials 23 (3) (2011) 733-758. doi:10.1021/cm102419z.

[63] G Witte, C Wöll. Growth of aromatic molecules on solid substrates for applications in organic electronics. Journal of Materials Research 19 (7) (2004) 1889-1916. doi:10.1557/JMR.2004.0251.

[64] JS Wilson, AS Dhoot, AJAB Seeley, MS Khan, A Köhler, RH Friend. Spin-dependent exciton formation in π-conjugated compounds. Nature 413 (6858) (2001) 828-831. doi:10.1038/35101565.

[65] SR Forrest. Exciton formation statistics under electrical injection in organic semiconductor thin films. Journal of Luminescence 110 (4) (2004) 378-383. doi:10.1016/j.jlumin.2004.08.035.

[66] M-K Lee, M Segal, ZG Soos, J Shinar, MA Baldo. Yield of singlet excitons in organic light-emitting devices: a double modulation photoluminescence-detected magnetic resonance study. Physical Review Letters 94 (13) (2005) 137403. doi:10.1103/PhysRevLett.94.137403.

[67] M Reufer, MJ Walter, PG Lagoudakis, AB Hummel, JS Kolb, HG Roskos, U Scherf, JM Lupton. Spin-conserving carrier recombination in conjugated polymers. Nature Materials 4 (4) (2005) 340-346. doi:10.1038/nmat1354.

[68] J Wang, A Chepelianskii, F Gao, NC Greenham. Control of exciton spin statistics through spin polarization in organic optoelectronic devices. Nature Communications 3 (2012) 1191. doi:10.1038/ncomms2194.

[69] DS McClure. Triplet-singlet transitions in organic molecules. Lifetime measurements of the triplet state. The Journal of Chemical Physics 17 (10) (1949) 905-913. doi:10.1063/1.1747085.

[70] A Köhler, H Bässler. Triplet states in organic semiconductors. Materials Science and Engineering: R: Reports 66 (4-6) (2009) 71-109. doi:10.1016/j.mser.2009.09.001.

[71] S Bogatko, PD Haynes, J Sathian, J Wade, J-S Kim, K-J Tan, J Breeze, E Salvadori, A Horsfield, M Oxborrow. Molecular design of a room-temperature maser. The Journal of Physical Chemistry C 120 (15) (2016) 8251-8260. doi:10.1021/acs.jpcc.6b00150.

[72] DM Pozar. Microwave engineering (4th ed.). Hoboken, NJ: Wiley; 2011.

[73] KA Sandberg. Theoretical investigations of spin-spin interactions: semiempirical estimations of exchange parameters of diradical metalloporphyrin cations and zero field splitting parameters for conjugated diradicals, Ph.D. thesis, North Carolina State University, Raleigh, NC (1998).

[74] J de Jong, C MacLean. The zero-field splitting of cyano-substituted benzenes in their lowest excited triplet states. Journal of Magnetic Resonance 11 (3) (1973) 373-380. doi:10.1016/0022-2364(73)90063-2.

[75] SS Kim. The triplet state of azulene in phenazine crystals by electron paramagnetic resonance. The Journal of Chemical Physics 68 (1) (1978) 333. doi:10.1063/1.435464.

[76] B Smaller. Electron and nuclear magnetic interaction in triplet states of various organic phosphors. The Journal of Chemical Physics 37 (7) (1962) 1578-1579. doi:10.1063/1.1733340.

[77] JS Brinen, MK Orloff. Zero-field splitting in phosphorescent triplet states of aromatic hydrocarbons III. Correlation between D and the triplet state energy. Chemical Physics Letters 1 (7) (1967) 276-278. doi:10.1016/0009-2614(67)80018-6.

[78] J de Jong. Zero-field splitting of methyl-substituted phosphorescent aromatic molecules in their lowest triplet states. Journal of Magnetic Resonance 9 (1) (1973) 185-189. doi:10.1016/0022-2364(73)90174-1.

[79] M Deguchi, D Suzuki, R Ito, M Matsumoto, M Yagi. Time-resolved electron paramagnetic resonance of the lowest excited triplet state of phenazinium cation. Spectrochimica Acta Part A: Molecular and Biomolecular Spectroscopy 61 (6) (2005) 1147-1151. doi:10.1016/j.saa.2004.06.033.

[80] M Kinoshita, N Iwasaki, N Nishi. Molecular spectroscopy of the triplet state through optical detection of zero-field magnetic resonance. Applied Spectroscopy Reviews 17 (1) (1981) 1-94. doi:10.1080/05704928108060401.

[81] C Bräuchle. Symmetry distortions and pseudo-Jahn-Teller effect of peri- and cata-hexabenzocoronene in their triplet states as obsevered by ODMR techniques. Chemical Physics 67 (1) (1982) 97-109. doi:10.1016/0301-0104(82)88062-2.

[82] T Ikeyama, T Azumi. Triplet spin-sublevel selectivity of the S1-T1 intersystem crossing in trans-stilbene. Chemical Physics Letters 212 (6) (1993) 551-554.

[83] BR Henry, W Siebrand. Spin-orbit coupling in aromatic hydrocarbons. The radiative triplet lifetime of naphthalene. Chemical Physics Letters 3 (2) (1969) 90-92.

[84] AB Zahlan. The triplet state. Cambridge University Press, Cambridge, UK, 1967.

[85] J Köhler, ACJ Brouwer, EJJ Groenen, J Schmidt. Single molecule electron paramagnetic resonance spectroscopy: hyperfine splitting owing to a single nucleus. Science 268 (5216) (1995) 1457-1460. doi:10.1126/science.268.5216.1457.

[86] MA El-Sayed, WR Moomaw, JB Chodak. The mechanism of the S1 ⇝ Tx intersystem crossing process in aromatic hydrocarbons from PMDR. Chemical Physics Letters 20 (1) (1973) 11-16.

[87] WR Dawson, JL Kropp. Radiative and radiationless processes in aromatic molecules. Coronene and benzcoronene. The Journal of Physical Chemistry 73 (3) (1969) 693-699.

[88] M de Groot, JH Van der Waals. Paramagnetic resonance in phosphorescent aromatic hydrocarbons II: determination of zero-field splitting from solution spectra. Molecular Physics 3 (2) (1960) 190-200. doi:10.1080/00268976000100221.

[89] G-C Wang, MA Winnik, HJ Schäfer, W Schmidt. Phosphorescence and photochemical studies of some benzophenone dicarboxylic acid esters. Journal of Photochemistry 33 (3) (1986) 291-296. doi:10.1016/0047-2670(86)87091-5.

[90] MR Wright, RP Frosch, GW Robinson. Phosphorescence lifetime of benzene. An intermolecular heavy-atom effect, a deuterium effect, and a temperature effect. The Journal of Chemical Physics 33 (3) (1960) 934-935. doi:10.1063/1.1731296.

[91] JS Brinen, JG Koren, WG Hodgson. ESR and phosphorescence spectra of the triplet states of phenyl s-triazines and phenyl benzenes. The Journal of Chemical Physics 44 (8) (1966) 3095-3099. doi:10.1063/1.1727185.

[92] J Zuclich, J von Schütz, A Maki. Direct measurement of spin-lattice relaxation rates between triplet spin sublevels using optical detection of magnetic resonance. Molecular Physics 28 (1) (1974) 33-47. doi:10.1080/00268977400101501.

[93] M Orrit, J Bernard. Single pentacene molecules detected by fluorescence excitation in a p-terphenyl crystal. Physical Review Letters 65 (21) (1990) 2716.

[94] J Zuclich. Triplet-state electron spin resonance of the aromatic amino acids and proteins. The Journal of Chemical Physics 52 (7) (1970) 3586. doi:10.1063/1.1673527.

7

Microscopic Particle Manipulation via Optoelectronic Devices

Xiaolu Zhu and Yifei Yang

Additional information is available at the end of the chapter

Abstract

The optoelectronic tweezers (or optically induced dielectrophoresis (DEP)) have showed the ability to parallelly position a large number of colloidal microparticles without any template. The microparticles can be trapped and driven by the dielectrophoretic forces induced by the optical micropatterns in OET devices. In this chapter, the design and fabrication of flat optoelectronic devices (FOD) and hybrid optoelectronic device (HOD) are described. In the typical FOD, the manipulation modes including filtering, transporting, concentrating and focusing controlling regimes are experimentally demonstrated and analyzed. The controllable rotation of self-assembled microparticle chains in FOD has also been investigated, and a method incorporating the optically induced electrorotation (OER) and AC electroosmotic (ACEO) effects is numerically and experimentally implemented for manipulating microparticle chains. Based on the above research of FOD, a hybrid DEP microdevice HOD is conceptually and experimentally proposed. The HOD integrates with metallic microelectrode layer and the underneath photoconductive layer with projected optical virtual electrodes. FOD and HOD hybrid device enables the active driving, large-scale patterning and local position adjustment of microparticles. These techniques make up the shortcoming of low flexibility of traditional metallic microelectrodes and integrate the merits of both the metal electrode-induced and microimage-induced DEP techniques.

Keywords: optoelectronic device, dielectrophoresis, optically induced electrorotation, AC electroosmosis

1. Introduction

The manipulation technique based on dielectrophoresis (DEP) [1–3] is easy-to-use and requires no moving parts, thus can achieve massively paralleled and non-contact manipulation for particles or cells [4, 5]. It is becoming an important enabling technology in filtering [6–8], concentrating [9–11], transporting [12–14] and flow focusing [15, 16] for microparticles. However, these DEP technologies require design and fabrication of sophisticated electrode structures, which lack flexibility and could not provide real-time alterable electrode-patterns for diverse manipulations. Therefore, how to improve the electrodes' flexibility and reconfigurability has become the key concern in the extended application of DEP. The proposal of optically induced DEP (ODEP) [17, 18] based on photoconductive effect in an optoelectronic device offered a potential option for the design of highly flexible and real-time reconfigurable optical microelectrodes. ODEP-based optoelectronic device offers more opportunity to develop miscellaneous particle manipulating techniques in biomedical microsystems, although ODEP device could not fully supersede the traditional physical electrode array.

Usually, the sample pretreatment process in microfluidic analysis system should include functional units capable of filtering, concentrating, transporting and focusing microparticles to be applicable to different sample handing requirements [9, 19, 20]. Filtration is essentially a separation technology, such as the insulative DEP approach for particle separation [21] and the DEP method for characterizing and distinguish stem cells [6]. Once the target particles are isolated after the filtration or separation, the concentration of the target particle population is sometimes low and requires the concentrating steps for the following detection. By now various DEP concentrations of viruses, bacteria have been reported in the literature [22–24]. The function of transporting microparticles nearly runs through the whole process in the microanalysis system, and biological/chemical applications [14, 25–27]. Besides, as an important approach for the continuous single-particle detection and counting, focusing particles into a straight line are also applied in microfluidic devices. The hydrodynamic forces [28, 29] and electrokinetic force [30, 31] are usually employed to realize flow focusing. However, the above sample handling approaches require fabricating complex multi-functional electrode array or structures to achieve versatile manipulation of the micro/nano bio-particles, and the previous work also has difficulties in providing real-time alterable electrode patterns to accommodate different manipulation requirements. So, the promotion of DEP technology in the application of the micro/nano manipulation faces some restrictions. In recent years, Chiou et al. [18, 32] reported parallel manipulation of single particles using a variant of DEP mechanism—ODEP mechanism which offers the ability to trap single particles parallelly and massively. ODEP has also been used to discriminate normal oocytes at a certain frequency [33]. However, the manipulation regimes still need more systematic study. Therefore, this chapter will experimentally verify and analyze the optical pattern and controlling regimes for filtering, transporting, concentrating and focusing based on ODEP.

In addition to the above manipulation modes for microparticle or cells, the ODEP-based optoelectronic device could serve as a more powerful tool to spatially manipulate and regulate the posture, orientation and position of microparticle chains or other microstructures. The previous methods for manipulating micro-objects usually need templates, physical or chemical reaction process involving more extra techniques [34]. Many of the created particle chains or closely linked chains are immobilized, and their relative positions cannot be stably controlled. The ODEP (also called optoelectronic tweezers, OETs) has showed the ability to parallelly position colloidal particles without any template [18, 35]. The microparticles can be trapped and driven by the DEP forces induced by the optical micropatterns in optoelectronic devices [36, 37]. However, the rotation of assembled microparticle chains by OETs has rarely been investigated, and the degree of freedom for varying the postures, positions or orientations of the microparticle chains still has limitations. To demonstrate the promoted ability of optoelectronic devices, this chapter will experimentally demonstrate the method incorporating optically induced electrorotation (OER) and alternating current electroosmotic (ACEO) effects, for the formation and motion control of microparticle chains.

By comprehensively analyzing the traditional physical electrode array and ODEP-based optoelectronic device, we further developed a hybrid optoelectronic device (HOD) that could integrate the merits of both the traditional physical electrode array and ODEP-based optoelectronic device. The ODEP-based optoelectronic device offers a flexible and real-time dynamic manipulating scheme but needs more external optic devices; the traditional physical electrode array offers a more stable working status and higher integration but lack the ability of real-time particle manipulation. In Section 5 of this chapter, HOD is conceptually proposed, and its prototype is designed and fabricated. This type of hybrid device enables the active driving, large-scale patterning and locally position adjustment of microparticles.

2. Design and fabrication of optoelectronic devices

2.1. The design and fabrication of flat optoelectronic devices

The flat optoelectronic devices (FOD) consisted of several layers, including indium tin oxide (ITO) and photoconductive layers. To fabricate photoconductive layer of FOD chip (as shown in **Figure 1(a)**), 50-nm doped hydrogenated amorphous silicon (n + *a*-Si:H), 1.5-μm intrinsic *a*-Si:H and 25-nm silicon carbide (SiC*x*) passivation layer were consecutively deposited by plasma-enhanced chemical vapor deposition (PECVD) method on the bottom ITO-coated glass at 200°C. The n+ *a*-Si:H was deposited from a gas ratio of 1% PH_3 in SiH_4, and then, the intrinsic *a*-Si:H was deposited from a gas mixture containing SiH_4 and H_2. Afterwards, the

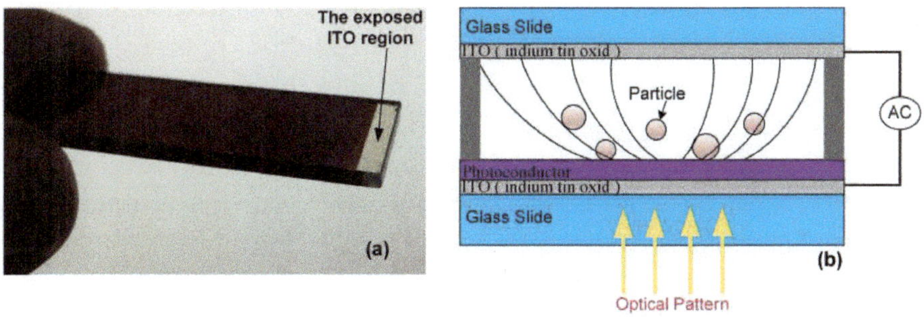

Figure 1. Schematic diagram of the FOD chip based on ODEP mechanism. (a) Photograph of the photoconductive substrate; (b) structure diagram of the ODEP microdevice.

SiC*x* layer was deposited by a SiH_4, CH_3 and N_2 mixture. After that, the ITO layer for bias connections was exposed via reactive ion etch (RIE), as shown in **Figure 1a**. The photoconductive layer of the microdevice consisted of n +*a*-Si:H layer, intrinsic *a*-Si:H layer and SiC*x* layer (as shown in **Figure 1b**). The n + *a*-Si:H layer was used as both a transitionary and adhesive layer from the intrinsic *a*-Si:H layer to the bottom ITO-coated glass. During chip packaging, firstly, a double-faced adhesive tape with a thickness of ~ 100μm was pasted on the four edges of the photoconductive surface to form a microchamber with a scale of ~15 mm × 8 mm. Then a droplet containing microsphere particles was dripped into the microchamber using a pipette. Lastly, the top ITO-glass slide was placed onto the roof of the microchamber to make the microchamber sealed as shown in **Figure 1b**. In the whole ODEP device, a 100-μm insulating spacer was used to maintain the gap between the upper ITO film and lower photoconductive layer (see **Figure 1b**). The alternating current (AC) voltage was supplied between the top and bottom ITO layers.

For the fabricated FOD chip as shown in **Figure 1**, the photoconductor has a high electric impedance (about 1×10^{-6} S/m, for example) prior to illumination, so at this time the majority of the applied voltage drops across the photoconductive layer, which causes a weak, uniform electric field inside the liquid layer. At this state, no ODEP force generates. When a projected light source hits the amorphous silicon layer (photoconductive layer), electron-hole pairs are excited, thus decreasing the impedance of the photoconductive layer by thousands of times (~5000 times, for example), and the electrical conductivity of the photoconductor increases to ~5×10^{-3} S/m. At this state, the virtual electrodes are formed in the photoconductor, and the voltage drop across the liquid layer in the illuminated area is much higher than that in the dark area. Consequently, the bright-dark pattern induces a non-uniform electric field inside this device. The dense electric filed lines appear at the illuminated areas, and the sparse field lines appear at the dark areas as shown in **Figure 1b**. Thus, a DEP force acting on microparticles can be induced in the liquid chamber(or called fluidic chamber), and the time-averaged DEP force acting on the spherical microparticle suspended inside liquid chamber can be mathematically described as follows [38]

$$\langle \mathbf{F}_{\mathrm{DEP}} \rangle = \pi R^3 \varepsilon_m \mathrm{Re}\left[\frac{\tilde{\varepsilon}_p - \tilde{\varepsilon}_m}{\tilde{\varepsilon}_p + 2\tilde{\varepsilon}_m}\right] \nabla E^2 \quad (1)$$

where R is the particle radius, E is the amplitude of electric field, $\tilde{\varepsilon}_p$ and $\tilde{\varepsilon}_m$ are the complex permittivities of the microparticle and suspending medium, respectively, $\tilde{\varepsilon}_p = \varepsilon_p - j\sigma_p/\omega$, and $\tilde{\varepsilon}_m = \varepsilon_m - j\sigma_m/\omega$. Here, σ_p and σ_m are ohmic electrical conductivities of the microparticle and liquid medium, respectively; ε_p and ε_m are the permittivities of the microparticle and liquid medium; ω is angular frequency of the electric field or the applied AC voltage. In Eq. (1), the expression in brackets (i.e., $(\tilde{\varepsilon}_p - \tilde{\varepsilon}_m)/(\tilde{\varepsilon}_p + 2\tilde{\varepsilon}_m)$) is referred to as the Clausius-Mossotti (CM) factor that is denoted as f_{CM}. When $\mathrm{Re}[f_{CM}] > 0$, positive DEP occurs and particles move toward high field regions; when $\mathrm{Re}[f_{CM}] < 0$, negative DEP occurs and particles move toward weak field regions. The particles in solution also undergo the opposing Stokes' drag [39], as follows

$$F_{\mathrm{drag}} = -6\pi\eta RV \quad (2)$$

where V is the particle velocity, and η is the dynamic viscosity of the liquid.

2.2. The design and fabrication of hybrid optoelectronic device (HOD)

The hybrid optoelectronic device (HOD) has a more compact structure, integrating another layer of physical electrode array above the photoconductive layer. In another word, the HOD (several cm^2) integrates with photoconductive layer and metal microelectrodes including the quadrupole electrodes and spiral electrodes as shown in **Figures 2** and **3**. **Figure 2** is the schematic design diagram of this microdevice. This HOD has a hamburger-like structure. The upper plate was transparent ITO conductive glass. The lower plate included the photoconductive layer (or called photoconductor) and the bottom glass slide. The photoconductive *a*-Si:H layer was on the bottom ITO film, and the insulating silicon nitride (or silicon carbide)

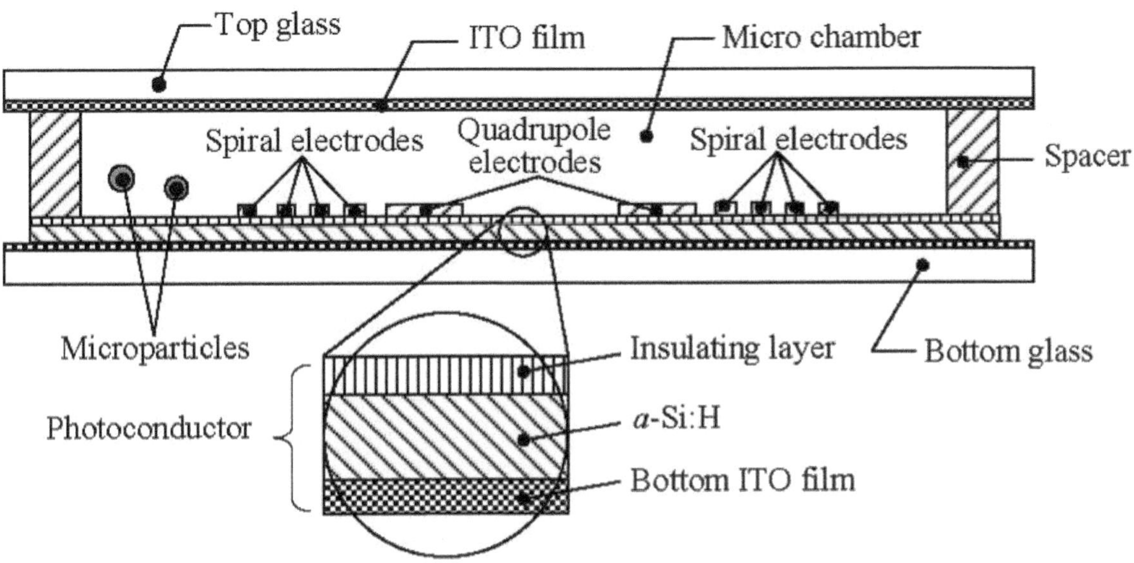

Figure 2. Schematic representation of HOD cross section along the line of AA′ in **Figure 3**. The photoconductor consists of insulating layer, *a*-Si:H layer and the bottom ITO film. Only a few representative electrodes are drawn for simplicity.

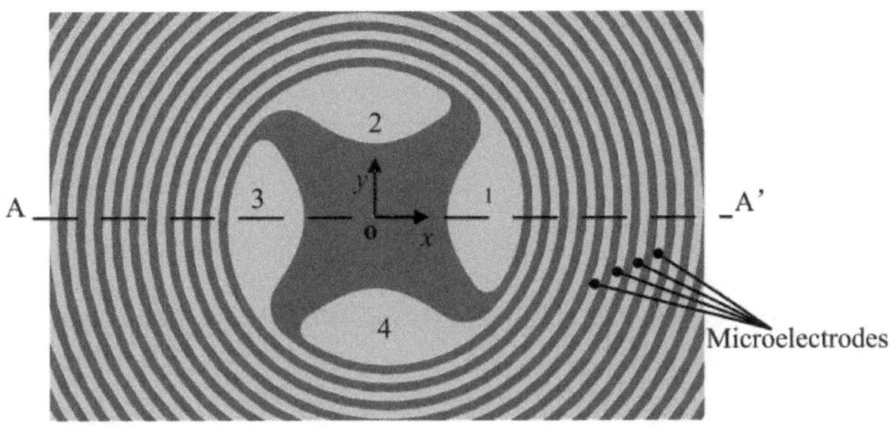

Figure 3. The enlarged drawing of quadrupole electrodes and spiral electrode bars with a width of 25 μm above the photoconductor. The coordinate system is shown, and the direction of z axis complies with right-handed screw rule [40].

layer and metal microelectrode layer were consecutively fabricated above the *a*-Si:H layer. The microfluidic chamber was designed and fabricated for suspending and manipulating particles. The photoconductor layer consisted of a 25–nm-thick insulating silicon nitride layer—to keep the insulation between the metal electrode layer and *a*-Si:H layer, a 1–2-μm-thick *a*-Si:H layer (include intrinsic *a*-Si:H layer and n + *a*-Si:H layer), and a bottom ITO conductive film (50 nm thick) in sequence. The potential of each successive signal could be set as phase shifted by a further multiple of 90° to produce quadrature signals or phases of 0°, 180°, 0°, 180° to produce traditional non-uniform field. It integrated metallic microelectrodes and the underneath photoconductive layer with LED-illuminating microspot.

The microelectrode layer contains of four parallel spiral electrodes as presented in **Figure 2**. When the four spiral electrode elements are energized with sinusoidal voltages with relative phase differences, a travelling electric field is generated and travels along the radial direction of the spiral array. A rotating electric field is also generated in the central free-space region denoted as central region. The spiral electrodes have the advantages of equivalent driving force to the particles at an arbitrary peripheral position. The quadrupole electrodes are located in the central area of spiral electrode array. The central region is surrounded by the quadrupole electrodes as shown in **Figure 3**. The geometrical shape of central region somewhat influences the distribution of the electric field around, and thus the sharp geometry should be avoided.

The experimental platform was established by assembling commercial optics, projector and an upright microscope. A digital micromirror device (DMD)-based projector (SHARP XG) was used as both the light source (via its dc lamp) and the interface between DMD driver circuit and the computer. The DMD forms a light image (or called optical pattern) according to the output signal of the PC's external monitor port. The optical image at the output of the projector lens was collimated, collected and reflected through the reflector at 45 degree to the horizontal and upward directed into a 10X driven lens from the lower location. The optical image was focused by the object lens and finally projected onto the photoconductive layer of ODEP device.

3. Manipulation modes and controlling regimes in FOD

3.1. Filtering

Usually, the impurities or non-target stuff (such as some types of microparticles) in the raw sample should be filtered firstly in the pre-treatment for the sample. The filtering experiment was implemented via an optical line scanning as shown in **Figure 4**. In the filtering experiment, with an applied 20 V_{pp} at 600 kHz, the normal pollen grains with diameters of 17–21 μm and most of the 30-μm diameter polystyrene particles were driven by the optical line with a width of ~70 μm from the upper area (**Figure 4a-b**) toward the bottom area of the view (**Figure 4c-d**). In contrast, the degenerative pollen grains and few polystyrene particles were left behind and stayed in their original positions. Therefore, the filtering function was able to be realized. This separation phenomenon is probably because the particles left behind were difficult to be polarized relative to the normal ones, which resulting in a much smaller DEP force for the abnormal ones. As long as the optical line velocity did not exceed the maximum synchronous velocity

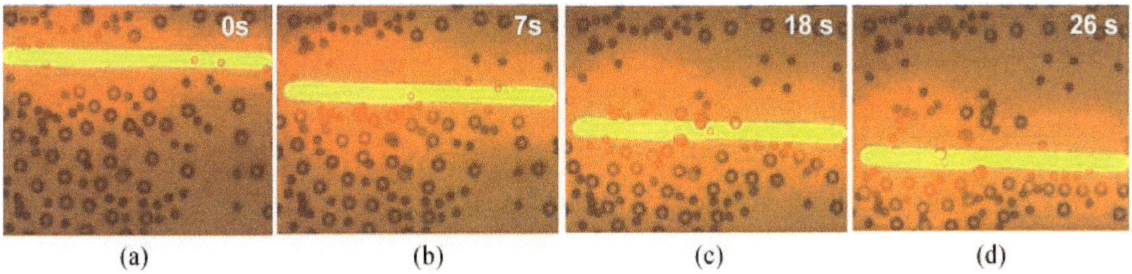

Figure 4. Snapshots of the filtering process in which a scanning light line moved from the top (a) to the bottom (d) of the viewing frame. Degenerative pollen grains and few polystyrene particles were left behind the optical line (b-d), while the normal particles were kinetically actuated by the scanning optical line.

(MS-velocity) of the normal particles [41] (i.e., maximum filtering velocity), the filtering function could be achieved. If the optical line velocity exceeded the MS-velocity, all particles would be left behind the scanning optical line, which made the filtering unavailable. This filtering experiment demonstrated the feasibility of the ODEP filtering.

3.2. Transporting

DEP can trap single particles and then transport the particles to desired positions, and the moving paths can be programed and controlled. As shown in, under the AC voltage of 20 V_{pp} and 1 MHz, two 30-μm diameter particles (**Figure 5a**) were trapped individually and moved along the programmed motion paths (two particles were marked by numbers 1 and 2 shown in **Figure 5**). These two particles were able to move simultaneously according to their own routings. The optical patterns can be controlled by the ODEP controlling software developed by our group. The polystyrene particles were always trapped by two optical rings those have an external diameter of ~150 μm and an interior diameter of ~40 μm during the transporting. Particle 1 and particle 2 kept moving and following the kinetic motion of the optical rings from the initial positions at the upper area (**Figure 5a**) to the lower area (**Figure 5h**) of the snapshot. The two particles moved simultaneously and always kept away from each other during the transporting, as shown in **Figures 5b–g** until they exchanged their positions and arrived at the terminative destinations.

3.3. Concentrating

In many biochemical applications, the reaction is very weak between a single bioparticle and other reagent, but a concentrated population of particles can react with other reagent intensively with much more obvious phenomena. Thus, when the raw sample was filtered and transported to the target locations, the selected particles would need concentration for the following detecting operations. The experiment presented in **Figure 6** serves as a typical example for presenting the concentrating regime of ODEP.

With an applied voltage of 20 V_{pp} at 800 kHz, at an arbitrary area (**Figure 6a**), the microlight pattern with a ring-shape can be used to collect a population of particles and can move the trapped particle population to an arbitrary location. The microparticles stayed inside the optical ring with a uniformly patterned array during the concentrating process (**Figure 6a-f**), and

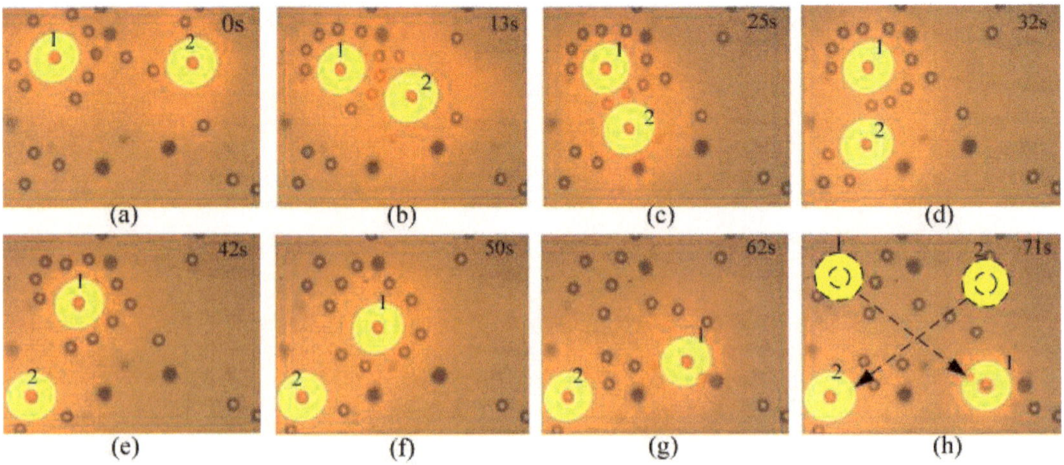

Figure 5. Snapshots of the transporting of two individual polystyrene particles parallelly. Particle 1 and particle 2 were driven from the upper area (a) toward the lower region (h) of the view field and the two particles switched their positions during the downward transporting process over time (c-g). The optical rings drawn by dashed lines in (h) indicate the original positions of the two particles [39].

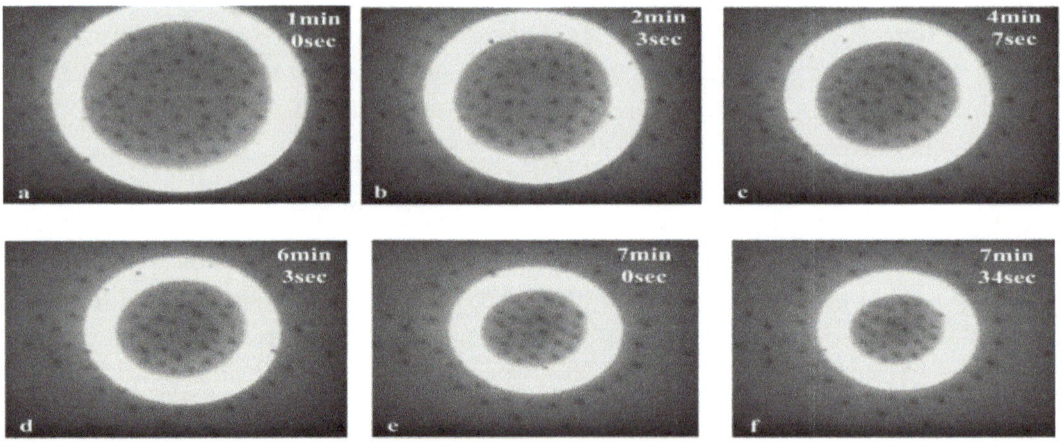

Figure 6. Concentration procedure for the microparticles with a diameter of 4 μm (20 Vpp, 800 kHz). (a) The initial distribution of microparticles. (b-e) The concentrating process of particles over time. (f) The final distribution of concentrated particles.

the distances between adjacent particles always kept the same. At this voltage magnitude, due to the dipole-to-dipole interaction between adjacent particles, the distance between adjacent particles is difficult to eliminate. The distances between particles were obviously larger than the particle diameter (**Figure 6**). If the voltage increased, the ODEP force would increase, which seems to drive the particles more concentrated. However, it is not that case when the voltage was larger than 20 V_{pp} in this experiment, because the increased voltage also enforced the interactional repelling force between particles.

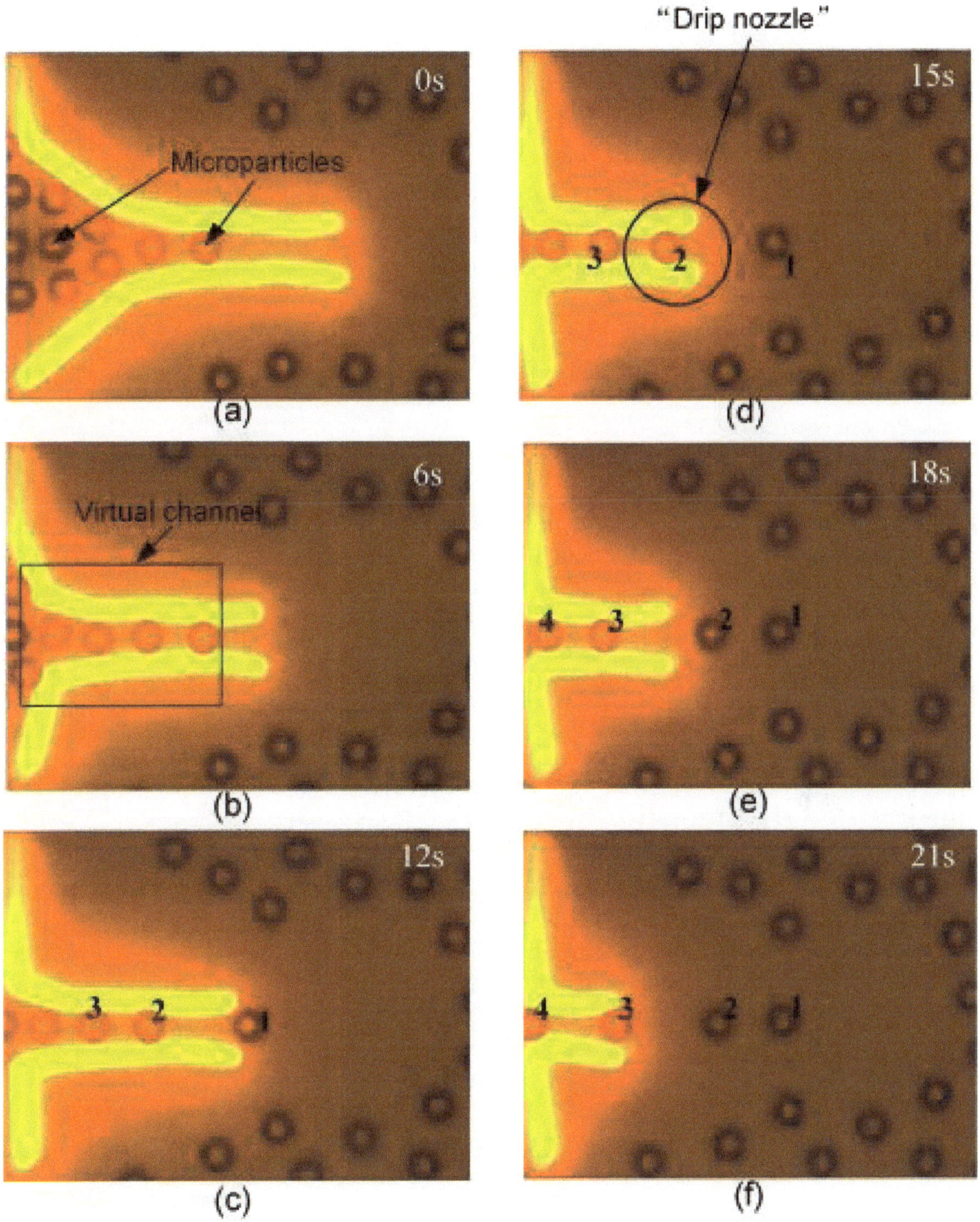

Figure 7. Microparticles are focused into a narrow particle line and ejected one-by-one by optical virtual channel and "drip nozzle." The micropattern projected onto the photoconductive layer to generate a negative ODEP force. Original state (a) is the initial spatial distribution of microparticles. While the optical virtual channel and "drip nozzle" deformed continuously, the random-located particles were gradually forming a straight particle line as shown in (b–d). Finally, the particles formed a linear array, and there were obvious spacing distances between particles as shown in (e–f).

3.4. Focusing

In many biochemical analyses, focusing is another important postprocessing mode apart from the procedures of concentrating particles. Focusing is mainly utilized to continuously and automatically count and detect individual particles. The ability to focus and eject particles based on ODEP is demonstrated in **Figure 7**.

In this study, a "drip nozzle" formed by the optical micropattern projected on the photoconductive layer was used to focus the microparticles in a straight sample flow and eject particles one by one. The optical lines forming the "drip nozzle" were set with a width of around the diameter of microparticles (**Figure 7**). The geometrical shape of optical pattern was real-time controlled so that the "drip nozzle" could continuously deform according to the current distribution of microparticles. The particles with 30-μm diameter moved out from the "drip nozzle" under the repelling ODEP force. This relative motion could also be regarded as the linear flow of particles relative to optical patterns. With an applied 24 V_{pp} at 600 kHz, the microparticles underwent negative ODEP force which always drove particles away from the virtual electrodes (optical micropattern) and thus kept particles horizontally swam in the virtual channel and then swam out of the "drip nozzle" accompany by the deformation of the optical virtual channel and "drip nozzle." Finally, the microparticles were focused into a straight particle line and had obvious spacing distances between particles 1 and 4 (**Figures 7e–f**). In this operation mode, there was no substantial fluidic flow to assist the focusing of the microparticles. The focusing efficiency grew when the optical line width increased but less than twice the particle diameter.

4. Manipulation of self-assembled microparticle chains by electroosmotic flow-assisted electrorotation in FOD

4.1. Background for microparticle chain manipulation

The microscopic particle chain is a common form of assembled particles, which involves in the colloidal, biological, electronic, photonic, magnetic fields. It plays an important role in these fields, as it serves as a link between the nanoscale world and meso- or macroscale objects. Currently, the methods for manipulating microparticle chains have become crucial for developing techniques on patterning magnetic nanoparticle arrays [42], forming regular superstructures by nanoparticles [43], constructing strings of particles [44], and creating multicellular assemblies [45] and live colloidosome structures [46]. In order to flexibly manipulate particle chains, various methods were developed, including the physical methods based on electric field, geometric constraint, dipole-dipole interactions or surface roughness direction [47] and the chemical methods based on linker molecules, oriented aggregation or non-uniform stabilizer distributions [34]. In this study, the formation and rotational manipulation of microparticle chains in real time are experimentally demonstrated by using optically induced electrorotation (OER) and ACEO flow in FOD.

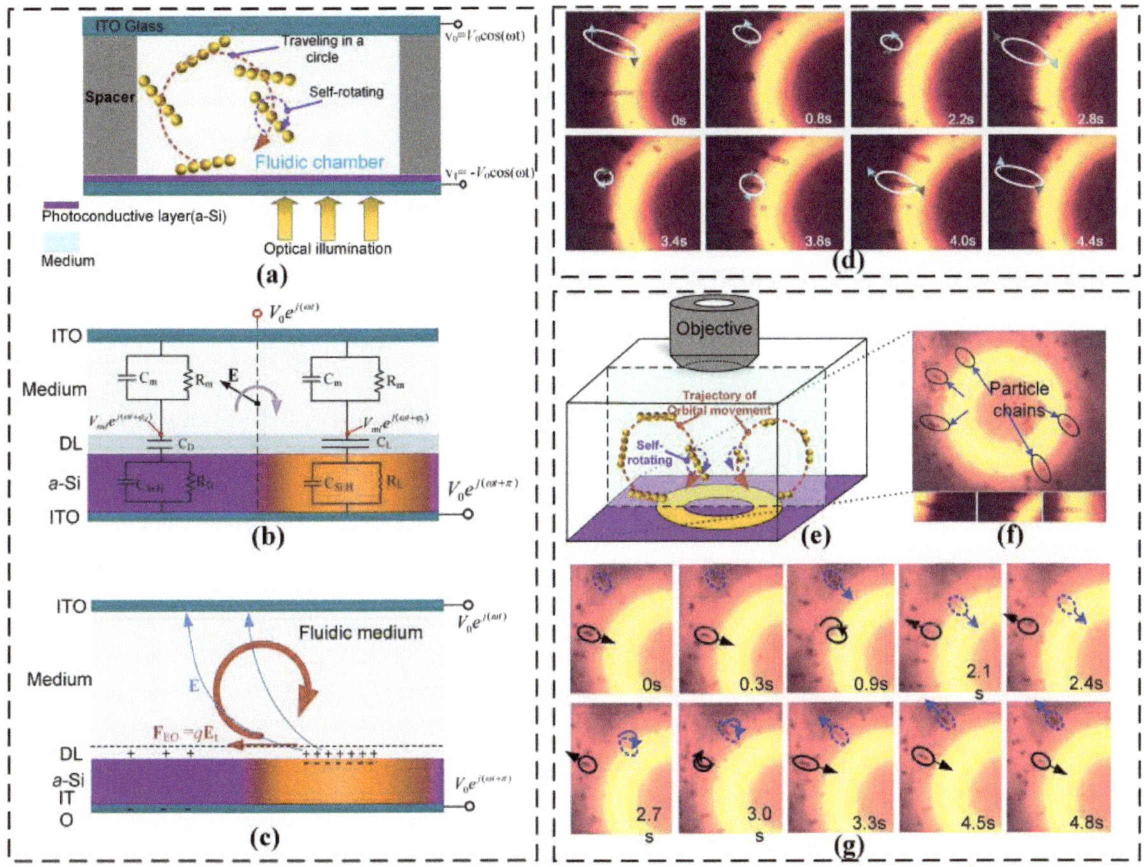

Figure 8. Schematically illustration of the OET device configuration and the principle for generating rotating electric field and ACEO flows. (a) Conceptual illustration of the self-rotation and circular travelling of an entire particle chains. The motion trajectories of the particle chain around the edge of illuminated area are indicated by dashed curve lines. (b) Device structure and mechanism for generating the rotating electric field around the border between dark and light regions. ω is the angular frequency, t is time, and V_0 is amplitude of AC voltage. (c) Electroosmotic fluid roll formed around the border between light and dark regions. (d–g) Experimental results of self-rotation and orbital travelling of an entire particle chains [48].

4.2. Experiment setup and working principle

The structure of the OET device and the scheme for rotating particle chains are conceptually illustrated in **Figure 8**. The equivalent circuit of the OET device is shown in **Figure 8b**. The device is composed by the upper and lower indium tin oxide (ITO) slides, photoconductive layer (hydrogenated amorphous silicon, *a*-Si:H) with a thickness of t_a and a relative permittivity ε_{ar}, and a fluidic chamber with a height of t_m and the applied AC voltages. The electrical conductivity of the photoconductor layer at unilluminated and illuminated areas is denoted as σ_{al} and σ_{ad}, respectively. The microfluidic chamber contains fluidic medium with a relative permittivity ε_{mr} and an electrical conductivity of σ_m. The micro-particles are suspended uniformly in the fluidic chamber at the initial time. The potentials applied at the upper and lower ITO layers.

The principle of the light-dark binary image induced rotating electric field (e-field) is indicated in **Figure 8a–b**. The a-Si:H layer has different impedances on light and dark sides, which leads to the amplitude and phase difference in the electric potentials at upper surface ($V_0 e^{j\omega t}$), photoconductive layer/medium interface in light area ($V_{ml} e^{j(\omega t+\varphi_l)}$) and dark area ($V_{md} e^{j(\omega t+\varphi_l)}$). In a particular frequency range, these three AC potentials can create a rotating e-field in fluidic medium around the border between light and dark regions, which will lead to self-rotating of the one-dimensional particle chains due to the interaction between the dipole moment of particle chain and rotating e-field.

According to equivalent circuit of the OET device shown in **Figure 8b**, and assuming that the impedance of conjunction between the light and dark regions in the a-Si:H layers is very large indicating no current flows across the border, the potential phasor at the bottom surface of fluidic chamber denoted as Φ_i is given by

$$\Phi_i = V_0 e^{j\omega t}\left(\frac{Z_a + Z_{DL} - Z_m}{Z_a + Z_{DL} + Z_m}\right) = V_{mi}\, e^{j(\omega t+\varphi_i)} \tag{3}$$

In (3), the subscript i in voltage amplitude V_{mi} and initial phase φ_i and can be replaced by "l" or "d" indicating the light or dark condition for the optical projection onto photoconductor, respectively. The impedance of photoconductor per unit area at light and dark regions is, respectively, expressed as follows

$$Z_{a\_l} = \frac{R_{a\_l}}{1 + j\omega C_a R_{a\_l}} \tag{4}$$

$$Z_{a\_d} = \frac{R_{a\_d}}{1 + j\omega C_a R_{a\_d}} \tag{5}$$

Here, the $C_a\left(=\varepsilon_{ar}\varepsilon_0/t_a\right)$ is the capacitance of a-Si:H layer per unit area, $R_{a\_l} = t_a/\sigma_{al}$ and $R_{a\_d} = t_a/\sigma_{ad}$ are the resistances of illuminated and unilluminated a-Si:H layer per unit area, respectively. Here, the double layer (DL) forms at medium/a-Si:H interface due to the interfacial polarization as shown in **Figure 8b**. The characteristic thickness of DL (t_{DL}) can be estimated by Debye length [38]. The detailed information could be found in Ref. [49]. The microparticle chain can be self-assembled under an e-field **E**, and then, a polarization could be induced in the particle chain. The effective dipole of the entire particle chain (denoted as **Pc**) forms after the establishment of the field **E** with a time delay, because it takes finite time for the dipole moment to completely form. When the e-field **E** rotates, **Pc** will follow it and rotates with a phase lag. The particle chain experiences a rotational torque as follows [50, 51],

$$\Gamma = \frac{1}{2}\mathrm{Re}[\,\mathbf{P}_c \times \mathbf{E}^*\,] \tag{6}$$

where $\mathbf{P}_C$ is the complex phasor of the effective dipole moment of a particle chain, and E^* is conjugated complex phasor of the e-field acting on particle chains. More information on effective dipole moment of a particle chain could be found in Refs. [49, 52].

ACEO flow is also induced at appropriate frequencies in the OET or ODEP device. ACEO flow is the motion of fluid due to an interaction between an e-field and the DL induced at medium/a-Si:H interface as shown in **Figure 8c**. The tangential e-field mainly arises around the edges of optical micropatterns. Therefore, the ACEO flow mainly occurs around the optical micropatterns projected on the photoconductive layer. ACEO flow lets the particle chains traveling in a nearly circular trajectory as shown in **Figure 8c**. After applying the voltage, counterions accumulate in the DL mainly within the illuminated region. Due to the distortion of electrical field induced by the optical pattern, a tangential component of e-field arises round the medium/a-Si:H interface. As a result, the ions in DL are subjected to the tangential electrostatic force along the a-Si:H surface. This force is then transmitted to the fluid and drive the boundary liquid to flow with a slip time-averaged velocity that can be calculated by using the Helmholtz–Smoluchowski equation as in Refs. [53, 54]

$$\langle u_{EO} \rangle = -\frac{1}{2}\frac{\varepsilon_{mr}\varepsilon_0}{\eta_b}\mathrm{Re}\left[(\varphi - \varphi_s)(-\tilde{E}_t^{\,*})\right], \tag{7}$$

where ϕ is the potential phasor just outside the DL, ϕ_s is the potential at the non-slip plane close to the solid surface of the DL. $\tilde{E}_t^*$ is the conjugated complex phasor of the tangential e-field just outside the DL. η_b is dynamic viscosity of the bulk solvent. $\eta_b = 1 \times 10^{-3}$ Pa·is for aqueous medium. The electrorotation of particle chains can be solely operated at relative higher frequencies because the ACEO flow decreases significantly at higher frequencies. Appropriately tuning down the frequency can make the ACEO flow available and then assist the self-rotating particle chains to travel along circled trajectory.

In experiment, the polystyrene (PS) microparticles did not form particle chains with a low voltage, but would start to form particle chains when the voltage increases to a threshold value. The PS particles with a diameter of 4 μm self-assembled into chains within the fluidic medium when the applied voltage was increased to 15 V_{pp}. The kinetic motion of particle chains could be enhanced with the increased voltage until 28 V_{pp}. The self-rotation of particle chain around its geometric center was obtained inside fluidic medium at 700 kHz because the ACEO flow can approximately be dismissed at these relatively higher frequencies. As shown in **Figure 8d**, the two ends of particle chains exchanged their positions after 2.8 s and had accomplished the first half-cycle of the rotation. Then, the particle chains continued to self-rotate through the other half-cycle, and it took only 1.6 s. The rotation rate of particle chain varied with time within a single period.

At lower frequencies, the particle chains circularly travelled around the center of fluid roll generated by the optically induced ACEO effect. As indicated in **Figure 8e**, the ACEO flow assisted particle chain rotation can be decomposed into two component motions: the self-rotation and orbital movement of the entire particle chain. **Figure 8f** shows the optically projected ring-shaped micropattern and the formed microparticle chains around the optical ring. The ACEO flow assisted microparticle chain rotation at 480 kHz is shown in **Figure 8g**. The particle chains were moving closer to the edge of optical pattern and turned around through the vertical planes after a short time (less than one second or a few seconds). Then, the particle chains were moving away from the optical ring along the circular path. They continued to turn around at the other corner of the circular path and then moved closer to optical ring again. The orbital moving of particle chains was not uniform motion and the vertical plane in which the circularly orbital movement occurred was approximate parallel to the diametrical dimension of optical ring. The complex movement of particle chains exhibited a three-dimensional (3D) movement of distributed particle chains that had more degrees of freedom than the 2D motion. These results offer more opportunities for spatially controlling the position, posture and orientation of particle chains for the applications regarding electrics, mechanics, optics and biodevices.

5. Manipulation of microparticles in a prototypical HOD

5.1. Integrated design and fabrication of metal electrode array and photoconductive layer

The virtual electrode is actually the microlight image in which we project in vertical direction on the photoconductive layer. It can precisely locate single microparticles and dynamically drive them. On the silicon nitride layer, we fabricated the planar microelectrode structure by photolithography. The microlight pattern could be designed according to specific experimental or application requirement, and it may be, for example, a linear or a spiral microelectrode array. In practical operations, we usually imposed low voltage to prevent insulating layer from breaking down by strong electric field. The spacer to mechanically support the fluidic chamber can be fabricated by polydimethylsiloxane (PDMS) or just double-sided tapes. The electric pads of microelectrodes were led out and connected to external PCB by gold wires. Then, the connectors on PCB were further connected to the external signal generator. The HODs were finally fabricated as shown in **Figure 9**.

The microelectrode array was fabricated by the depositing about 2000nm-thick layer of aluminum or gold on top of the insulating layer through electron beam evaporation. Then, the spiral pattern (**Figure 9**) was produced by standard or "lift-off" lithography process. In the standard lithography process, the corrosion rate needs to be strictly controlled in order to prevent possible damage to the insulating layer as well as the photoconductive layer. In our study, we had also sputtered gold over the silicon carbide layer (insulating layer) and then fabricated microelectrodes. Sputtering process and standard lithography process had a certain impact on the layer-to-layer electrical connections. In general, after the deposition by electron beam evaporations, the insulating property between the aluminum layer and the substrate is better. This usually could meet the insulating requirement of layers in this device. On the other hand, depositing gold by sputtering method onto the photoconductive film usually involved a relatively higher sputtering power or higher temperature in vacuum chamber (typically the sputtering involved a much higher temperature than the electron beam evaporation). These process features may result in that the *a*-Si:H layer was locally damaged within very tiny scale, which may enhance the partial defects in the thin film. More importantly,

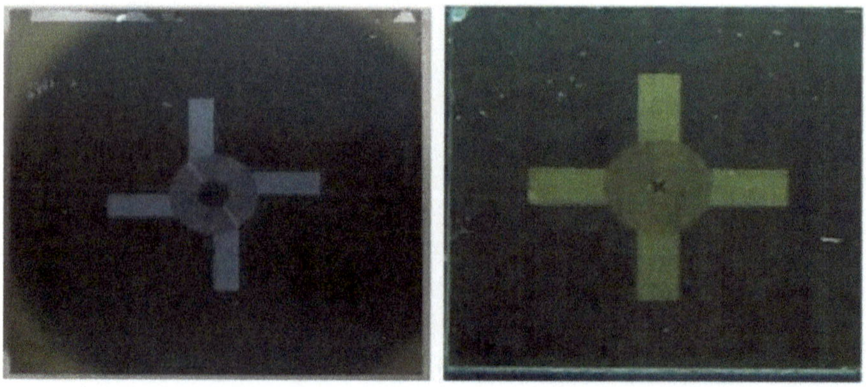

Figure 9. The prototypical photos of the substrates with metal microelectrodes and photoconductive layers for HOD.

the sputtering process also could form the Schottky contact between the metal layer and the below semiconductor layers. Consequently, the time-averaged magnitude of local AC impedance between the metal layer and bottom ITO layer was smaller. The measurement results showed a low resistance of a few hundred ohms to thousands of ohms between the metal layer and the bottom ITO layer.

5.2. Experiment result of the distribution of particles when light pattern is off

In this experiment, a 10 V peak-to-peak (V_{pp}) AC bias with a frequency of 1 MHz is applied between the top and bottom ITO electrodes, and the light pattern is not projected onto the HOD. The metal electrodes might be connected to the bottom ITO layer by a very low resistance through the Schottky contact between the metal and photoconductive layer and/or through the local defects of *a*-Si:H layer, when the AC voltage was applied. Therefore, most of the applied voltage drops across the liquid layer (for the resistance between the metal electrodes, and bottom ITO electrode is very small). However, in the gap region between the metal electrodes, most of the voltage drops across the photoconductive layer, only minority of the voltage drops across the liquid layer. Therefore, most of the particles moved to the gap region between metal electrodes by the negative DEP, as the white dotted circles indicated in **Figure 10A** and **B**. In addition, a few particles remained on the metal electrodes, as shown in **Figure 10A**. The single particle on metal electrode was initially near the center region of the metal electrodes where the DEP is very small, and it was already under the balanced state from the very beginning. That is why it remained on the electrodes instead of being driven to the gap region. The Schottky contact formed at the interface of the gold layer and photoconductive layer, the electrons flew from the photoconductive layer to the metallic layer. It enhanced the short connection effect between the bottom ITO film and the metal electrodes.

5.3. Experiment result of the particle's assembly under the assistance of light pattern

Before the microlight pattern was projected onto the HOD, the particles had distributed along the trend of spiral electrodes (**Figure 10**). The particles could be actuated by moving the microlight pattern when the light pattern (the diameter was nearly 120μm in this experiment)

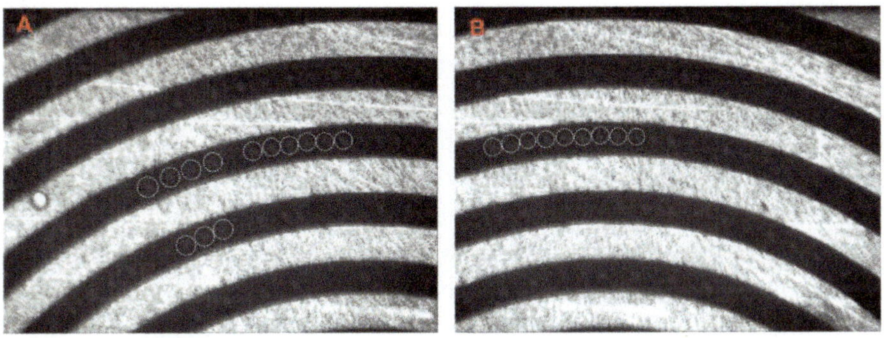

Figure 10. Generation and distribution of particle array or chains when ac voltage (10 Vpp, 1 MHz) was applied between the top and bottom ITO layers. (A) The arrayed microparticles in the gap of electrodes and a few ones on the metal electrodes. (B) The arrayed microparticles all in the gap of electrodes. The metallic electrode array was not energized by external voltages. Particles diameter is 10 μm. The widths of the electrodes and gap are both 20 μm.

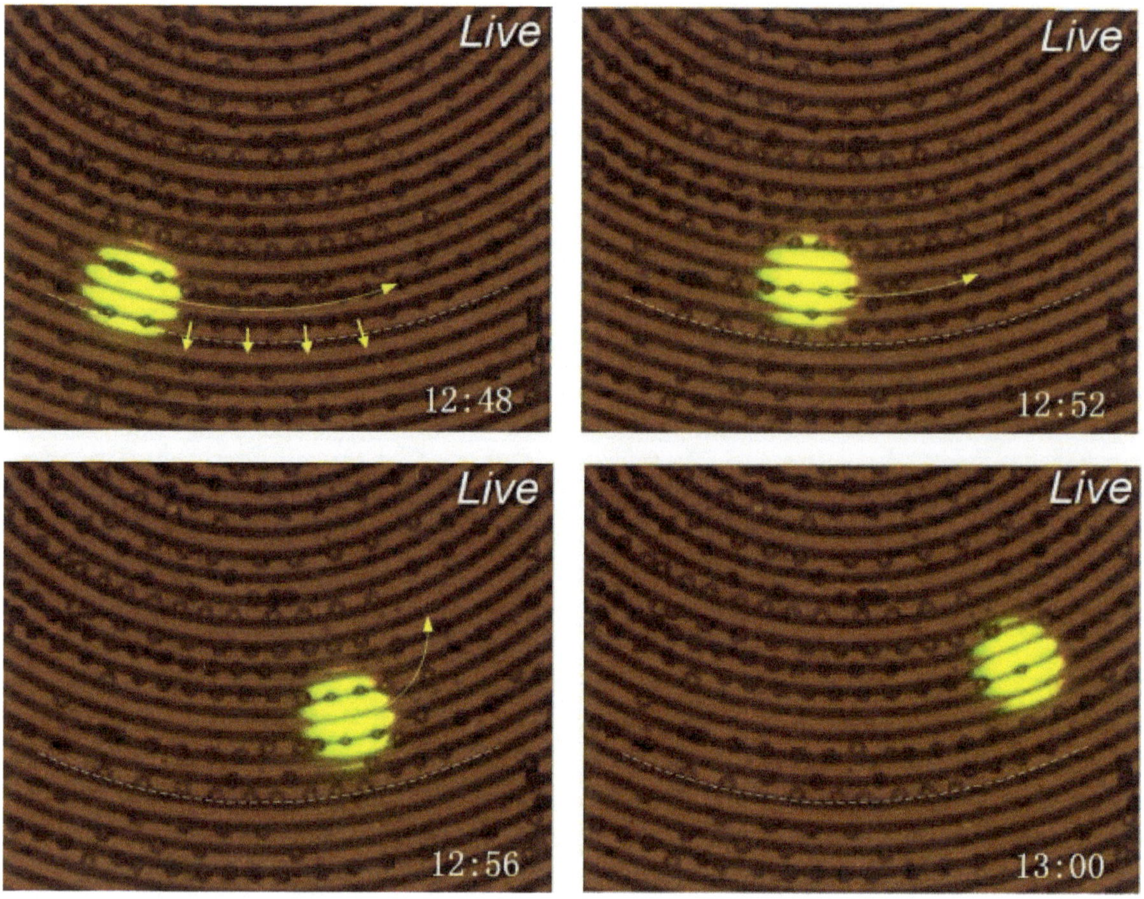

Figure 11. Particle manipulation in typical HOD. The curve arrow indicated the moving track of the light spot. Particles along spiral electrode and away from the white dotted arc were gradually moving closer to the position marked by the dotted arc due to the repulsive force from the light spot. The AC voltage between the top and bottom ITO layers is 5 V_{pp} and 1 M Hz. The particle diameter was 30 μm, and the widths of the electrode and gap were both 20 μm.

was projected as shown in **Figure 11**. The light pattern produced by LED can meet the needs of the power for operation. The spatial interval between the particles can be manipulated by adjusting the magnitude of the voltage applied between the upper and lower ITO layers, and the increase in voltage leads to larger separation between the particles. The metallic electrode array was not electrically energized. As shown in **Figure 11**, the DEP forces acting on microparticles were generated due to the participation of both the planar metal electrodes and microlight pattern in the electric field generated by the voltage applied between the upper and lower ITO layers. The curve arrow indicated the moving track of the light spot. Particles along spiral electrode and away from the white dotted arc were gradually moving closer to the position marked by the dotted arc due to the repulsive force from the light spot. This HOD can meet the need of the large-scale particles manipulation requiring the particles to form specific geometrical array. In this device, incorporation of metal microelectrode array was more suitable for the situation that the required separation between particles was small, and the tracks of particle distribution are more elaborate than the virtual electrode produced by light projection system. That is because microlight pattern needs to be produced by using even more complex and higher cost optical device, and the Gaussian distribution of light pattern border, the diffraction and the scattering may also limit the light pattern to have the smaller, finer and sharper geometry. This combination of metal microelectrode and light-induced virtual electrode in HOD is appropriate for the situation that the area for arranging particles is compact, and the particular geometric form (like curve shape distribution) is expected. This technique may also have the potential for manipulating large number of cells and be incorporated into the self-organization scheme for constructing multicellular structure [55, 56], in order to develop biomaterials applied in regenerative medicine.

6. Summary

In this chapter, *firstly*, the design and fabrication of flat optoelectronic devices (FOD) and hybrid optoelectronic device (HOD) were elucidated. *Secondly*, the scanning filtering, multiparticle parallel transportation, particle concentration and the focusing for single particle queue were experimentally demonstrated and analyzed. *Thirdly*, the self-electrorotation and circularly orbital movement of the microparticle chains in real time are experimentally demonstrated by FOD. The rotating electrical field for implementing the optically induced electrorotation (OER) of particle chains can be generated by utilizing the different impedances of photoconductor layer at light and dark areas. The optically induced ACEO flow was simultaneously produced accompanying the OER effect. This non-contact manipulation method for rotating particle chains has the potential for spatially regulating the posture, orientation and position of microparticle chains or microstructures. *Fourthly*, The hybrid device HOD enables rapid large-scale patterning and locally position adjustment of single microparticles. This technique makes up the shortcoming of lacking flexibility possessed by metallic microelectrodes and takes advantages of both electrode-based DEP and microimage-based DEP technologies.

Acknowledgements

This work was supported by grants from the National Natural Science Foundation of China (51505127), the Natural Science Foundation of Jiangsu Province (BK20161197), the Changzhou Sci&Tech Program (CE20165029), and the Fundamental Research Funds for the Central Universities of China (2015B04414). Parts of this chapter are reproduced from authors' recent conference publication [40, 48].

Author details

Xiaolu Zhu* and Yifei Yang

*Address all correspondence to: zhuxiaolu@hhu.edu.cn

College of Mechanical and Electrical Engineering, Hohai University, China

References

[1] Medoro G, Nastruzzi C, Guerrieri R, Gambari R, and Manaresi N. Lab on a chip for live-cell manipulation. IEEE Design & Test of Computers. 2007; 24(1): 26-36.

[2] Vahey MD and Voldman J. An equilibrium method for continuous-flow cell sorting using dielectrophoresis. Analytical Chemistry. 2008; 80(9): 3135-3143.

[3] Wang X-B, Huang Y, Gascoyne PRC, and Becker FF. Dielectrophoretic manipulation of particles. IEEE Transactions on Industry Applications. 1997; 33(3): 660-669.

[4] Gascoyne PRC and Vykoukal JV. Dielectrophoresis-based sample handling in general-purpose programmable diagnostic instruments. Proceedings of the IEEE. 2004; 92(1): 22-42.

[5] Small WR, Stoyanov SD, and Paunov VN. Scaffold free fabrication of linear multicellular assemblies by dielectrophoretic hydrogel trapping technique. Biomaterials Science. 2013; 1(9): 996-1002.

[6] Flanagan LA, Lu J, Wang L, Marchenko SA, Jeon NL, Lee AP, and Monuki ES. Unique dielectric properties distinguish stem cells and their differentiated progeny. Stem Cells. 2008; 26(3): 656-65.

[7] Iliescu C, Tresset G, and Xu G. Dielectrophoretic field-flow method for separating particle populations in a chip with asymmetric electrodes. Biomicrofluidics. 2009; 3(4): 044104-10.

[8] Eguchi M, Imasato H, Yamakawa T, and Ieee, Separation of blood cells by employing dielectrophoresis and traveling-wave electric fields. 2012 World Automation Congress; 2012.

[9] Schonfeld F, Griebel A, Konrad R, Rink S, and Karlsen F. Development of a [mu]-concentrator using dielectrophoretic forces. Journal of the Association for Laboratory Automation. 2002; 7(6): 130-134.

[10] Yokokawa R, Manta Y, Namura M, Takizawa Y, Le NCH, and Sugiyama S. Individual evaluation of DEP, EP and AC-EOF effects on lambda DNA molecules in a DNA concentrator. Sensors and Actuators B-Chemical. 2010; 143(2): 769-775.

[11] Cheng IF, Chang HC, Chen TY, Hu CM, and Yang FL. Rapid (<5 min) identification of pathogen in human blood by electrokinetic concentration and surface-enhanced raman spectroscopy. Scientific Reports. 2013; 3: 2365. DOI: 10.1038/srep02365

[12] Crews N, Darabi J, Voglewede P, Guo F, and Bayoumi A. An analysis of interdigitated electrode geometry for dielectrophoretic particle transport in micro-fluidics. Sensors and Actuators B: Chemical. 2007; 125(2): 672-679.

[13] Dos Santos MVP, Lima LPB, Mayer RA, Beron F, Pirota KR, and Diniz JA. Dielectrophoretic manipulation of individual nickel nanowires for electrical transport measurements. Journal of Vacuum Science & Technology B. 2015; 33(3): 031804.

[14] Hossan MR, Dutta P, Dillon R, and ASME. Numerical investigation of DC dielectrophoretic particle transport. ASME Fluids Engineering Division Summer Meeting—2014, Vol 2: Fora; 2014.

[15] Fu LM, Yang RJ, Lin CH, Pan YJ, and Lee GB. Electrokinetically driven micro flow cytometers with integrated fiber optics for on-line cell/particle detection. Analytica Chimica Acta. 2004; 507(1): 163-169.

[16] Yao B, Luo GA, Feng X, Wang W, Chen LX, and Wang YM. A microfluidic device based on gravity and electric force driving for flow cytometry and fluorescence activated cell sorting. Lab on a Chip. 2004; 4(6): 603-607.

[17] Jamshidi A, Pauzauskie PJ, Schuck PJ, Ohta AT, Chiou PY, Chou J, Yang PD, and Wu MC. Dynamic manipulation and separation of individual semiconducting and metallic nanowires. Nature Photonics. 2008; 2(2): 85-89.

[18] Chiou PY, Ohta AT, and Wu MC. Massively parallel manipulation of single cells and microparticles using optical images. Nature. 2005; 436(7049): 370-372.

[19] Tsutsui H and Ho CM. Cell separation by non-inertial force fields in microfluidic systems. Mechanics Research Communications. 2009; 36(1): 92-103.

[20] Cheng IF, Chang HC, Hou D, and Chang HC. An integrated dielectrophoretic chip for continuous bioparticle filtering, focusing, sorting, trapping, and detecting. Biomicrofluidics. 2007; 1(2): 021503.

[21] Hawkins BG, Smith AE, Syed YA, and Kirby BJ. Continuous-flow particle separation by 3D insulative dielectrophoresis using coherently shaped, dc-biased, ac electric fields. Analytical Chemistry. 2007; 79(19): 7291-7300.

[22] Yeo WH, Lee HB, Kim JH, Lee KH, and Chung JH. Nanotip analysis for dielectrophoretic concentration of nanosized viral particles. Nanotechnology. 2013; 24(18):185502. DOI: 10.1088/0957-4484/24/18/185502

[23] Maruyama H, Kotani K, Masuda T, Honda A, Takahata T, and Arai F. Nanomanipulation of single influenza virus using dielectrophoretic concentration and optical tweezers for single virus infection to a specific cell on a microfluidic chip. Microfluidics and Nanofluidics. 2011; 10(5): 1109-1117.

[24] Cho YK, Kim TH, and Lee JG. On-chip concentration of bacteria using a 3D dielectrophoretic chip and subsequent laser-based DNA extraction in the same chip. Journal of Micromechanics and Microengineering. 2010; 20(6): 1023-1035.

[25] Sen AK and Sajeesh P. Electrokinetic transport and separation of droplets in a microchannel. Microfluidics and Nanofluidics. 2014; 17(1): 97-106.

[26] Wang CH, Wang XZ, and Jiang ZD. Dielectrophoretic driving of blood cells in a microchannel. Biotechnology & Biotechnological Equipment. 2011; 25(2): 2405-2411.

[27] Khoshmanesh K, Nahavandi S, Baratchi S, Mitchell A, and Kalantar-Zadeh K. Dielectrophoretic platforms for bio-microfluidic systems. Biosensors & Bioelectronics. 2011; 26(5): 1800-1814.

[28] Zhang JM, Aguirre-Pablo AA, Li EQ, Buttner U, and Thoroddsen ST. Droplet generation in cross-flow for cost-effective 3D-printed "plug-and-play" microfluidic devices. RSC Advances. 2016; 6(84): 81120-81129.

[29] Pittermannova A, Ruberova Z, Zadrazil A, Bremond N, Bibette J, and Stepanek F. Microfluidic fabrication of composite hydrogel microparticles in the size range of blood cells. RSC Advances. 2016; 6(105): 103532-103540.

[30] Song YX, Wang CF, Li MQ, Pan XX, and Li DQ. Focusing particles by induced charge electrokinetic flow in a microchannel. Electrophoresis. 2016; 37(4): 666-675.

[31] Li D, Lu XY, Song YX, Wang JS, Li DQ, and Xuan XC. Sheathless electrokinetic particle separation in a bifurcating microchannel. Biomicrofluidics. 2016; 10(5): 1-13.

[32] Park S, Pan C, Wu T-H, Kloss C, Kalim S, Callahan CE, Teitell M, and Chiou EPY. Floating electrode optoelectronic tweezers: light-driven dielectrophoretic droplet manipulation in electrically insulating oil medium. Applied Physics Letters. 2008; 92(15): 151101.

[33] Hwang H, Lee D-H, Choi W, and Park J-K. Enhanced discrimination of normal oocytes using optically induced pulling-up dielectrophoretic force. Biomicrofluidics. 2009; 3(1): 14103.

[34] Kitching H, Shiers MJ, Kenyon AJ, and Parkin IP. Self-assembly of metallic nanoparticles into one dimensional arrays. Journal of Materials Chemistry A. 2013; 1(24): 6985-6999.

[35] Hwang H and Park JK. Optoelectrofluidic platforms for chemistry and biology. Lab Chip. 2011; 11(1): 33-47.

[36] Zhu XL, Yin ZF, and Ni ZH. Dynamics simulation of positioning and assembling multi-microparticles utilizing optoelectronic tweezers. Microfluidics and Nanofluidics. 2012; 12(1-4): 529-544.

[37] Yang YJ, Mao YF, Shin KS, Chui CO, and Chiou PY. Self-locking optoelectronic tweezers for single-cell and microparticle manipulation across a large area in high conductivity media. Scientific Reports. 2016; 6: 22630. DOI: 10.1038/srep22630

[38] Morgan H and Green NG. AC electrokinetics: colloids and nanoparticles. Microtechnologies and Microsystems Series, 2. 2003, Philadelphia, PA: Research Studies Press.

[39] Zhu XL, Yin ZF, Gao ZQ, and Ni ZH. Experimental study on filtering, transporting, concentrating and focusing of microparticles based on optically induced dielectrophoresis. Science China-Technological Sciences. 2010; 53(9): 2388-2396.

[40] Zhu XL, Yi H, Ni ZH, and IEEE. Novel design of multiphase optoelectronic microfluidic device for dielectric characterization of single biological or colloidal particles. 2009 ICME International Conference on Complex Medical Engineering. 2009. 542-547.

[41] Zhu XL, Yi H, and Ni ZH. Frequency-dependent behaviors of individual microscopic particles in an optically induced dielectrophoresis device. Biomicrofluidics. 2010; 4(1) :013202. DOI:10.1063/1.3279788

[42] Schelhas LT, Farrell RA, Halim U, and Tolbert SH. Directed self-assembly as a route to ferromagnetic and superparamagnetic nanoparticle arrays. Advanced Functional Materials. 2014; 24(44): 6956-6962.

[43] Zhou F, Xie MX, and Chen DY. Structure and ultrasonic sensitivity of the superparticles formed by self-assembly of single chain janus nanoparticles. Macromolecules. 2014; 47(1): 365-372.

[44] Termonia Y. Entropy-driven self-assembly of nanoparticles into strings. Colloids and Surfaces A: Physicochemical and Engineering Aspects. 2014; 447: 23-27.

[45] Brandy M-L, Cayre OJ, Fakhrullin RF, Velev OD, and Paunov VN. Directed assembly of yeast cells into living yeastosomes by microbubble templating. Soft Matter. 2010; 6(15): 3494-3498.

[46] Fakhrullin RF, Brandy M-L, Cayre OJ, Velev OD, and Paunov VN. Live celloidosome structures based on the assembly of individual cells by colloid interactions. Physical Chemistry Chemical Physics. 2010; 12(38): 11912-11922.

[47] Kraft DJ, Ni R, Smallenburg F, Hermes M, Yoon K, Weitz DA, van Blaaderen A, Groenewold J, Dijkstra M, and Kegel WK. Surface roughness directed self-assembly of patchy particles into colloidal micelles. Proceedings of the National Academy of Sciences of the United States of America. 2012; 109(27): 10787-10792.

[48] Zhu X. Electroosmotic flow assisted rotation of self-assembled microparticle chains using optoelectronic tweezers. in IEEE International Conference on Nano/micro Engineered and Molecular Systems. 2015.

[49] Zhu X. Manipulation of self-assembled microparticle chains by electroosmotic flow assisted electrorotation in an optoelectronic device. Micromachines. 2015; 6(9): 1387-1405.

[50] Jones TB. Electromechanics of Particles. 1995, New York: Cambridge University Press.

[51] Jones TB. Basic theory of dielectrophoresis and electrorotation. IEEE Engineering in Medicine and Biology Magazine. 2003; 22(6): 33-42.

[52] Jiang ZH, He XD, Han JC, Zhang BM, and Du SY. Fractal characterization of the dipole moments of dielectric particle chains. Journal of Electrostatics. 1998; 44(1-2): 47-51.

[53] Ramos A, Morgan H, Green NG, Gonzalez A, and Castellanos A. Pumping of liquids with traveling-wave electroosmosis. Journal of Applied Physics. 2005; 97(8): 084906.

[54] Lyklema J. Fundamentals of Interface and Colloid Science, Volume II: Solid–Liquid Interfaces. Vol. 2. New York: Academic Press; 1995.

[55] Zhu X, Gojgini S, Chen T-H, Teng F, Fei P, Dong S, Segura T, and Ho C-M. Three dimensional tubular structure self-assembled by vascular mesenchymal cells at stiffness interfaces of hydrogels. Biomedicine & Pharmacotherapy. 2016; 83: 1203-1211.

[56] Zhu X and Yang H. In-silico constructing three-dimensional hollow structure via self-organization of vascular mesenchymal cells. in *The 16th International Conference on Nanotechnology (IEEE NANO 2016)*. 2016. Sendai, Japan: IEEE.

8

Surface-Barrier Photodiodes with Transparent Electrodes for High-Performance Detection in the UV-NIR Spectrum

Oleksandr Malik and
Francisco Javier De la Hidalga-Wade

Additional information is available at the end of the chapter

Abstract

The aim of this chapter is to present a short review in the field of design, fabrication technology, and testing of high-efficiency surface-barrier photodiodes with electrodes based on thin-film transparent conducting oxides (TCO) such as tin-doped indium oxide (ITO) and fluorine-doped tin oxide (FTO). Most of this review is based on our own results obtained and reported during the last 30 years. Besides a brief description of the low-cost spray pyrolysis deposition technique, mainly used in our work for deposition of the TCO films on a semiconductor surface, structural, morphological, and optoelectronic properties of these TCO films are discussed. As an example, a successful application of these TCO films is shown and used in high-efficiency surface-barrier photodiodes for a wide spectral range, from near ultraviolet (UV) to near infrared (NIR), and fabricated on different semiconductor substrates such as traditional Si, wide energy band ZnS, and GaP compound semiconductor substrates. The possible use of the Si surface-barrier structures as radiation-resistant detectors and gamma radiation detectors is discussed. The properties of high-efficiency surface-barrier photodiodes based on a perspective ternary semiconductor compound, $Hg_3In_2Te_6$ mercury indium telluride (MIT), for detection of 1.3 μm and 1.55 μm radiation for applications in fiber optics, are also reported.

Keywords: surface-barrier photodetectors, thin film, tin-doped indium oxide, fluorine-doped tin oxide, spray pyrolysis

1. Introduction

Physics of surface-barrier (SB) devices is based on the well-known Schottky barrier, named after Walter H. Schottky, which is a potential energy barrier for carriers formed at a metal-semiconductor

(M-S) junction [1]. In the design of SB photodetectors, the main function of the barrier is to separate carriers photogenerated inside the semiconductor substrate. Such photodiodes are very simple device structures; because a p-n junction is not necessary, it does not present optical losses in the highly doped p-layer and also present a very high speed of response on modulated optical signal. The SB structure is very useful for fabrication of optoelectronic devices on semiconductors, in which a p-n junction cannot be created due to doping troubles. At the same time, in order to use M-S structures as radiation detectors, the metal electrode needs to be extremely thin (<15 nm) to avoid the radiation absorption losses and also must be chemically resistant to prevent device degradation in time. A thinner metallization leads to devices susceptible to degradation. Among all the metals, only some of them such as Au, Pt, and Ni are suitable for this application. Moreover, the high reflectivity of these metallic layers demands the use of antireflection coatings. The electrical properties of SB photodetectors based on M-S structures in many cases can be enhanced by the introduction of a very thin (<3 nm) insulating layer between the metallic film and the semiconductor (M-I-S structures). The presence of this insulator layer reduces the number of localized states at the semiconductor interface and hence serves to reduce interface carrier recombination. Its presence can also reduce significantly thermionic emission current because of an increase in the potential barrier for majority carriers [2]. At that, the thickness of the insulating layer must not reduce the transport of minority carriers from the semiconductor to the metal.

For best photoelectrical properties of SB radiation detectors based on M-S and M-I-S structures, the metallic opaque layer can be changed by a thin film based on some transparent conducting oxide (TCO), such as tin-doped indium oxide (ITO) or fluorine-doped tin oxide (FTO), which are widely used for optoelectronic applications. Such films can be deposited on a semiconductor surface by different technological methods as thermal vacuum deposition, e-beam deposition, sputtering, chemical vapor deposition, sol-gel deposition, and spray pyrolysis [3]. Spray pyrolysis is a low-cost and simple technological method used successfully since 1976 for the fabrication of efficient solar cells based on metal oxide (ITO, FTO)-silicon structures [4–11]. In such structures, the TCO films work as an active metal-like transparent conducting electrode. Thickness tuning of the TCO films allows for its use as an effective antireflection coating for reducing radiation losses.

In this chapter we pay attention to application of spray-deposited ITO and FTO films for fabrication of high-efficiency SB photodetectors for applications in a wide spectral region. The material presented in this chapter is based on original results reported in this field during the last 35 years; hence, this review will be very useful for a large number of readers.

2. Spray pyrolysis processing

Spray pyrolysis [12] is a process in which a thin film is deposited by spraying a solution on a heated surface, where the constituents react to form a chemical compound. The chemical reactants are selected such that the products other than the desired compound are volatile at the temperature of deposition. Usually, two types of processing are used: the process in which the droplet resides on the surface as the solvent evaporates, leaving behind a solid that may further react in the dry state and the process where the solvent evaporates before the droplet reaches the surface and the dry solid impinges on the surface, where decomposition occurs.

Three sprayed solution were used for photodetectors fabrication. The ITO films were deposited from solution contents 13.5 mg of $InCl_3 \times 4H_2O$ dissolved in 170 ml of water and ethylic alcohol (1:1) mixture, with an addition of 5 ml of HCl to prevent hydrolysis. The different ratios of Sn/In in the solution were controlled by adding in the solution a calculated amount of tin chloride ($SnCl_4 \times 5H_2O$) [13]. As alternative, the ITO layers can be deposited by spraying a solution of Indium (III) acetate, $In(OOCCH_3)_3$ dissolved in ethanol at 25°C by adding Tin(IV) chloride, $SnCl_4 \times 5H_2O$, and HCl. Ratio Sn/In in the solution varied by calculated amount of tin chloride. A glass atomizer was designed in order to produce small-size droplets. The spraying was conducted using compressed air. Substrates were placed on heater having fixed temperature of 460 ± 2°C. The heater was covered with polished carbon disk having a high thermal conductivity and chemical resistance in order to assure a uniform temperature of the substrates. The substrate temperature was controlled with a thermocouple. The optimum distance from the atomizer to the substrate and the compressed air pressure were 25 cm and 1.4 kg/cm^2, respectively. In order to prevent the deposition of the films under nonequilibrium thermal conditions, due to a possible fast cooling of the substrate surface with the stream of the precursor and the compressed gas, periodical cycles of the deposition, with durations of 1 sec and intervals of 10 sec, were employed.

Fluorine-doped tin oxide (FTO) films [14] were deposited from solution contents a 0.2 M solution of tin chloride ($SnCl_4 \times 5H_2O$) dissolved in methanol. A small amount of NH_4F dissolved in water was added in the starting solution for the preparation of the precursor with 0–1 F/Sn molar concentrations. A few drops of HCl were added to the precursor solution in order to prevent the hydrolysis of the tin chloride due to the presence of water. The flow rate of the precursor during the deposition process was 10 ml/min. Other deposition parameters were the same as for the ITO films.

For fabrication of the M-S or M-I-S structures, the ITO and FTO films were deposited on chemically prepared surface of semiconductor substrates. A Corning glass or sapphire substrates placed on the heater together with the semiconductor substrates were used as referents for the measurement of the films' properties. The structural characterization of the ITO films was conducted using a Bruker D8 Advanced X-ray diffractometer (XRD), with Cu K_α radiation (λ = 0.15406 nm). The morphological features of the ITO films were examined using the Ambios USPM atomic force microscope (AFM). Keithley's Series 2400 Source Measurement Unit and the Ecopia HMS-5000 Hall Effect Measurement System were used for electrical characterization. The optical measurements were carried out using the F20 Filmetrics and the Agilent 8453 spectrometers.

3. Properties of the metal oxide films

Both, the ITO and FTO films deposited with conditions described above were microcrystalline. X-ray diffraction (XRD) spectra for films with optimal doping concentration are shown in **Figure 1**.

The optimal doping concentration for the ITO and FTO films means the value of doping at which is obtained the lowest sheet resistance $R_s = \rho/d$, where ρ and d are the film resistivity and thickness,

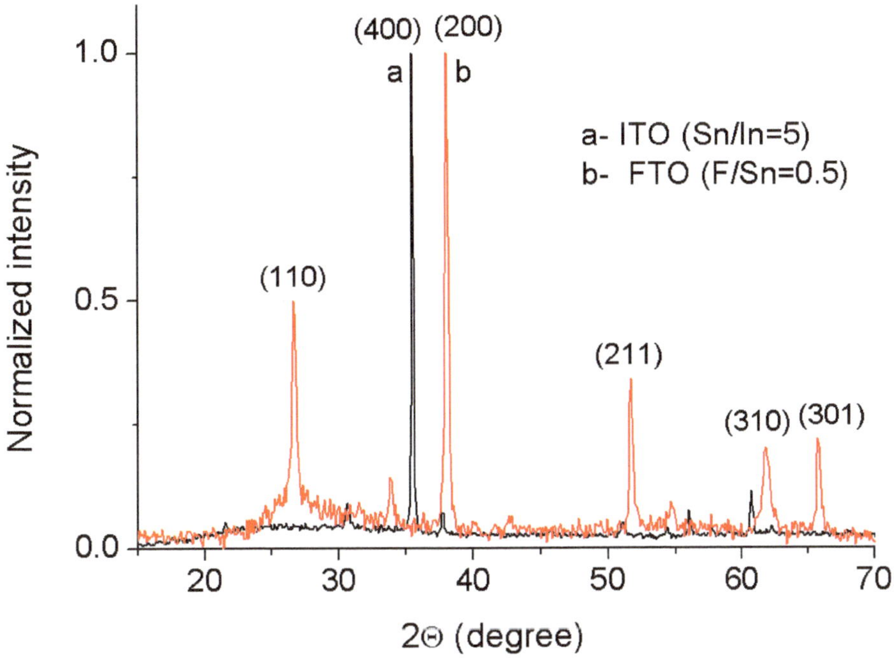

Figure 1. XRD patterns for sprayed ITO and FTO films.

respectively. At the same time, the value of transmittance (T) is important. For practical applications of the TCO films as transparent conducting electrodes, the fabrication process needs to be optimized for obtaining a better balance between R_s and T. However, a simultaneous obtaining of maximum transmission and conductivity is not possible to achieve in most cases. Haacke [15] proposed the figure of merit (FOM) $\phi_{TC} = T^{10}/R_s$ for comparison of films with different thickness and fabrication histories. A higher FOM value indicates a higher performance; this characterizes a film showing simultaneously a low sheet resistance (R_s) and a high transparency (T). However, the value of one of these parameters can compromise the value of the other. To avoid influence of an interference effect in the thin films, the average integrated value of the films transmittance (T_{int}) at wavelengths in the 400–700 nm spectral range is better to be used for the FOM value calculation. **Table 1** presents structural, electrical, and optical parameters of the ITO and FTO films deposited at 460°C from the solutions containing the optimal doping concentration.

The parameters obtained allow to conclude that high-quality FTO and ITO films can be fabricated by a simple and inexpensive spray deposition method. Our recently published works [13, 14] give more detailed information about the dependence of the film properties on the doping concentration as well as a comparison of our results with the ones published in literature.

FTO and ITO are degenerated wide band gap semiconductors. Films of these materials can be used as transparent conducting electrodes in ultraviolet (UV) SB photodetectors. In nominally doped semiconductors, the Fermi level lies between the conduction and valence bands. As the doping concentration is increased, electrons populate states within the conduction band which pushes the Fermi level higher in energy, and in the case of degenerate level of doping, the Fermi level lies inside the conduction band. At that, using transmission/reflection spectroscopy, the

Parameters	Spray deposited FTO film	Spray deposited ITO film
Ratio F/Sn in solution	0.5	–
Ratio Sn/In in solution	–	5
Preferred grain orientation	200	400
Grain size (nm)	120	110
Specific resistance (10^{-4} Ω cm)	1.8	2
Carrier concentration (10^{20} cm^{-3})	16.0	11.5
Mobility ($cm^2V^{-1}s^{-1}$)	21.0	28
Integrated transmittance (T_{int})	0.84	0.85
FOM (10^{-3} Ω^{-1})	38	20

Table 1. Comparison properties of the FTO and ITO films deposited by spray pyrolysis from the solutions with optimal doping concentration (ratio of F/Sn and Sn/In).

optical energy gap (E_{opt}) can be measured. This value is not coinciding with value E_g that is a distance between the valence and conduction bands as is shown in **Figure 2**. The difference ΔE is named as Burstein-Moss shift [16]. In the case of a degenerate semiconductor, an electron from the top of the valence band can only be excited into conduction band above the Fermi level (which now lies in conduction band) since all the states below the Fermi level are occupied states. Pauli's exclusion principle forbids excitation into these occupied states. The value of ΔE is proportional to the carrier concentration. Burstein-Moss shift for FTO and ITO films, characteristics of which are shown in **Table 1**, is 0.5–0.8 eV. At that, the value of optical energy gap determined from analysis of transmittance (**Figure 3**) for the FTO and ITO films is 4.3 eV and 4.8 eV, respectively. From **Figure 3** we made an important conclusion that FTO films are more suitable for UV SB photodetectors.

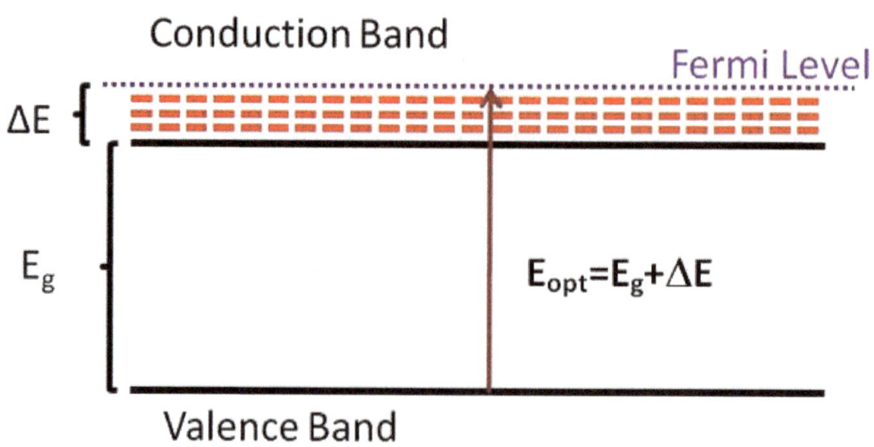

Figure 2. Illustration of the Burstein-Moss shift in degenerated semiconductors.

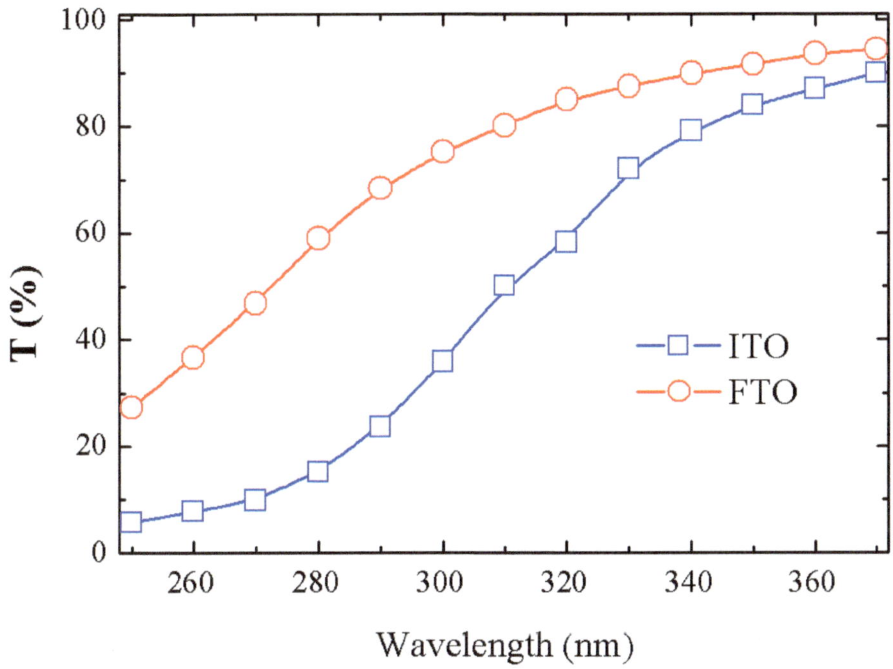

Figure 3. Transmittance of FTO and ITO films in the UV spectral range.

4. Surface-barrier photodetectors for the UV spectral range

There are many applications (biomedical, astronomy, flame safeguard systems, and others) that require the measurement of blue or UV radiation in the presence of visible light. The spectral selectivity of photodiodes depends on a band gap of semiconductor which is shown in **Figure 4**.

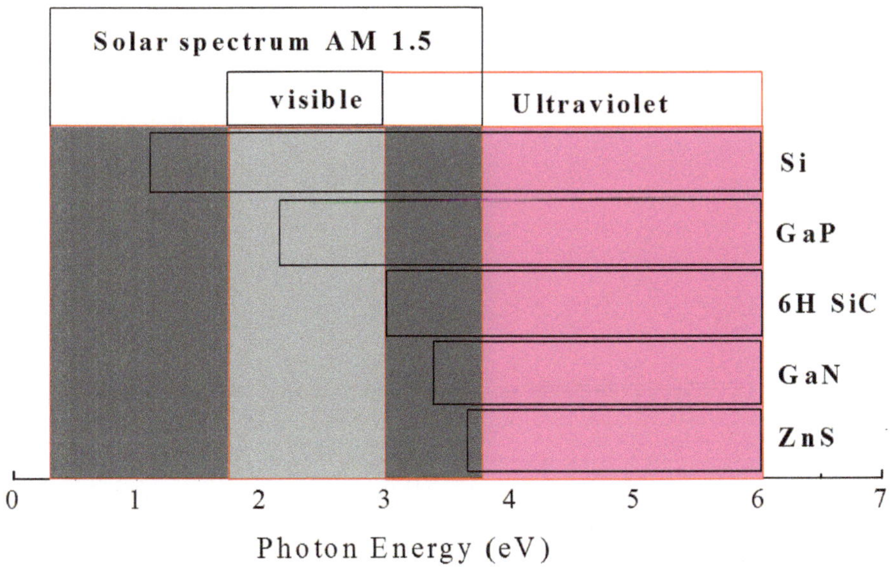

Figure 4. Semiconductors of interest for visible, UV, and solar-blind photodetectors.

In this direction, researchers are paying attention to large-band-gap semiconductor compounds such as GaP, SiC, and GaN in contrast to Si-based photodiodes in which the selectivity can be provided with additional blocking filters or dielectric coating to fit them with the spectral region of interest.

4.1. Solar-blind UV photodetectors based on monocrystalline ZnS

The large-band-gap (3.6 eV) zinc sulfide (ZnS) greatly reduces the sensitivity to photons in the terrestrial solar radiation, and it is an attractive material for developing solar-blind photodetectors.

N-type ZnS monocrystalline wafers were used for the fabrication of ITO-ZnS and FTO-ZnS [17, 18] UV SB photodetectors, with active areas ranging from 5 to 100 mm^2. FTO and ITO films have been deposited at the top surface of the monocrystals by spray technique at conditions described in Section 2. The backside ohmic contact to the ZnS was made as described in [17]. The current-voltage characteristic of FTO-ZnS photodetector with an area of 1 cm^2 is shown in **Figure 5**.

Figure 6 shows the spectral response of the FTO-ZnS photodetector [18] in comparison with characteristic of ITO/ZnS [17] and SiC [19] UV photodetectors.

In contrast to ITO-ZnS detectors, the maximum spectral sensitivity peak of the FTO/ZnS photodetectors shifted from 330 to 290 nm, due to the higher transparency of the FTO layer in UV, when compared to the transparency of the ITO layer (**Figure 3**). The combination of a low dark current and a high sensitivity originates a low value for the noise-equivalent power (NEP), ranging from 10^{-14} to 10^{-12} W/Hz$^{1/2}$, that is a function of the area of the sensors. In **Figure 7**, the

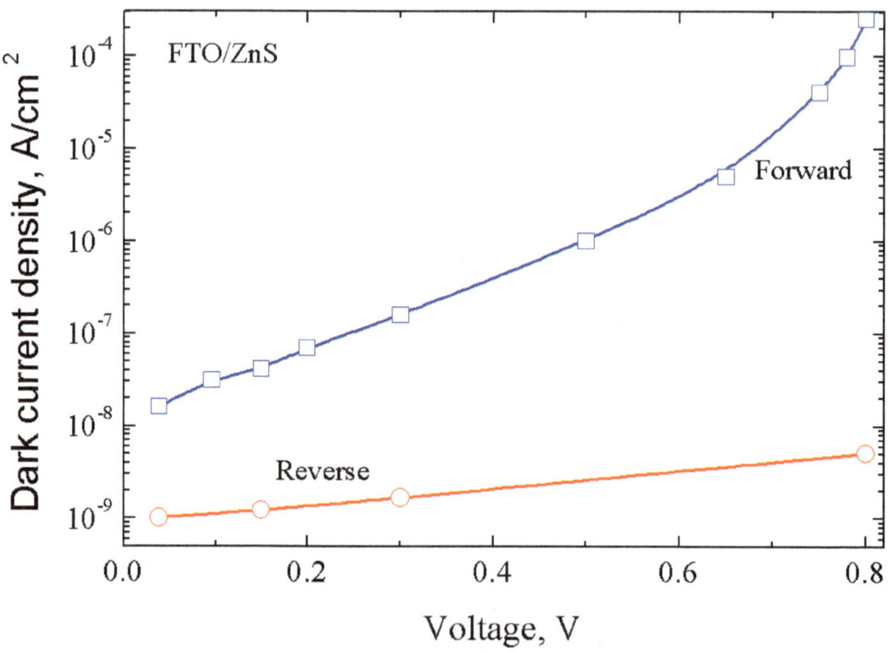

Figure 5. Current-voltage characteristic of FTO-ZnS photodetector with an area of 1 cm^2.

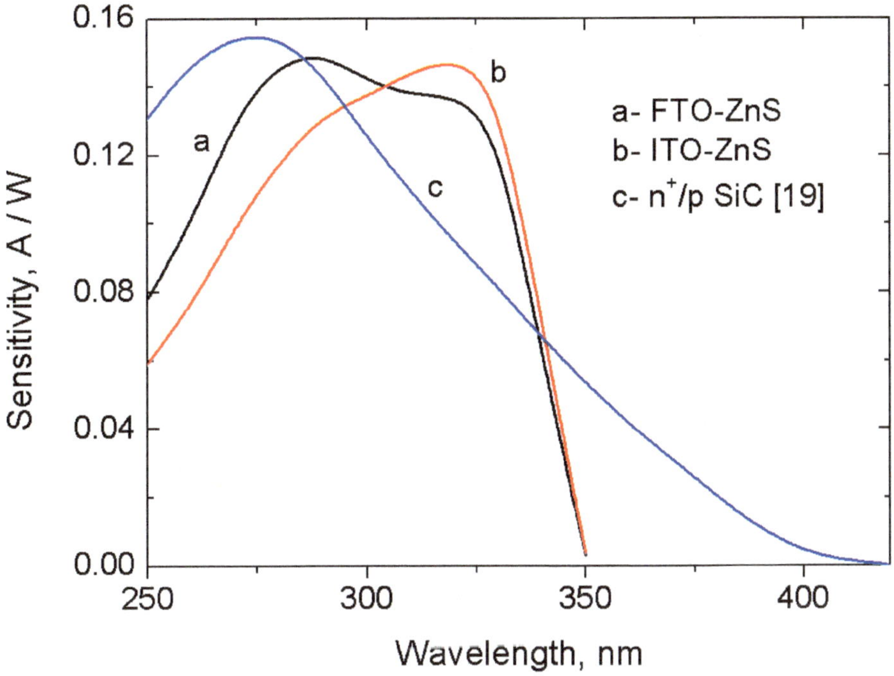

Figure 6. Spectral sensitivity of FTO-ZnS and ITO-ZnS photodetectors in comparison with a SiC UV photodetector [19].

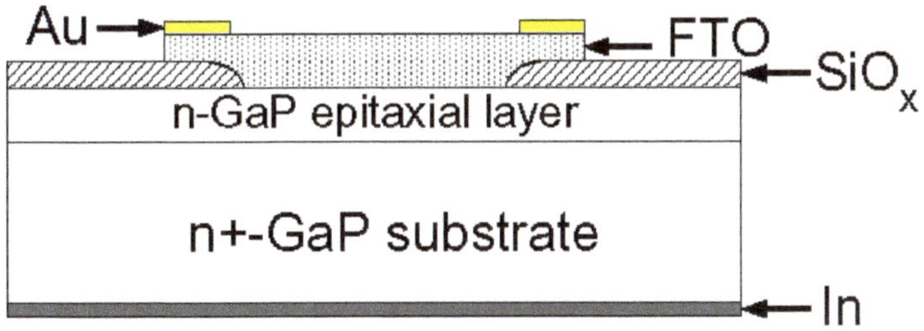

Figure 7. Schematic cross section of the FTO/n-n$^+$ GaP SB photodetector.

dependence of the photocurrent on incident light power is shown. A linear dependence of photocurrent on incident light power ranging from 10^{-8} to 10^{-3} W was found.

Thus, SB UV FTO-ZnS photodetectors show excellent characteristics as solar-blind UV detectors with high, 66%, external quantum efficiency, and can be used for different technical applications without losing their photoelectrical parameters under direct solar irradiation.

4.2. UV photodetectors based on epitaxial GaP

Our research results about SB photodetectors with transparent TCO electrode deposited by spray pyrolysis technique on gallium phosphide (GaP) with a band gap 2.27 eV were reported during the last 20 years [20–24]. Here, a brief resume of these works in comparison with reported by other authors is done.

A simple and low-cost pyrolysis technique has been used to form the high transparent conducting FTO or ITO electrode on prepared surface of GaP n-n$^+$ epitaxial structures with 10–12 μm thick epitaxial layers. The estimated electron density in the epitaxial layer is around 10^{16} cm^{-3}. The back ohmic contact was formed by indium diffusion. **Figure 7** shows the schematic cross section of a photodiode based on the FTO-GaP heterostructure.

The typical dark current-voltage (I-V) characteristic of the FTO-GaP photodiode with an active area of 3 mm^2 measured at room temperature is shown in **Figure 8**.

The results show excellent junction properties of FTO-GaP photodiodes. The equivalent shunt resistance R_{sh}≈100 GΩ was evaluated at a bias ±50 mV, and this resistance is a main source of the

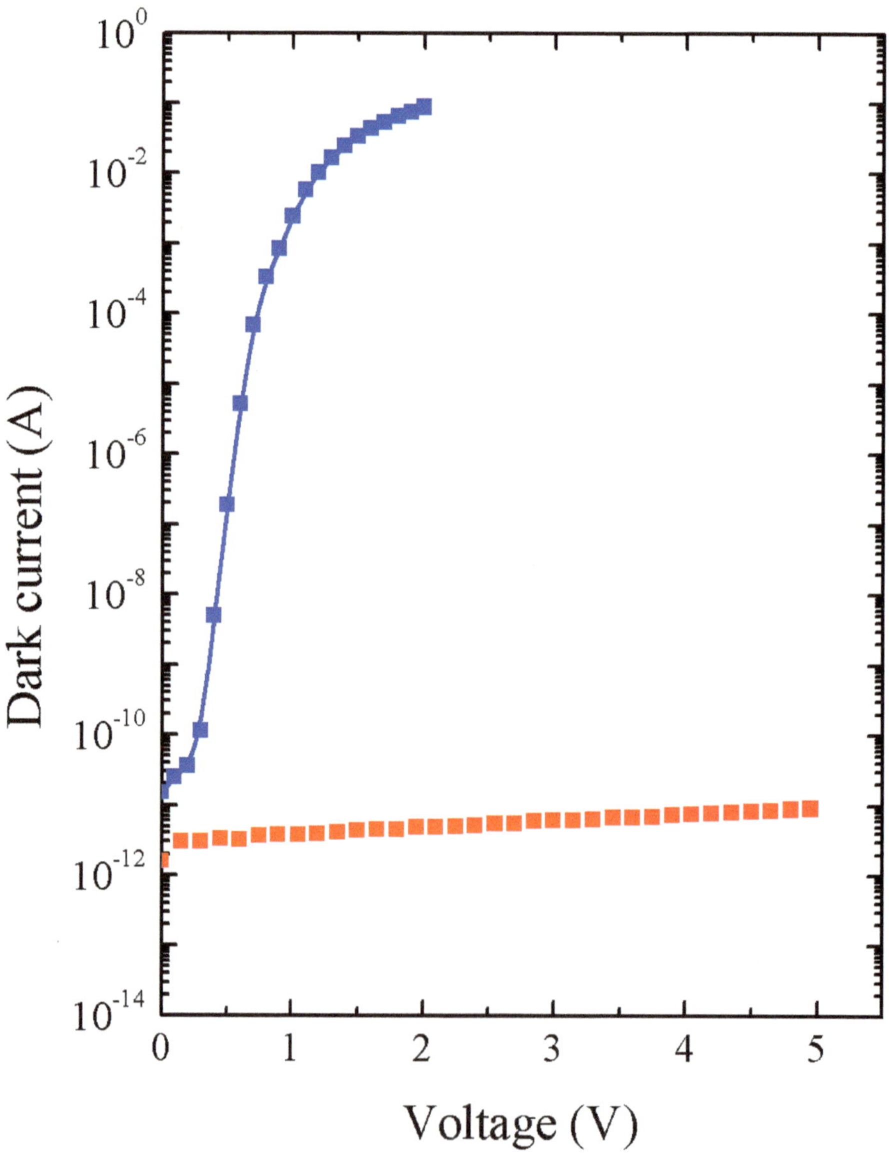

Figure 8. The typical I-V characteristics of FTO-GaP photodetector.

device noise, which limits the detectivity in short-circuit current mode. The FTO film resistance causes the non-exponential current increasing at forward bias higher than 0.6 V. A saturation current of $I_o \approx 10^{-15}$ A and an ideality factor $n = 1.1$ were determined from the exponential region of the forward I-V characteristic. The donor concentration $N_D = 3.8 \cdot 10^{16}$ cm^{-3} and the built-in potential $V_{bi} = 1.33$ V in the GaP epitaxial layer were determined from the C-V characteristics at 10^3 Hz. The thickness of FTO film is an important factor for spectral sensitivity of the developed photodetectors due to the possibility using this film as an effective antireflection coating. The spectral sensitivity of FTO-GaP SB photodetectors having different thicknesses of the FTO electrode as well as ITO-GaP photodetector with 65 nm thick ITO film is shown in **Figure 9**.

The typical performance parameters of the photodiodes with TCO layer are summarized in **Table 2**. The maximal spectral sensitivity depends on the thickness of the TCO film.

Maximal sensitivity of commercial GaP photodetectors based on Au-GaP Schottky barriers (G1962, EDP-365-0-2.5, JEP 5-365) fabricated by well-known companies Hamamatsu, Roithner LaserTechnik, and electro optical components (EOC) lies in the range of 0.07–0.12 A/W with electric parameters that do not exceed as reported here. More close parameters were rep orted for Ag-GaP photodetectors [25]; maximal spectral sensitivity of which was reported as 0.12 A/W.

High-efficient FTO-GaP photodetectors [23] have been used in the design of the radiometric heads of the UVR-21 UV radiant power meter [26]. The UVR-21 UV (**Figure 10**) is an easy-to-use, handheld power meter with unique future intended to make the calibrated measurements of UV-A, UV-B, and UV-C radiation. At the heart of the URV-21 is spectral-filtered

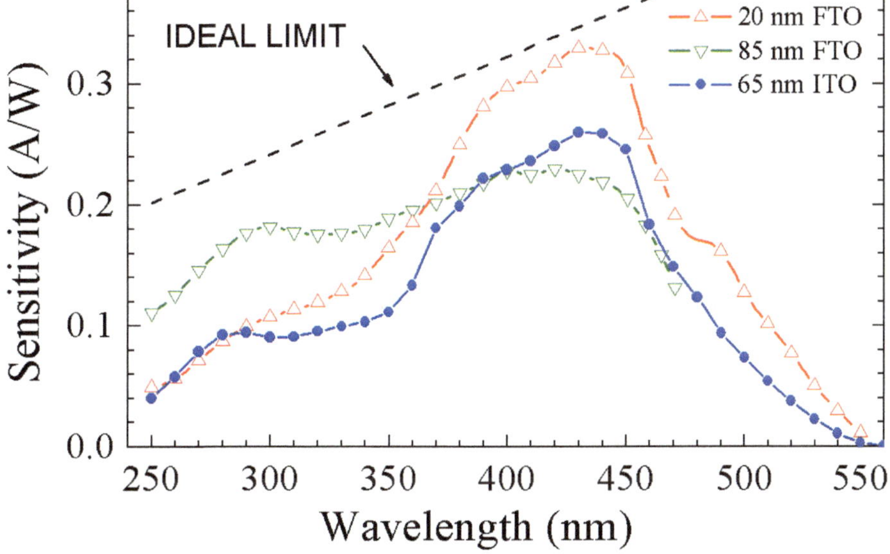

Figure 9. The spectral sensitivity of FTO-GaP and ITO-GaP SB photodetectors for different thicknesses of the transparent conducting electrode.

Electrode material	FTO	ITO
Active area (mm^2)	3	3
Terminal capacitance (nF)	2	1.5
Dark current at V = 1V (pA)	2	20
Shunt resistance (GΩ)	100	30
Peak sensitivity wavelength (nm)	439	435
Peak sensitivity (A/W)	0.2–0.26	0.2–0.26
Peak quantum efficiency (%)	57–74	57–74
Noise-equivalent power ($W/Hz^{1/2}$)	1.7×10^{-15}	3×10^{-15}

Table 2. Comparison of the properties of photodetectors with FTO and ITO electrodes deposited by spray pyrolysis.

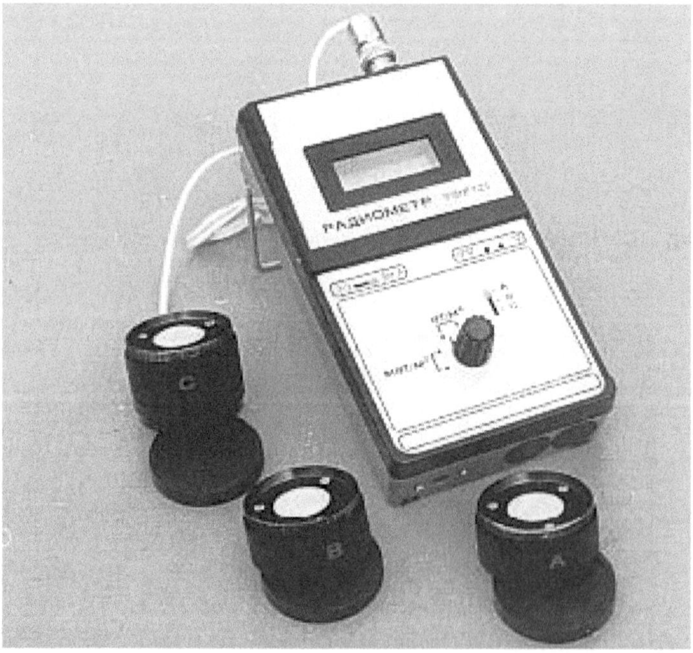

Figure 10. A handheld UV radiant power meter UVR-21 with FTO-GaP–based radiometric heads.

FTO-GaP-based radiometric heads. The interference filters to select requirement spectral band were fabricated on quartz substrates by vacuum deposition. The interchangeable UV radiometric heads allow measuring UV-A, UV-B, and UV-C radiation in the range from 10^{-3} to 10^{3} Wm^{-2}. A rechargeable NiCd battery life is 6–8 hours. Specification limit of measurement relative error is less than 5%.

The performance data of UV radiometric heads are shown below.

Spectral range (nm)		
UV-C radiometric head	220	280
UV-B radiometric head	280	315
UV-A radiometric head	135	400
Irradiance range (Wm^{-2})	10^{-3}	10^{3}
Linearity error (%)		<1
Response uniformity (%)		<1
Measurement error (%)		<5
Cosine error (%) at zenith angle		
30°		3
60°		7
80°		15

5. Silicon surface-barrier photodetectors for visible and near-infrared spectral range

The first results about silicon photodiodes were published more than 60 years ago, and the first detailed investigations of the silicon surface-barrier diodes with thin gold electrode have been published in 1962 [27, 28] by researches from the USA. The first superior electric and photoelectric characteristics of SnO_2-Si heterojunction diodes fabricated by a simple process of spraying a solution containing a tin compound were published in 1975 [29] by authors from Japan.

In this chapter, a brief review of our original results [30, 31] in the field of different types of SB silicon photodetectors with transparent conducting FTO and ITO electrodes is shown.

5.1. Fabrication, properties, and applications of FTO-Si surface-barrier photodetectors

The main technological fabrication steps of FTO-Si photodetectors are shown in **Figure 11**.

Photodetectors were fabricated on the base of high resistivity (1–2 kΩ cm) n-type oxidized silicon wafers with phosphorous-doped backside. Because the photolithographic patterning of the FTO film is difficult due to a high resistance of FTO film to acids, lift-off method was chosen for patterning. On the silicon dioxide layer previously by spray pyrolysis was deposited ITO or ZnO film that sufficiently easy can be etched in HCl solution. The active area of photodetectors after lift-off technological step was 50 mm^2. Current-voltage characteristic of photodetectors with 50 mm^2 area is shown in **Figure 12**.

5.1.1. Γ-Radiation detectors

The Γ-radiation detectors were designed on the base of the FTO-Si photodetectors with single crystal $Bi_4Ge_3O_{12}$ scintillator optically matched on the FTO electrode by silicon-based metal-organic lubricant. The optical matching of the photodetector with the scintillator is illustrated in **Figure 13**.

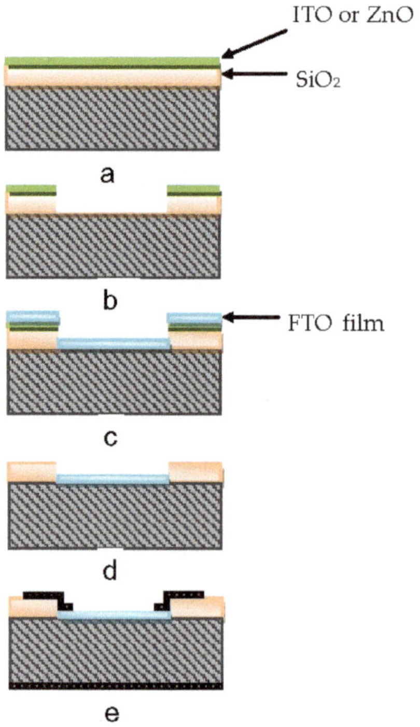

a- Oxidized high resistivity n-type silicon with doped backside and sprayed ITO or ZnO film on the SiO_2 layer.

b- After photolithography.

c- After FTO film deposition by spray pyrolysis method.

d- After removal of ITO or ZnO film by etching in HCl solution (lift-off)

e- After deposition of the metallic contacts on FTO film and the silicon backside.

Figure 11. Technological fabrication steps of the FTO-Si photodetectors. (a) Oxidized high-resistivity n-type 350 μm thick silicon with doped backside and sprayed ITO or ZnO film on SiO_2 layer. (b) After photolithography. (c) After FTO film deposition by spray pyrolysis method. (d) After removal of ITO or ZnO film by etching in HCl solution (lift-off). (e) After deposition of the metallic contacts on FTO film and the silicon backside.

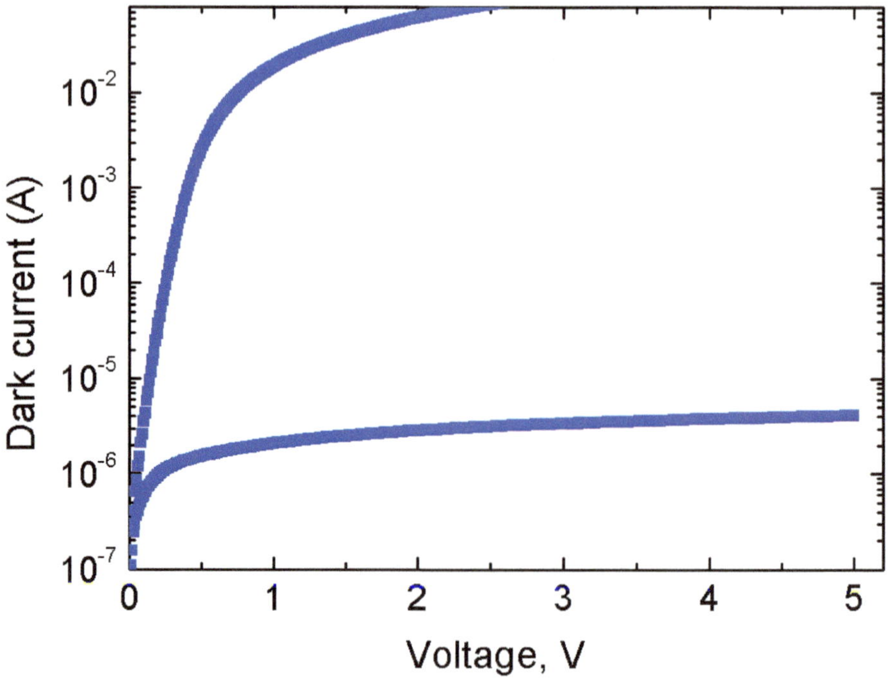

Figure 12. Current-voltage characteristic of FTO-Si photodetectors with 50 mm^2 area.

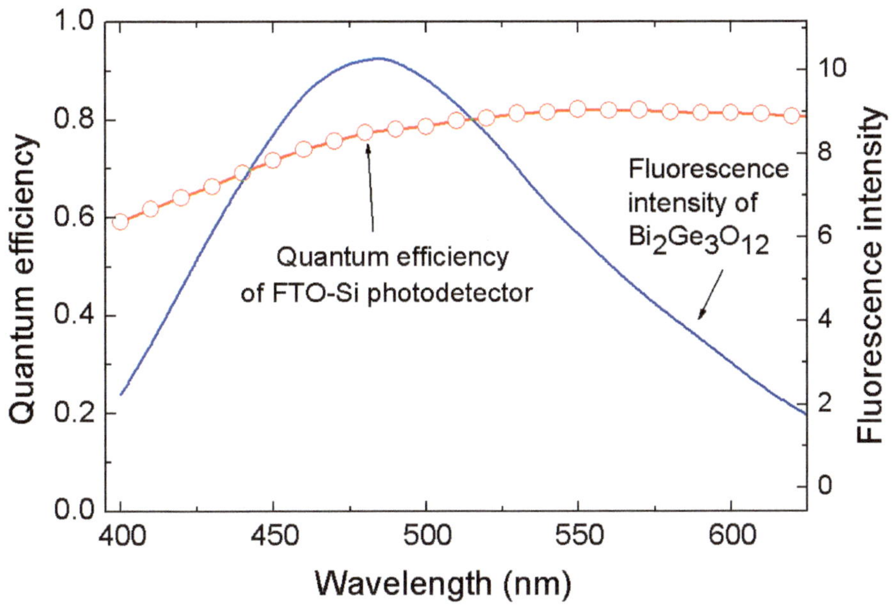

Figure 13. Illustration of optical matching of the photodetector with the scintillator.

Cooling of photodetectors up to 176 K reduces the dark current to a value around 10^{-10} A, while fluorescence lifetime of the scintillator (~ 3 ms) increases 10 times. Registration of low-level photocurrent produced by the photodetector under irradiation by γ-particles from ^{137}Cs was made by using a cooled preamplifier with noise of 250 e^- RMS.

Figure 14 shows recorded pulse height spectra when the photodetector with $Bi_4Ge_3O_{12}$ scintillator is irradiated from a137 Cs (Cs-137) source of γ-rays having energy of 661.5 keV. The registered signal is related to the main photon peak radiation from metastable nuclear isomer ^{137m}Ba (Ba-137). The energy resolution (FWHM/peak) is 16.5%. The FTO-Si photodetector with low noise amplifier was cooled to 176 K and operated at 5 V reverse voltage bias.

Figure 15 shows recorded pulse height spectra when the photodetector with $Bi_4Ge_3O_{12}$ scintillator is irradiated from 22Na (Na-22) source of γ-rays. The registered signals are related to two emission peaks, due to electron-positron annihilation (E_γ = 511 keV) and due to nuclear energy transition (E_γ = 1.27). The FTO-Si photodetector with low noise amplifier was cooled to 250 K and operated at 40 V reverse voltage bias.

Our silicon SB detectors of γ-rays fabricated by a simple and low-cost spray pyrolysis technique operate at low-voltage bias, and this is an advantage in comparison with much more expensive solid-state detectors based on cadmium telluride or mercuric iodide operating at reverse biases of order 100 V. Not cooled silicon detectors of γ-rays based on commercial Hamamatsu photodiodes (S1790-01) show low energy resolution of about 4% with 122 keV γ-rays from ^{97}Co radiation spectra [32].

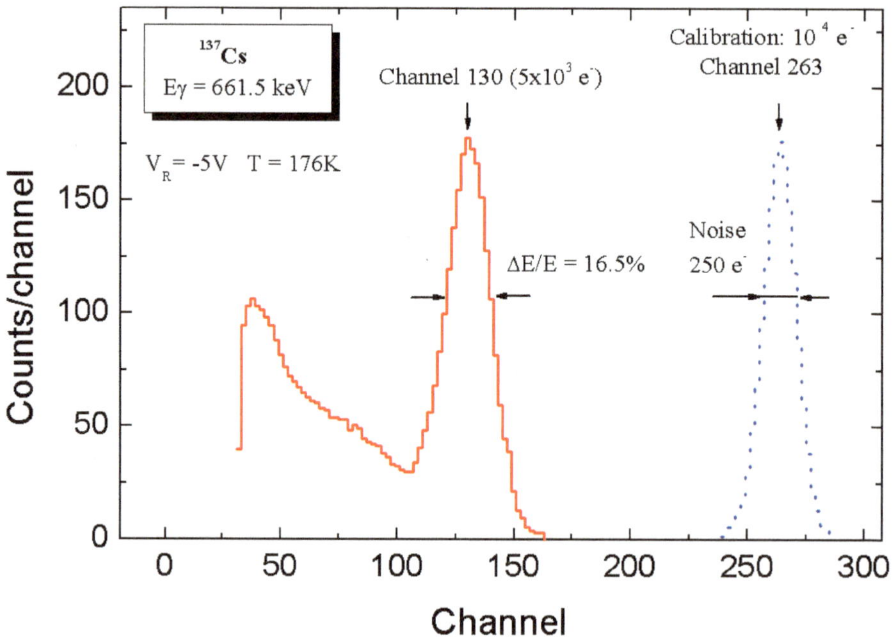

Figure 14. The pulse height spectra when $Bi_4Ge_3O_{12}$ scintillator is irradiated from 137Cs source of γ-rays having energy of 661.5 keV.

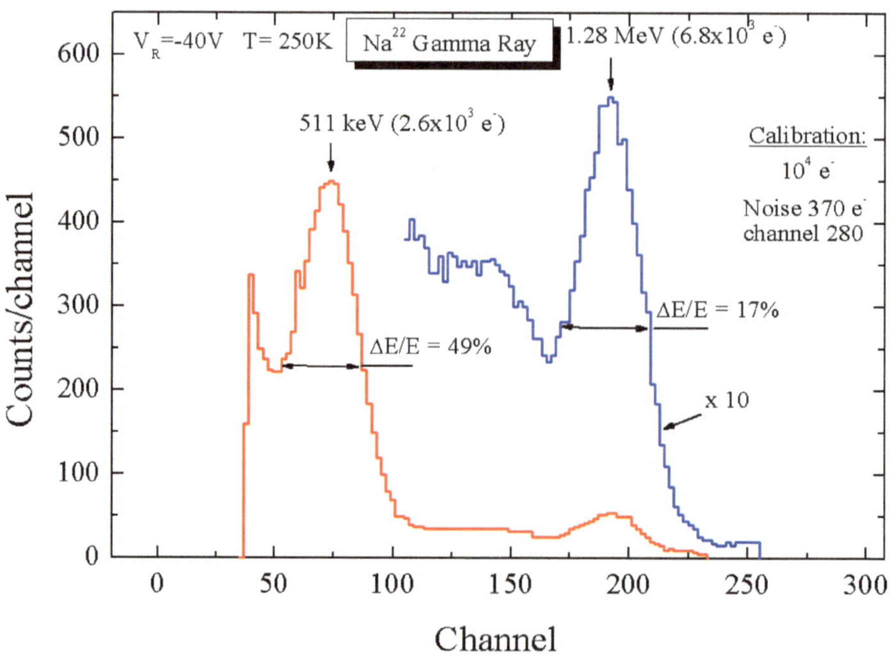

Figure 15. The pulse height spectra when $Bi_4Ge_3O_{12}$ scintillator are irradiated from 22Na source of γ-rays.

5.1.2. Radiation-resistant surface-barrier silicon photodetectors

Many technical applications require radiation-tolerant photodetectors that operate under high-energy particle irradiation, in particular under irradiation by reactor neutron at high flux of fast neutrons [33]. Commercial photodiodes such as the Hamamatsu Photonics photodiode S1723 was tested with respect to fast neutron radiation [34]. Significant changes of the leakage current and of the noise at a fluency of neutrons as low as 2.5×10^{10} n cm^{-2} were found. Damage also reduces the spectral sensitivity of the silicon photodiodes at long wavelengths due to a dramatic reduction of the minority lifetime.

Fast and radiation-resistant surface-barrier ITO-Si photodetectors have been fabricated using silicon epitaxial n-n$^+$ structures with 8 or 25 μm thick undoped epitaxial layer. The epitaxial layers with donor concentration 10^{13} cm^{-3} were grown by chemical vapor deposition on high-doped (10^{19} cm^{-3}) 300 μm n$^+$-silicon substrate. The silicon dioxide layer was grown on the surface of the epitaxial layer. An active device area (3–5 mm^2) was defined by chemically etching SiO_2 film. The ITO film as transparent conducting electrode with thickness of about 85 nm was deposited by spray pyrolysis method. A Cr/Ni metallic contact was deposited on the backside of n$^+$ substrate. The front metallic contact to ITO layer was created by the lift-off photolithography technique. After probing, the wafer with numerous fabricated devices was diced, and the chips were mounted on T0-8 body using silver epoxy.

The idea of the radiation-resistant photodiodes is based on existence of built-in electric field inside of the epitaxial silicon layer due to its nonuniform auto-doping as a result of the solid-state diffusion of antimony from the heavy-doped substrate during the growing process [35]. If built-in electric field is overlapped with the electric field inside the depletion layer, photogenerated minority carrier will be separated by a drift mechanism, and its recombination in defects formed by radiation will be negligible. Such photodetector will be high resistant to neutron flux even at zero bias.

An inhomogeneous impurity distribution of donors (N_D) inside the epitaxial layer estimated with capacitance-voltage (C-V) measurements [36] is shown in **Figure 16**, where the experimental points were approximated by two exponential curves (dotted lines). The doping variation in the epitaxial layer causes appearance of the built-in electric field $E = (kT/q)(1/N_D)(dN_D/dx)$.

The current-voltage characteristics of the ITO-Si photodetectors fabricated using 25 μm epitaxial silicon structures is shown in **Figure 17**.

In photodetectors with 25 μm epitaxial layer operating without voltage bias, a strong built-in electric field due to auto-doping effect does not overlapped with the electric field in depletion region. At that, part of photogenerated minority carriers needs to cross a distance without electric field by diffusion inside the epitaxial layer. However, after exposition under a neutron flux of about 10^{14} cm^{-3}, lifetime of the minority carriers is reduced from 36 μs to 12 ns, and photogenerated carriers recombine in this neutral area. For radiation resistance, thinner epitaxial layer needs to be used. **Figure 18** shows the spectral sensitivity of ITO-Si photodetectors at zero bias voltage fabricated on 8 μm epitaxial layers under irradiation by fast neutrons with flux of 3×10^{14} cm^{-2}.

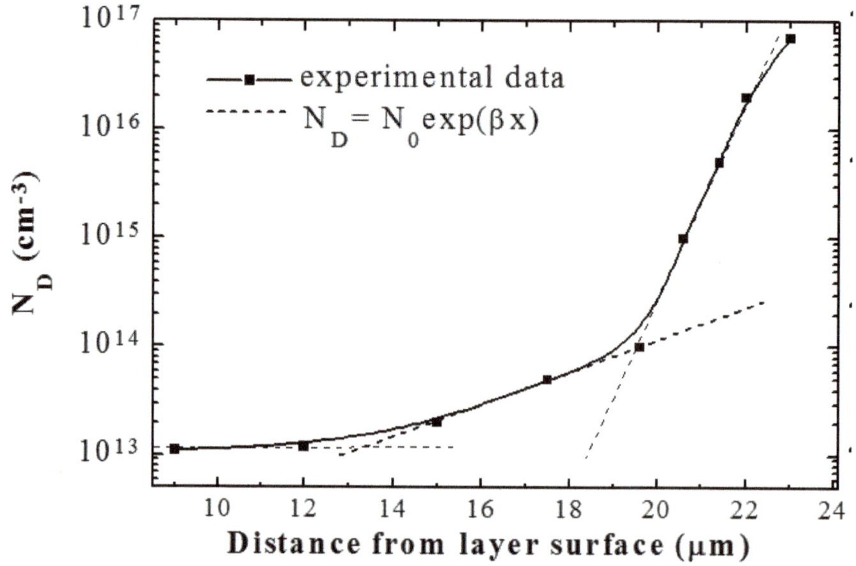

Figure 16. The donor concentration distribution inside the 25 μm epitaxial layer. Distance is shown from the epitaxial layer surface.

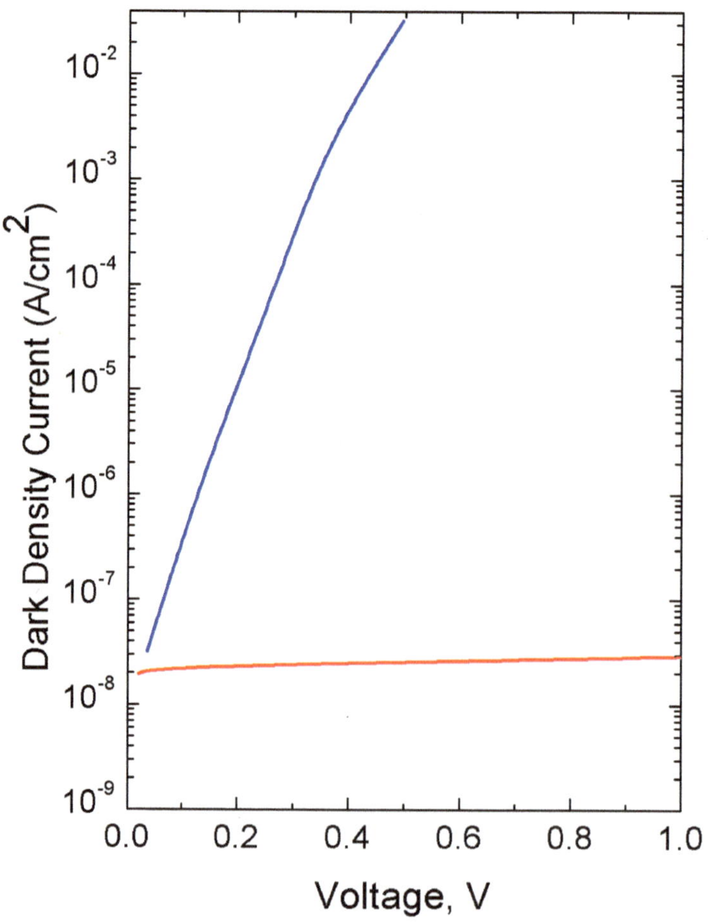

Figure 17. Current-voltage characteristics of ITO-Si photodetectors with an area of 3 mm^2.

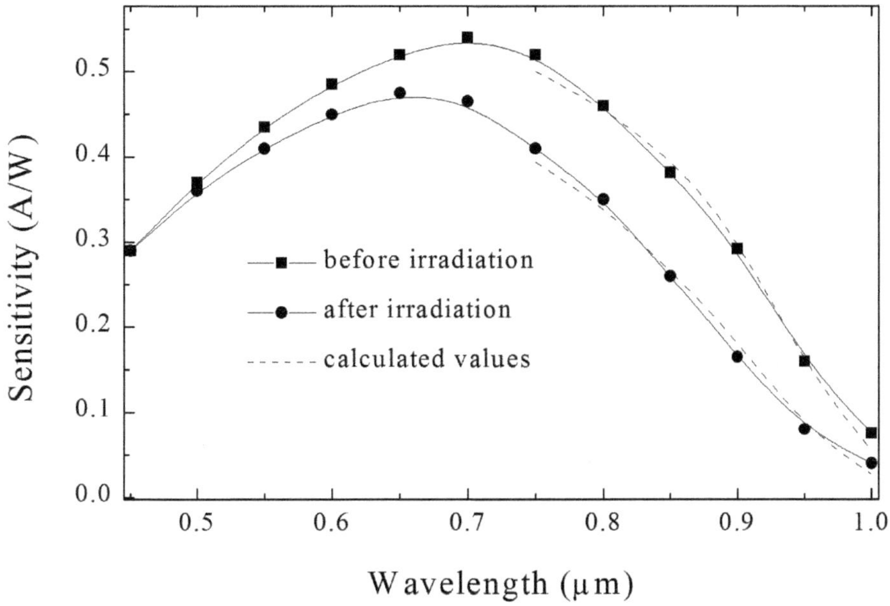

Figure 18. Spectral sensitivity of ITO-Si photodetectors at zero bias voltage fabricated on 8 μm epitaxial layers under irradiation by fast neutrons with a flux of 3×10^{14} cm^{-2}.

We suppose that ~20% reducing sensitivity of irradiated photodiodes in near-infrared spectral range is caused by lack of a contribution in photocurrent from carriers' photogenerated inside of heavy-doped silicon substrate. After neutron irradiation, such contribution is not significant in this spectral region due to a considerable shortage of the minority carrier lifetime. Thus, ITO-Si photodetectors with 8 μm epitaxial layers operating at zero voltage bias are extremely resistant to very big neutron flux. Photodetectors fabricated on the epitaxial layers with bigger thickness need to operate with the voltage bias. However, radiation damage produces an unacceptable level of the dark current because of dramatically reduction of the minority carrier lifetime. Devices with reported design can be used in optoelectronic equipment operating at hard radiation conditions without loss of their principal parameters.

6. Surface-barrier photodetectors for 1.3–1.55 μm

In this section, a new SB photodetector based on HgInTe-semiconductor compound. The development and improvement of fiber-optic communication system imply the use of high-speed and high-efficiency photodetectors. The transmission signal with a wavelength of 1.55 μm is especially important for long-distance communications. For this purpose, epitaxial layers of quaternary $In_xGa_{1-x}As_yP_{1-y}$ alloy with x = 0.57 and y = 0.95 are practically used for effective detector fabrication. Twenty-five years ago, the perspective of SB photodiodes based on $Hg_3In_2Te_6$ or mercury indium telluride (MIT) semiconductor has been shown [37–39]. The next years increasing interest in photodetectors based on this semiconductor [40–42]. The

$Hg_3In_2Te_6$ is a direct energy gap semiconductor with E_g = 0.72 eV at 300 K, close to germanium, and crystallizes in the zinc blend structure in which case 1/3 of all sites in the cationic sub-lattice are vacant. These stoichiometric vacancies with a concentration of about 10^{12} cm^{-3} determine unique properties of this semiconductor. It is inactive to dopants up to the doping concentration of 10^{20} cm^{-3} for fabrication p-n junctions, electrical properties of the MIT are indifferent to active defects [43], and material is high resistant to ionizing radiation [44]. The properties of the SB Au-MIT [40], Ni-MIT [41], and ITO-MIT [43] photodetectors were studied. In this section, electrical and optical properties of SB photodetectors on the base of the MIT single crystals with the ITO film as transparent conducting electrode deposited on chemically treated MIT surface are reported. Almost the same detectors in which the chemical treatment was changed on treatment in oxidizing RF plasma were patented recently in China [45], but their characteristics are not published.

The starting material for the photodetector fabrication was an *n*-type MIT mechanically and chemically polished wafer with carrier concentration of about 1×10^{13} cm^{-3}. The wafers were etched in 5% solution of Br_2 in N,N-dimethylformamide at the etching speed approximately of 0.3 μms^{-1}. The density of surface defects visible with a microscope was found as 30 cm^{-2}. After etching, wafers were boiled in azeotropic acetonitrile-based mixture and dried in N_2 atmosphere. The roughness of surface after the abovementioned polishing was determined as 0.03 μm.

The attempts to form a Schottky potential barrier on the MIT surface directly by the deposition of the ITO film lead to a weak rectifying nature of the ITO-MIT contact. The best rectifying characteristic has been observed if the ITO deposition was performed on a thin (0.02–0.03 μm) oxide layer chemically grown on the MIT surface. For chemical growth of oxide layer on the MIT surface, wafer was dipped on a few minutes in (1:1:1) solution prepared H_2O_2, N,N-dimethylformamide, and NH_4OH, previously heated to beginning of intensive foaming. The oxide thickness was measured using an ellipsometer. The chemically grown oxide composition was studied by X-ray photoelectron spectroscopy (XPS) [39]. Results of measurements are shown in **Figure 19**.

Analysis of XPS peaks corresponding the binding energies of the In $3d_{5/2}$, Te $3d_{5/2}$, and Hg 4f at 444.8 eV, 576.4 eV, and 100.7 eV, respectively, allows estimating the oxide composition: 40 at. % In_2O_3, 50 at. % TeO_2, and 10 at. % HgO. Almost the same oxide content was reported in Ref. [45], where the oxide layer was fabricated by RF plasma oxidation. The ITO film can be deposited on the oxide surface by spray pyrolysis technique or by sputtering.

In this work, we focused our attention on two SB photodetector types, fabricated on the thick (700 μm) and the thin (70 μm) MIT substrates. Typical current-voltage characteristic of ITO-oxide-MIT photodetectors with 3 mm^2 active area is shown in **Figure 20**.

Density of the dark current is almost the same as reported in [40, 41], but the spectral external quantum efficiency of the ITO-oxide-MIT photodetectors is more than 90% (**Figure 21**) and very differs from one reported for the Au-MIT [40] and Ni-MIT [41] Schottky-based photodetectors with 20–50% quantum efficiency.

Note that the quantum efficiency of ITO-MIT photodetectors with the ITO film deposited by sputtering on the MIT surface without specially created oxide layer is only about 60% [46].

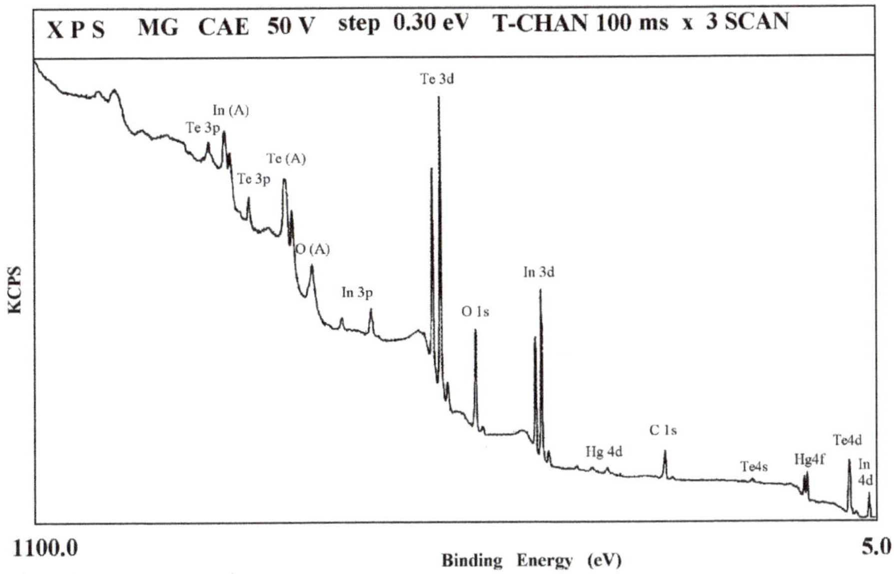

Figure 19. XPS spectrum of chemically grown oxide on the MIT surface.

The time response characteristic of the photodetectors fabricated using 70 μm thick MIT wafers measured at the reverse voltage bias of 120 V irradiated by laser pulse with 1.3 μm wavelength is shown in **Figure 22**.

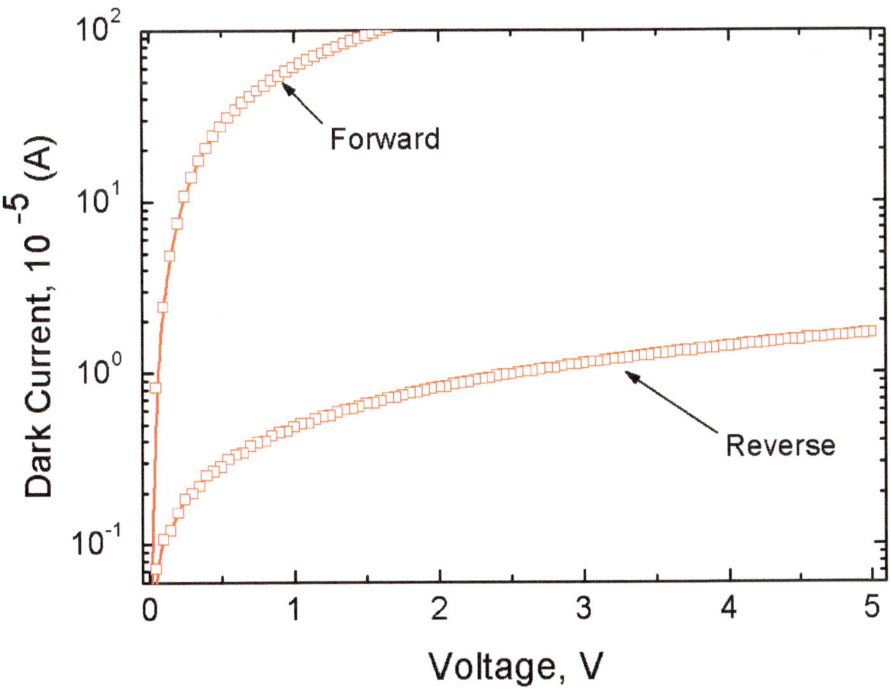

Figure 20. The current-voltage characteristic of ITO-oxide-MIT photodetectors with 3 mm² active area.

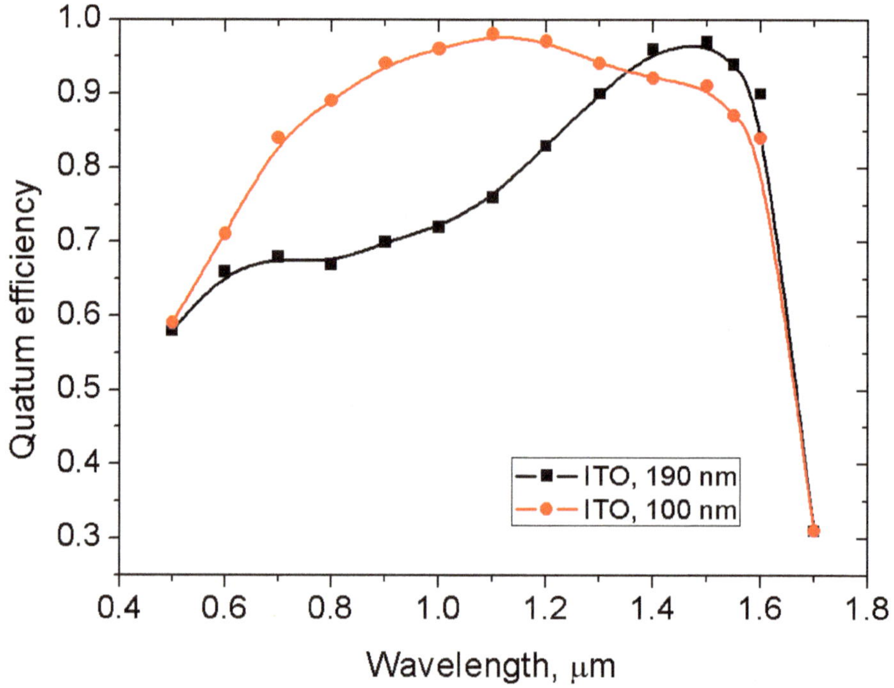

Figure 21. The spectral external quantum efficiency of the ITO-oxide-MIT photodetectors for two thicknesses of the ITO film.

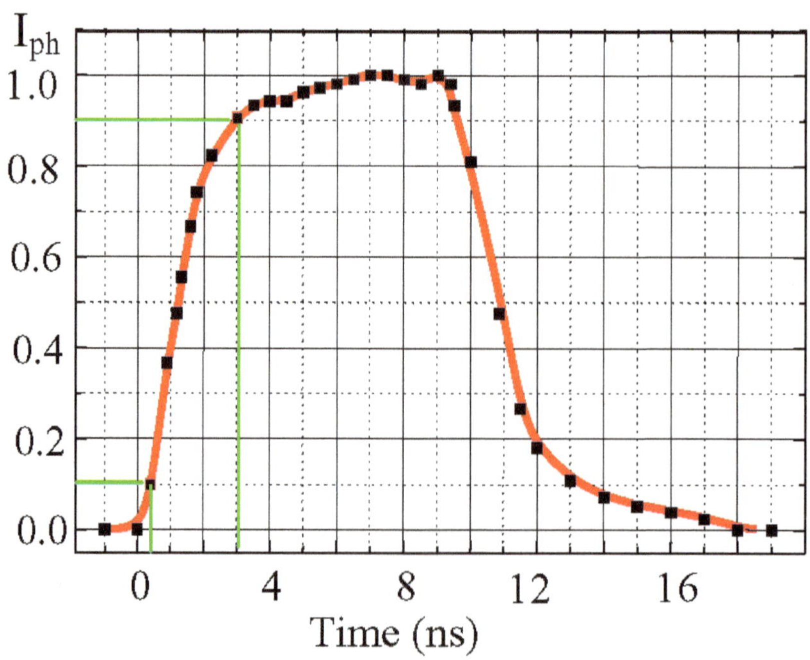

Figure 22. The time response characteristic of the 70 μm thick MIT photodetectors with 3 mm[2] area at a reverse voltage bias of 120 V irradiated by a laser pulse with 1.3 μm wavelength.

7. Conclusions

We presented the properties of surface-barrier photodiodes suitable for a wide spectral range and fabricated using a simple and low-cost spray deposition method that does not require expensive technological equipment. Of course, in this short report, it was not possible to describe in detail some technological aspects, such as the chemical treatment of the semiconductors used for the photodetectors. Readers can found further information in the works referenced in this chapter. The aim of this work was to review illustratively a simple technological method for the fabrication of high-efficiency surface-barrier photodiodes for applications in a wide spectral range, from UV to near infrared, using different semiconductor materials. This method has been successfully used for the fabrication of high-efficiency silicon solar cells whose characteristics have already been reported [47].

Author details

Oleksandr Malik* and Francisco Javier De la Hidalga-Wade

*Address all correspondence to: amalik@inaoep.mx

Electronics Department, Instituto Nacional de Astrofísica, Óptica y Electrónica (INAOE), Puebla, Mexico

References

[1] R. T. Tung. The physics and chemistry of the Schottky barrier height. Appl. Phys. Rev. 2014; 1: 1-55.

[2] S. J. Fonash. Solar cell device physics. Academic Press. NY 1981: 331 p. ISBN 0-12-261980-3.

[3] A. L. Dawar, J. C. Joshi. Review semiconducting transparent thin films: their properties and applications. J. Mater. Sci. 1984; 19: 1-23.

[4] J. DuBow, D. Burk, and J. Sites. Efficient photovoltaic heterojunctions of indium tin oxides on silicon. Appl. Phys. Lett. 1976; 29(8): 494-496.

[5] J. Manifacier, L. Szepessy. Efficient sprayed In_2O_3 n-type silicon heterojunction solar cell. Appl. Phys. Lett. 1977; 31; 459-462.

[6] T. Feng, A. K. Ghosh, and C. Fishman. Spray deposited high efficiency SnO_2-silicon solar cells. Appl. Phys. Lett. 1979; 35(3): 266-268.

[7] A. Malik, V. Baranyuk, and V. Manasson. Solar cells based on the SnO_2-SiO_2-Si heterojunction. Appl. Sol. Energy. 1979; 2: 83-84. ISSN: 0003-701x.

[8] A. Malik, V. Baranyuk, and V. Manasson. Improved model of solar cells based on In_2O_3/SnO_2-SiO_x-nSi structures. Appl. Solar Energy. 1980; 1:1-2.

[9] S. Ashok, P. Sharma, and S. Fonash. Spray-deposited ITO-silicon SIS heterojunction solar cells. IEEE Trans. Electron. Dev. 1980; ED-27(4): 725-730.

[10] O. Malik, F. J. De la Hidalga-W, C. Zúñiga-I, and G. Ruíz-T. Efficient ITO-Si solar cells and power modules fabricated with a low temperature technology: results and perspectives. J. Non Cryst. Solids. 2008; 354: 2472-2477.

[11] O. Malik, F. J. De la Hidalga-W, C. Zúñiga-I, and G. Ruíz-T. Efficient ITO-Si solar cells and power modules fabricated with a low temperature technology: results and perspectives. J. Non Cryst. Solids. Elsevier. 2008; 354: 2472-2477.

[12] J. B. Mooney, S. B. Radding. Spray pyrolysis processing. Ann. Rev. Mater. Sci. 1982; 12: 81-101.

[13] O. Malik, F. J. De la Hidalga-Wade. Comparison of tin-doped indium oxide films fabricated by spray pyrolysis and magnetron sputtering. Cryst. Res. Technol. 2015; 50(7): 516-521.

[14] O. Malik, F. J. De la Hidalga-Wade, and R. Ramírez Amador. Fluorine-doped tin oxide films with a high figure of merit fabricated by spray pyrolysis. J. Mater. Res. 2015; 30(13): 2040-2045.

[15] G. Haacke. New figure of merit for transparent conductors. J. Appl. Phys. 1976; 47: 4086-4089.

[16] E. Burstein. Anomalous optical absorption limit in InSb. Phys. Rev. 1954; 93: 632-633.

[17] A. Malik, A. Seco, E. Fortunata, and R. Martins. New UV-enhanced solar blind optical sensors based on monocrystalline zinc sulphide. Sens. Actuators A. 1998; 67: 68-71.

[18] V. Machnii, A. I. Malik, and N. Melnik. Solar blind photodiode based on ITO/ZnS heterostructures. Sov. Phys. Tech. Phys. 1990; 60: 146-147.

[19] D. Caputo, G. de Cesare, F. Irrera, and F. Palma. Solar-blind UV photodetectors for large area applications. IEEE Trans. Electron. Devices. 1966; 43: 1351-1356.

[20] A. I. Malik, G. G. Grushka. Optoelectronic properties of metal oxide-gallium phosphide heterojunctions. Sov. Phys. Semicond.. 1991; 25(10): 1017-1020. ISSN 0038-5700.

[21] A. Malik, A. Seco, E. Fortunato, R. Martins, B. Shabashkevich, and S. Piroszenko. A new high ultraviolet sensitivity FTO-GaP Schottky photodiode fabricated by spray pyrolysis. Semicond. Sci. Technol. 1998; 13: 102-107.

[22] Yu. Vygranenko, A. Malik, M. Fernandes, R. Schwarz, and M. Vieira. UV-Visible ITO/GaP photodiodes: characterization and modeling. Phys. Status. Sol. A. 2001; 185(1): 137-144.

[23] A. I. Malik, B. Shabashkevich, and S. Piroszenko. Photodiode for the ultraviolet spectral range. Patent of Ukraine N. 71544.

[24] O. Malik, F. J. De la Hidalga-Wade, C. Zuniga-Islas, and J. Abundus Patiño. UV-sensitive optical sensors based on ITO-gallium phosphide heterojunctions. Phys. Status Sol. C. 2010; 7(3-4): 1176-1179.

[25] A. N. Pikhtin, S. A. Tarasov, and B. Kloth. Ag-GaP Schottky photodiodes for UV sensors. IEEE Trans. Electron. Dev. 2003; 50(1); 215-217.

[26] V. C. Boutenko, Y. G. Dobrovolskiy, S. I. Piroszenko, B. Shabashkevich, V. G. Yoriev. Radiometer of power luminosity in the ultraviolet spectral range. Patent of Ukraine N. 82843.

[27] E. Ahlstrom. W. W. Gartner. Silicon surface barrier photocells. J. Appl. Phys. 1962; 33(8): 2602-2606.

[28] A. J. Tuzzolino, E. L. Hubbard, M. A. Perkins, and C. Y. Fan. Photoeffects in silicon surface-barrier diodes. J. Appl. Phys. 1962; 33(1): 148-155.

[29] H. Kato, J. Fujimoto, T. Kanda, A. Yoshida, and T. Arizumi. SnO_2-Si photosensitive diodes. Phys. Status Sol. A. 1975; 32: 255-261.

[30] A. Malik, R. Martins. Metal oxide/silicon heterostructures: new solution for different optoelectronic applications. MRS Proc. 1997; 487: 375-380.

[31] A. Malik, M. Vieira, and M. Fernandes. Surface-barrier Si-based photodetectors fabricated by spray pyrolysis technique. Philosophical Mag. B. 2000; 80(4); 781-790.

[32] C. Bilton, S. Hedges, P. R. Hobson, and D. C. Imrle. Low-cost silicon photodiodes for X-ray detection. J. Phys. E Sci. Instrum. 1988; 21: 809-811.

[33] M. Hasegawa, S. Mori, T. Ohsugi, H. Kojima, A. Taketani, T. Kondo, and M. Noguchi. Radiation damage of silicon junction detectors by neutron irradiation. Nucl. Instrum. Methods. 1989; A277: 395-400.

[34] W. Dabrowski, K. Korbel, and A. Skoczeń. Fast neutron damage of silicon PIN photodiodes. Nucl. Instrum. Methods Phys. Res. A. 1991; 301(2): 288-294.

[35] B. A. Joyce, J. C. Weaver, and D. J. Maule. Impurity redistribution processes in epitaxial silicon layers. J. Electrochem. Soc. 1965; 112(11): 1100-1106.

[36] P. J Baxandall, D. J Colliver, and A. F Fray. An instrument for the rapid determination of semiconductor impurity profiles. J. Phys. E Sci. Instrum. 1971; 4: 213-221.

[37] A. I. Malik, G. G. Grushka, and N. R. Tevs. High-efficiency measuring photodiodes based on mercury indium telluride. Sov. Tech. Phys. 1990; 35: 723-726.

[38] A. I. Malik, G. G. Grushka. Self-calibrated radiometric IR photodiode based on the defect semiconductor $Hg_3In_2Te_6$ for the spectral range 0.85-1.5 μm. Sov. Tech. Phys. 1990; 35: 1227-1231.

[39] A. I. Malik, M. Vieira, M. Fernandes, F. Macarico, and Z. Grushka. Near-infrared photodetectors based on HgInTe-semiconductor compound. Proc. SPIE. Photodetectors: Materials and Devices IV. 1999; 3629: 433-442.

[40] L. A. Kosyachenko, Yu. S. Paranchich, V. N. Makagonenko, V. M. Sklyarchuk. E. F. Sklyarchuk, and I. I. German. Electrical performance of HgInTe surface-barrier photodiodes. Tech. Phys.. 2003; 48(5): 647-650.

[41] L. A. Kosyachenko, I. S. Kabanova, V. M. Sklyarchuk, O. F. Sklyarchuk, and I. M. Rarenko. $Hg_3In_2Te_6$-based photodiodes for fiber optic communication. Phys. Status Solidi A. 2009; 206(2): 351-355.

[42] L. Zhang, X. L. Zhang, W. G. Sun, and Z. X. Lu. Proc. SPIE. 2011; 8193: 1-6.

[43] O. L. Maslyanchuk, L. A. Kosyachenko, I. I. German, I. M. Rarenko, V. A. Gnatyuk, and T. Aoki. Electrical and optical properties of $Hg_3In_2Te_6$ single crystals. Phys. Status Solidi C. 2009; 6(5): 1154-1157.

[44] O. G. Grushka, V. T. Maslyuk, S. M. Chupyra, O. M. Mysliuk, S. V. Bilichuk, and I. I. Zabolotskiy. The effect of irradiation with electrons on the electrical parameters of $Hg_3In_2Te_6$. Semiconductors. 2012; 46(3): 312-314. ISSN 10637826.

[45] Te-In-Hg photoelectronic detector chip manufactured method. Patent of China N. CN101060144A 2014.

[46] L. A. Kosyachenko, I. M. Rarenko, E. F. Sklyarchuk, I. I. German, and Sun Weiguo. Electrical characteristics of the ITO/HgInTe photodiodes. Semiconductors. 2006; 40(5): 554-557.

[47] O. Malik, F. J. De la Hidalga-W. Physical and technological aspects of solar cells based on metal oxide-silicon contacts with induced surface inversion layer. In: Application of solar energy, R. Rugescu Editor. 2013. Intech, Croatia. ISBN 978-953-51-0969-3.

9

Sputtered Indium Tin Oxide Films for Optoelectronic Applications

Oleksandr Malik and
Francisco Javier de la Hidalga-Wade

Additional information is available at the end of the chapter

Abstract

High-quality indium tin oxide (ITO) films have been fabricated using a DC sputtering technique in a pure argon atmosphere with a postannealing in an oxygen environment at atmosphere pressure. Structural, morphological, and electro-optical parameters of the ITO films were studied at different annealing temperatures for the films fabricated on two types of glass substrates, soda lime and alkali-free substrates. A comparative analysis shows that low-cost soda lime substrates are suitable for the fabrication of high-quality nanocrystalline ITO films after annealing them at 300°C. This result is of great importance for reducing the cost of thin film solar cells, in which ITO films serve as transparent conducting electrodes. We present a comparison of the properties of sputtered ITO films with those fabricated using a spray pyrolysis deposition technique, which is useful for some optoelectronic applications.

Keywords: thin film, indium tin oxide, DC sputtering, postannealing, soda lime glass, alkali-free glass

1. Introduction

Thin films of transparent conducting oxides (TCOs) have been widely used as a transparent electrode due to their low electrical resistivity and high transparency. As a degenerated semiconductor, ITO can be used as an antireflection layer in n^+-p optical sensors and as a flat panel displays including liquid crystal displays, organic light emitting diodes, and plasma displays [1]. Among the various TCOs, indium oxide films doped with tin (ITO) are widely transparent conducting electrodes in optoelectronic devices [3–6].

ITO thin films can be prepared using a wide variety of techniques such as thermal evaporation, chemical vapor deposition, electron beam evaporation, sol-gel, pulsed laser deposition, magnetron sputtering, and spray pyrolysis [2].

DC and RF magnetron sputtering using ceramic ITO targets is the most widespread method for deposition of ITO films on different substrates. Usually, such sputtering is carried out in a mixed argon-oxygen atmosphere. Normally, sputtering in a pure argon atmosphere is a more simple, cheap and highly controllable technique. However, a practical realization of this process using commercial ITO targets presents some difficulties. Commercial ceramic sputtering targets of ITO usually are fabricated from a mixture of high-purity In_2O_3 and SnO_2 powders (90:10 wt%), according to the solubility limits of Sn in In_2O_3 (12.4–15 mol% SnO_2) [7–9], using hydrothermal processes in the temperature range of 1300–1500°C. A key condition for obtaining ITO films presenting the best parameters is achieved when specific sintering techniques are used, which are also effective to tailor distinct properties of the target materials. X-ray diffractometer (XRD) peaks and the coloration of the ceramic target are references of the complete transformation to indium tin oxide. A thermal treatment in air or oxygen leads to a light-green coloring of the target, whereas a thermal treatment in reducing atmospheres (such as H_2, for instance) results in a black coloring [10]. This specific color of the target can be due to the presence of foreign phases such as metallic tin and/or blue-black tin oxide SnO due to the dissociative decomposition of SnO_2 into SnO + 1/2 O_2. At the same time, the unstable phase SnO can turn into Sn_3O_4 and pure Sn, and Sn_3O_4 will decompose into SnO_2 + Sn.

If only argon is used as the working gas for the sputtering, the deviation in stoichiometry of the target will be duplicated onto the film leading to a poor transparency [11–15]. Therefore, a certain amount of oxygen needs to be added into the working gas, otherwise, the films must be annealed in an air/oxygen atmosphere for obtaining the expected stoichiometry and high transparency; thus, a quasi-reactive deposition in argon mixed with reactive gas O_2 takes the place in contrast to sputtering in pure argon gas. Sputtering target obtained from AJA Corporation and used in this work had a black color that is not usual for light or dark-green color of the ITO targets sintered in an oxygen atmosphere. According to Ref. [15], the use of a reduction atmosphere during the sintering process leads to a black coloring of the target. Thus, the knowledge of the thermal history of the mixed powders, which determines their characteristics, is fundamental in order to determine the target quality, which in turn has a profound impact on the characteristics of the sputtered thin layers [16]. According to Refs. [17, 18], the ITO targets fabricated under laboratory conditions with sintering only in an air atmosphere at 1300°C, during 6 hours, present a light coloring close to that of the initial powders. The stoichiometry of the ITO target confirmed by XRD analysis allows for the fabrication of highly transparent ITO films in a pure argon atmosphere without oxygen adding and a postannealing procedure. Since the chemical content of the as-deposited films, fabricated in a pure argon atmosphere by using the AJA target, was far from stoichiometry and had a very low transparency, for this study, we used a thermal annealing in an oxygen atmosphere at different temperatures. The crystalline structure, the work function, and the optical and electrical properties of the ITO films were investigated.

Another objective of our work was to analyze comparatively the films fabricated with a post-annealing on two types of substrates, namely, low-cost soda lime and expensive alkali-free glass (AFG) substrates. Currently, the cost of soda-lime glass substrates is 20–30 times lower than the cost of alkali-free borosilicate substrates. The applicability of such inexpensive glass substrates in a large-scale thin film solar cells production is a significant factor for reducing the cost of thin film solar cells, in which the ITO films are an integral part of the solar cell structure as the transparent conducting electrode. Nevertheless, some reported results [19, 20] show that the presence of a large amount of alkali ions (Na^+, K^+) in the soda-lime glass substrates is dangerous for the ITO film parameters if high deposition or annealing temperatures are used in the fabrication process. This is due to the out-diffusion of ions from the substrate into the ITO film, leading to the possible formation of secondary phases such as Na_2SnO_3 or $Na_2Sn_2O_5$ observed in SEM images [21]. These disadvantages from known published results were minimized by intercalating a barrier of TiO_2-SiO_2 composite films [19], or SiO_2 and Al_2O_3 films [20] between the ITO film and the soda-lime glass substrate.

2. Experimental

ITO films with a thickness of 200 ± 10 nm were deposited simultaneously on both, soda lime glass (SLG) (Corning 2947 Plain Microscope Slides) and alkali-free glass (AFG) (Corning 1737 near zero alkali aluminosilicate glass) substrates, using the AJA ATC Orion 5 UHV DC Magnetron Sputtering System with 2 inches heads. SLG and AFG substrates differ on their Na_2O content, which are 14% and less than 1%, respectively. The ITO target (90 wt% In_2O_3, 10 wt% SnO_2) was manufactured by AJA Corporation. Only pure argon was used as the working gas; the detailed deposition conditions can be found in our earlier reports [22, 23]. The substrate surface was not heated and maintained at a temperature below 50°C. The films removed from the sputtering system were thermally annealed during 60 min under a constant flow of a pure oxygen, at temperatures in the range of 200–500°C using a 50°C step, in an improved annealing equipment with a more uniform oxygen flux than the one used in an early published work [22].

The structural characterization of the ITO films was carried out using a Bruker D8 advanced X-ray diffractometer (XRD), with CuK_α radiation (λ = 0.15406 nm). Morphological features of the ITO films were examined using the Ambios USPM atomic force microscopy (AFM). The Keithley's Series 2400 source measurement unit and the Acopia HMS-5000 Hall Effect Measurement System were used for the electrical characterization. Optical measurements were carried out in the 400–1100 nm wavelength range using an F20 Filmetrics spectrometer.

3. Comparison of sputtered ITO films on SLG and AFG substrates

As-deposited, the ITO films grown at room temperature on both glass substrates were amorphous and black colored. The dark color of the films can be explained by nonstoichiometric film contents due to the inclusions of metals and indium, and tin sub-oxides present in the used sputtering target. Crystallization of the films in the (222) and (400) planes (**Figures 1** and **2**) with preferred grains (222) orientation was detected after a thermal annealing in oxygen at a temperature above 200°C.

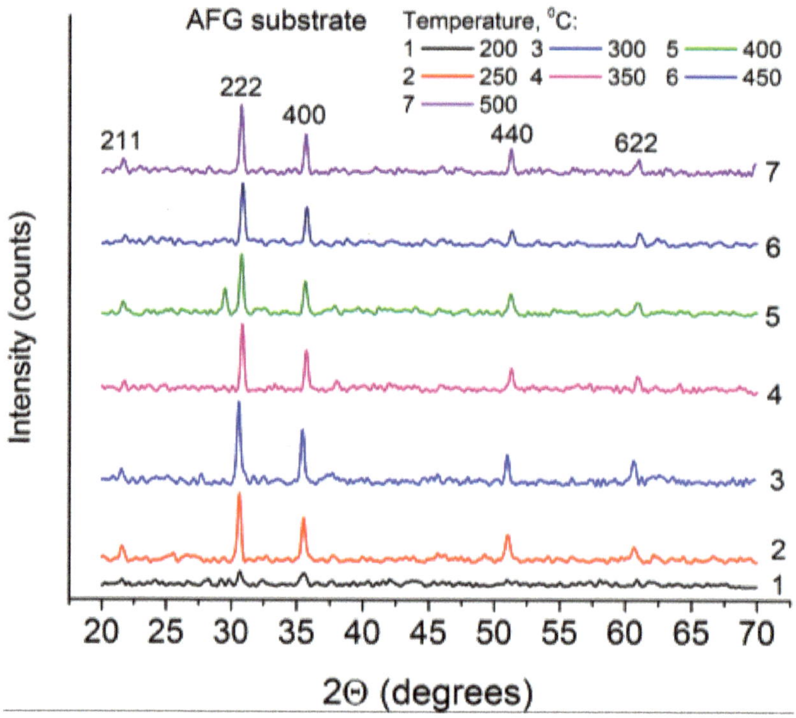

Figure 1. XRD diffraction patterns of the ITO films deposited on AFG substrates and annealed at different temperatures in an oxygen atmosphere.

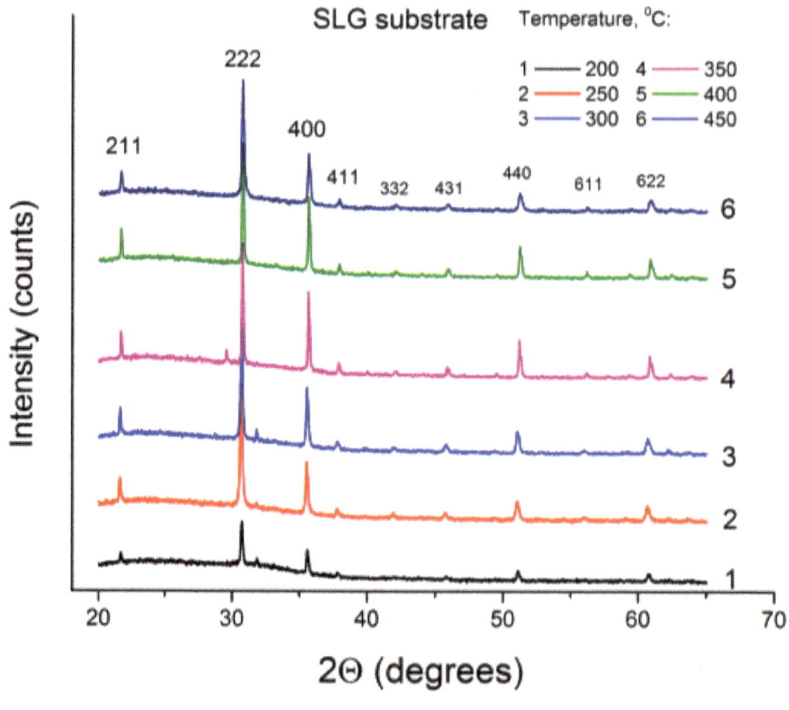

Figure 2. XRD diffraction patterns of the ITO films deposited on SLG substrates and annealed at different temperatures in an oxygen atmosphere.

The films deposited on both substrates and annealed at temperatures 200–450°C in the oxygen atmosphere are polycrystalline. Their XRD spectra (**Figures 1** and **2**) have a little difference in the height of peaks. The preferred grains orientation is (222). This result is consistent with the works presented by Raoufi and Alam [24] and Salehi [25], who obtained the same (222) predominant peak. The XRD peaks for the films deposited on SLG substrates are sharper than those from the films deposited on AFG substrates. The average crystallite size (s) was calculated from the (222) peak, as shown in **Figure 3**, using the well-known Scherrer formula [26].

The surface roughness of the films obtained from AFM analysis is shown in **Figures 4** and **5**. The substrate roughness serves as a reference.

From **Figures 4** and **5**, one can see that the roughness of the AFG substrates is significantly less than that for SLG substrates. However, the roughness of the ITO films deposited on the SLG substrates is less than that of the substrate roughness. We correlate this observed experimental fact with a planarization effect occurring during the film growing on the SLG substrate. A higher roughness of the SLG substrates serves as crystallization centers for the film growth that leads to a bigger crystallite size than that obtained for the ITO films deposited on ALG substrates. The reduction of the grains size of the ITO films deposited on the SLG substrates and annealed at temperatures above 350°C can be correlated with the beginning of the diffusion of Na^+ ions from the SLG substrate with the subsequent formation of secondary phases such as Na_2SnO_3 or $Na_2Sn_2O_5$ that inhibits the normal grain growth [20]. At the same time, the grains size of the ITO films on the AFG substrates is almost independent of the annealing temperature. **Figure 6** shows the dependence of the resistivity of the ITO films deposited on

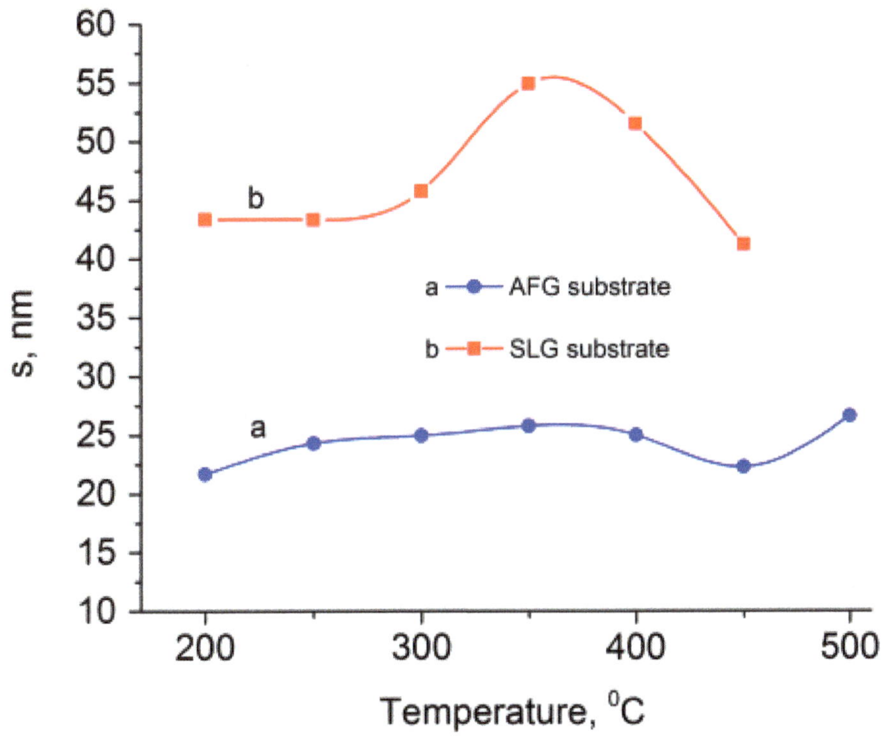

Figure 3. The average crystallite size calculated from the (222) peak.

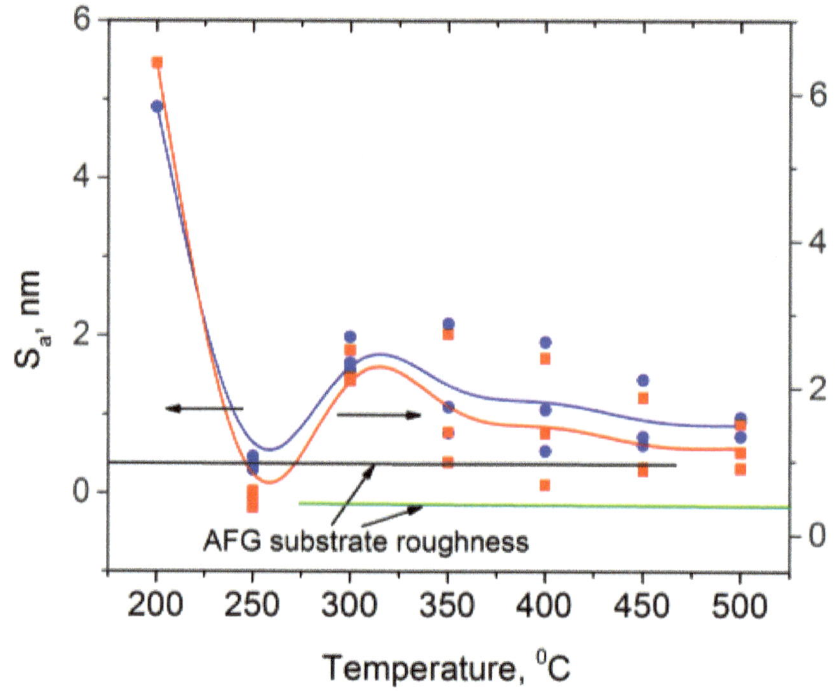

Figure 4. Average (S_a) and root mean square (S_q) roughness for the ITO film deposited on the AFG substrate and annealed in oxygen at different temperatures.

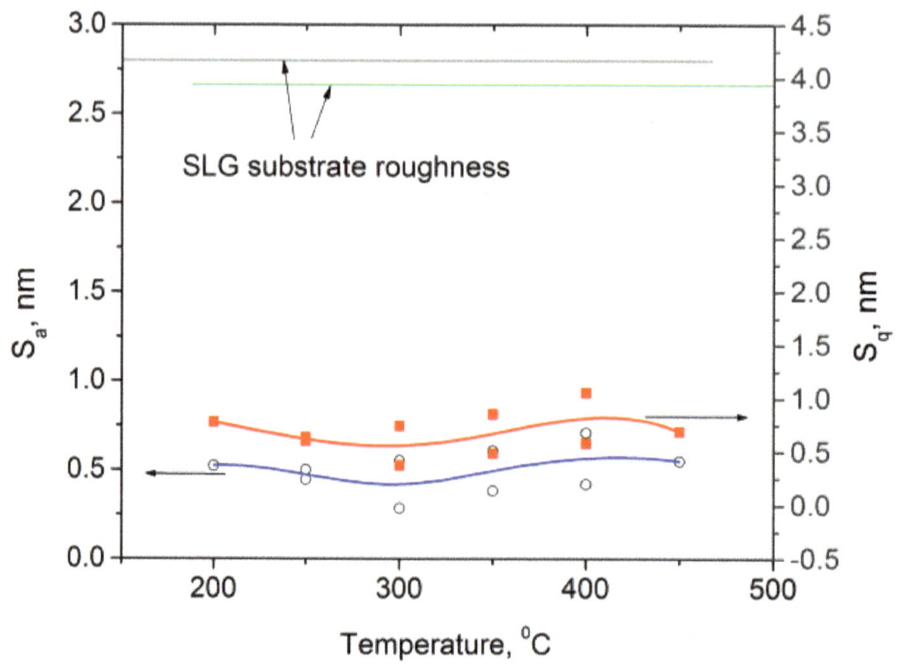

Figure 5. Average (S_a) and root mean square (S_q) roughness for the ITO film deposited on the SLG substrate and annealed in oxygen at different temperatures.

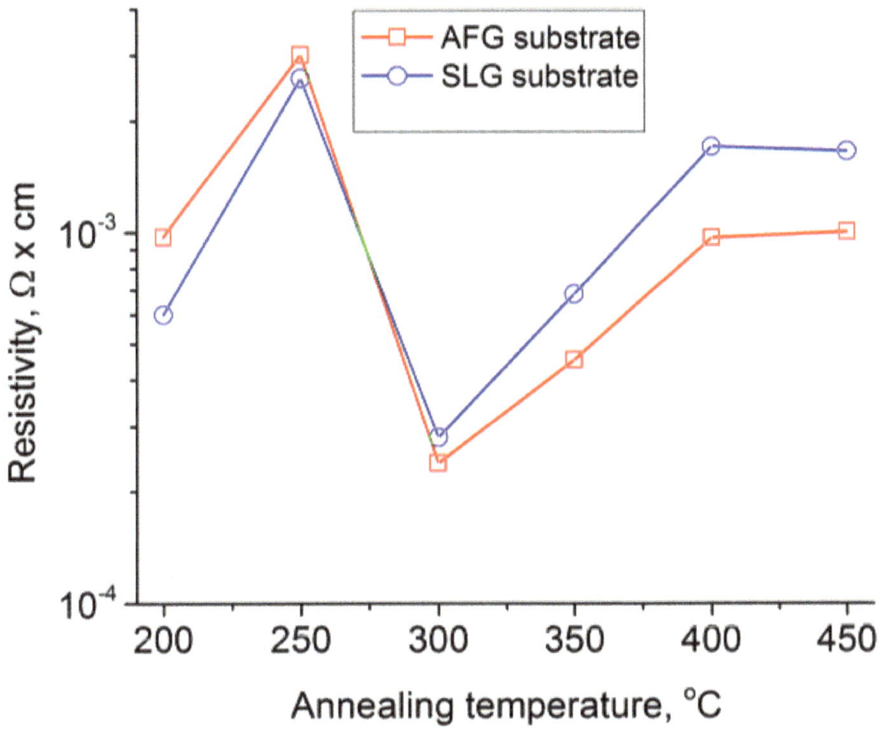

Figure 6. The resistivity dependence of the ITO films deposited on both types of substrates as a function of the annealing temperature in the oxygen atmosphere at different temperatures.

both types of substrates as a function of the annealing temperature in the oxygen atmosphere. The carrier concentration and carrier mobility obtained from Hall measurements are shown in **Figures 7** and **8**, respectively.

The as-deposited films are amorphous with a high concentration of nonstoichiometric contents such as metals and sub-oxides inclusions. This leads to a significant reduction of the resistivity in spite of the fact that the films were opaque and presented a black color. The film crystallization takes place at temperatures above 200°C. Higher annealing temperatures promote tin doping of the ITO films due to an effective replacement of the three-valence indium atoms by four-valence tin atoms in the crystal lattice of In_2O_3. A minimum value of the resistivity, (2–3) × 10^{-4} Ω×cm (**Figure 6**), is obtained at 300°C for both types of substrates. At higher annealing temperatures, an increasing resistivity is observed due to a reduction of oxygen vacancies that are doubly charged donor impurity in the ITO films; it is higher for the ITO films deposited on the SLG substrates, which can be due to the effect of alkali ions out-diffusion from the SLG substrate [19, 20]. The same explanation can be applied to the carrier concentration reduction (**Figure 7**) at annealing temperature above 300°C. The increment of the carrier mobility (**Figure 8**) with annealing temperatures for the ITO films deposited on both types substrates comes from an enhancement of the films structural morphology as well as a reduction of defects at the grain boundaries.

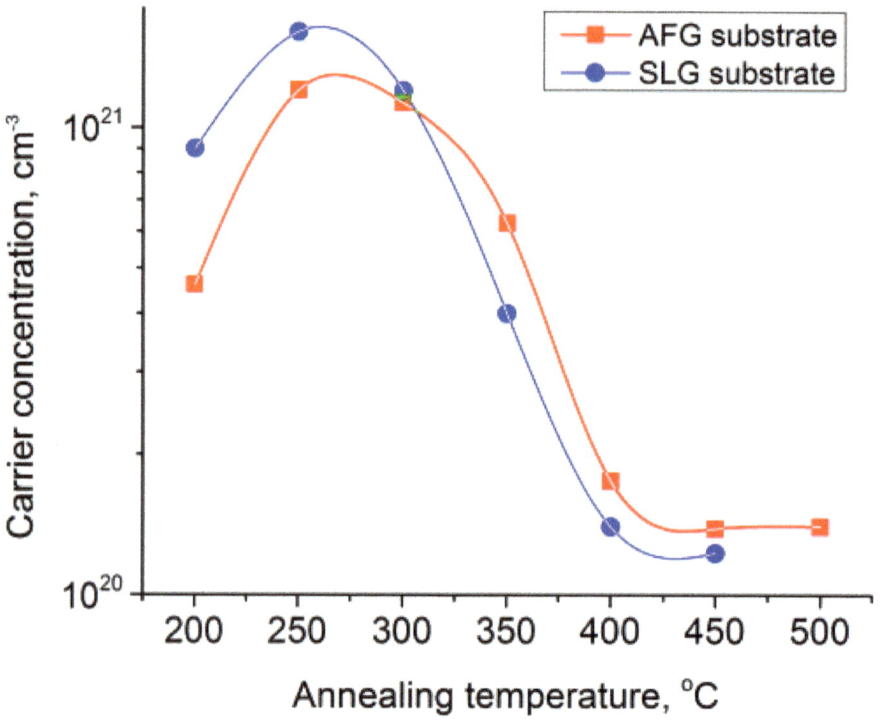

Figure 7. Dependence of carrier concentration on the annealing temperature for the ITO films deposited on AFG and SLG substrates.

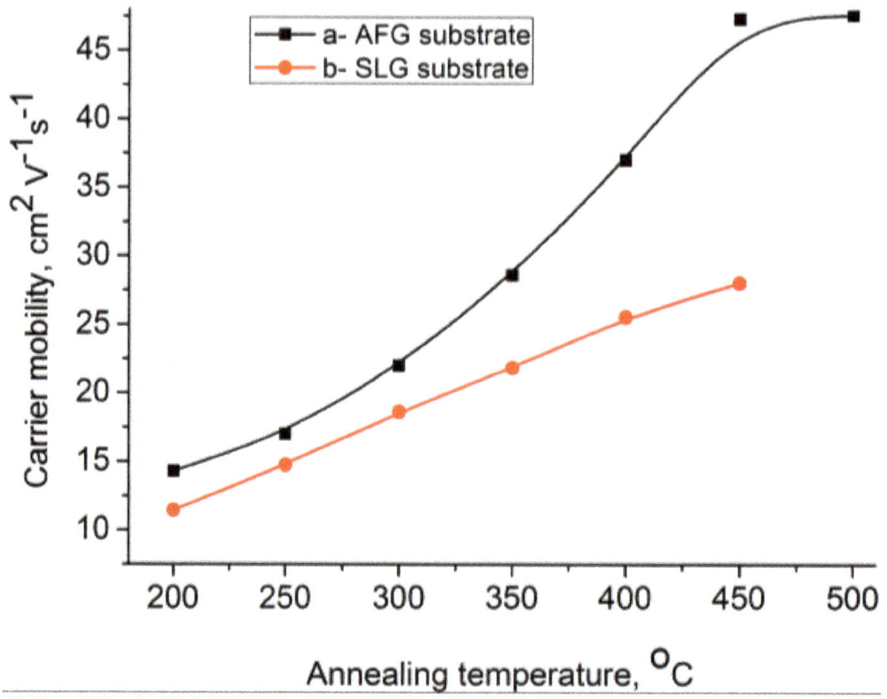

Figure 8. Dependence of carrier mobility on annealing temperature for the ITO films deposited on AFG and SLG substrates.

From the above-mentioned experimental results, we conclude that the highest structural and electrical parameters of the ITO films are obtained after a postannealing in the oxygen atmosphere at 300–350°C temperature range.

For practical applications of the ITO films as transparent conducting electrodes, the fabrication process needs to be optimized for obtaining a better balance between the transmittance (T) and the sheet resistance $R_s = \rho/d$, where ρ and d are the resistivity and the thickness of the film, respectively. However, a simultaneous obtaining of maximum transmission and conductivity is not possible to achieve in most cases. Haacke [27] proposed the figure-of-merit (FOM) $\phi_{TC} = T^{10}/R_s$ for comparison of films with different thickness and fabrication history. A higher FOM indicates a higher performance, this characterizes a film showing simultaneously a low sheet resistance (R_s) and a high transparency (T). However, the value of one of these parameters can compromise the value of the other. To avoid the influence of an interference effect in the thin films, the average integrated value of the films transmittance (T) at wavelengths in the range of 400–700 nm was used for the FOM value calculation. The dependence of integrated optical transmittance in visible spectral range (400–700 nm) of the ITO films deposited on AFG and SLG substrates is shown in **Figure 9** for different annealing temperatures.

Table 1 shows the structural, electrical parameters, the FOM for ITO films deposited on both glass substrates, AFG and SLG, after postannealing in the oxygen atmosphere at two different temperatures that are close to the optimal value.

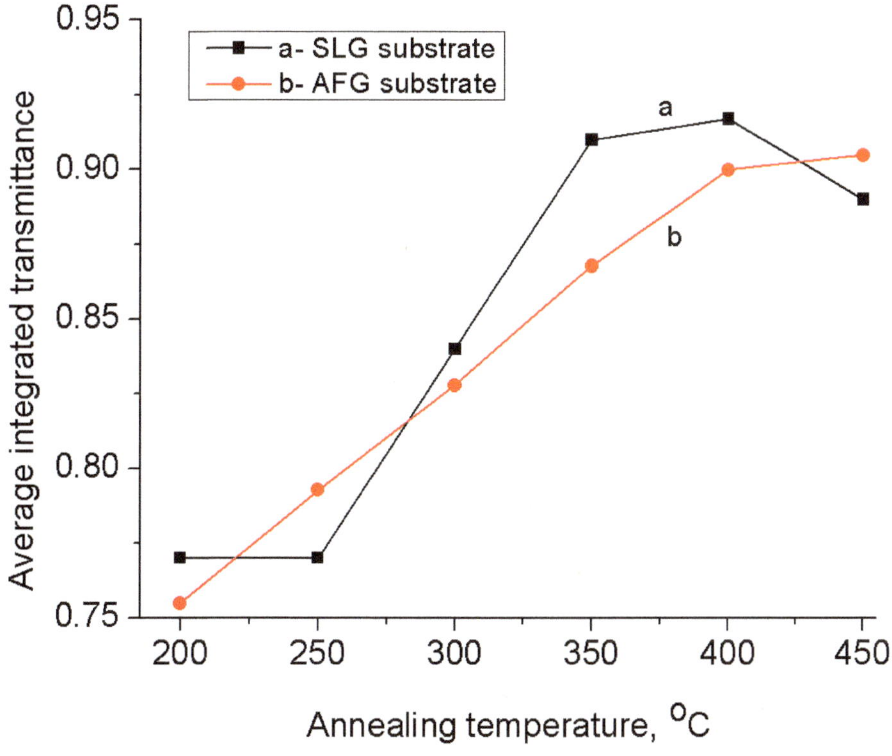

Figure 9. Dependence of the transmittance on annealing temperature for the ITO films deposited on SLG and AFG substrates.

Parameters/substrate	AFG	SLG	AFG	SLG
Annealing temperature (°C)	300	300	350	350
Substrate roughness (nm)	0.37	2.8	0.37	2.8
Ratio XRD peaks (222/400)	1.16	1.23	1.11	1.16
Grain size (nm)	25.0	45.0	25.0	55.0
ITO roughness (nm)	1.74	0.43	1.55	0.52
Specific resistance (10^{-4} Ω-cm)	2.2	2.8	4.2	6.8
Carrier concentration (10^{20} cm^{-3})	11.0	11.8	6.2	4.0
Mobility (cm^2 V^{-1} s^{-1})	25.0	17.5	28.0	23.0
Integral transparency	0.83	0.84	0.87	0.92
FOM (10^{-3} Ω^{-1})	14.1	12.5	11.8	12.8

Table 1. Comparison of properties of the ITO films deposited on AFG and SLG glass substrates after annealing in oxygen at 300 and 350°C.

From **Table 1**, the maximum $_{\phi TC}$ is observed for the films after postannealing in oxygen at 300– 350°C. At 300°C, where a minimum resistivity is observed, the films deposited on both substrates do not show a large difference in their FOM. Usually, ITO films present an FOM of the order of 10^{-3} Ω^{-1}. The highest known value of the FOM for commercially available sputtered ITO films is (14–17) × 10^{-3} Ω^{-1} [28, 29]. Also, we can compare the maximum value of the FOM obtained in this work (12 × 10^{-3} Ω^{-1}), with those reported for ITO films fabricated using other alternative methods: 6.6 × 10^{-3} Ω^{-1} by electron beam evaporation [30], 13 × 10^{-3} Ω^{-1} by sputtering [31], 13.6 × 10^{-3} Ω^{-1} by pulse laser deposition [32], 3.9 × 10^{-3} Ω^{-1} by spray pyrolysis [33], and 0.3 × 10^{-3} Ω^{-1} by sol-gel deposition [34].

We conclude that the ITO films sputtered in a pure argon atmosphere on low-cost SLG substrates, after annealing in oxygen at 300°C, are suitable for obtaining ITO films with high electro-optical parameters without the creation of a diffusion barrier between the substrate and the ITO film. The use of low-cost SLG substrates is an important factor for a significant reduction in the cost of thin film solar cells and modules for large-scale production. Moreover, a low annealing temperature allows for the fabrication of the ITO film on flexible plastic substrates. The ITO films, whose XRD spectra are shown in **Figure 10**, have been successfully fabricated on the Kapton substrate with annealing in oxygen at 300°C during 1 hour [23].

The incomprehensible preferred grains orientation (231) characteristics of a cubic film structure, or the (110) for hexagonal structure, need a further investigation. The electrical parameters of the 215-nm thick ITO films, deposited on Kapton plastic substrates at room temperature after annealing in an oxygen atmosphere during 1 hour at 300°C, are resistivity $\rho = 3.2 \times 10^{-4}$ Ω-cm, carrier concentration $n = 1.4 \times 10^{21}$ cm^{-3}, and mobility $\mu = 13.7$ cm^2/V-s.

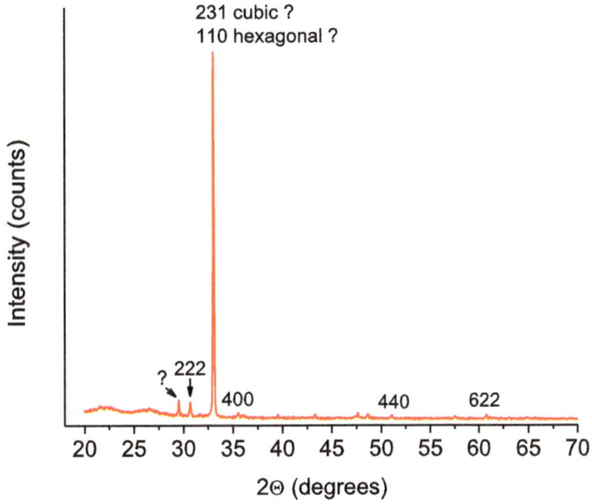

Figure 10. XRD spectra of the ITO film deposited on a Kapton substrate and annealed for 1 hour in oxygen at 300°C.

4. Comparison of sputtered and spray deposited ITO films

In this section, we present a comparison of the properties of the ITO films, fabricated using a two-step process for fabrication of films (sputtering an annealing), with those properties obtained using a one-step spray pyrolysis process [6, 23]; the second option is no expensive because it does not demand the use of sophisticated technological equipment. The spray pyrolysis is a stable and repetitive process very convenient for laboratory purposes and also for sufficiently large-scale production in the frame of small enterprises.

The spray pyrolysis technique was used for the deposition of 175–220 nm ITO films on an SLG substrate (Corning 1747). A glass atomizer was designed in order to produce small-size droplets. The SLG substrates were placed on a heater covered with a carbon disk in order to assure a uniform temperature, and the spraying was conducted using compressed air. The deposition rate was high at around 200 nm/min. For the ITO films deposition, 13.5 mg of $InCl_3 \times 4H_2O$ were dissolved in 170 ml of 1:1 water and ethylic alcohol mixture, with an addition of 5 ml of HCl to prevent hydrolysis. The different ratios of Sn/In achieved in the ITO films were controlled by adding in the solution a calculated amount of tin chloride ($SnCl_4 \times 5H_2O$). The substrate temperature was controlled using a thermocouple at 480 ± 5°C. The optimum distance from the atomizer to the substrate and the compressed air pressure were 25 cm and 1.4 kg/cm^2, respectively. We achieved a high deposition rate at around 200 nm/min. In order to prevent the deposition of the films under nonequilibrium thermal conditions, which can be due to a possible fast cooling of the substrate surface with the stream of the precursor and the compressed gas, periodical cycles of the deposition with durations of 1 s and intervals of 10 s were employed.

The X-ray diffraction patterns of the ITO films, prepared by spray pyrolysis at a substrate temperature of 480°C (**Figure 11**), show that these films are polycrystalline in nature for all the Sn/In ratios in the precursor and have a strong preferred columnar (400) grains orientation. Such preferred grains orientation significantly differs from grains orientation for sputtered and annealed ITO films (**Figure 1**). The average grain size estimated from full width at half maximum (FWHM) of the (400) peak using the Scherrer formula [26] is 160 nm, this is bigger than the one obtained in sputtered and annealed films as well as the average roughness, $S_a = 30$ nm, determined from the analysis of AFM measurements.

Figure 12 shows the thickness and the sheet resistance (R_s) of the spray deposited ITO films prepared from the precursors with different values of the Sn/In ratio. The minimum $R_s = 12$ Ω/square or resistivity $\rho = 2 \times 10^{-4}$ Ω-cm was obtained for the films deposited using the solution with Sn/In = 5%. The dependencies of the carrier concentration and mobility on the Sn/In ratio are shown in **Figure 13**. Both parameters present their maximum for Sn/In = 5%. The posterior reduction of these parameters for a Sn/In ratio above 5% can be explained as due to the limit of solubility of tin in indium oxide and a possible formation of neutral complexes of tin inside the indium oxide lattice.

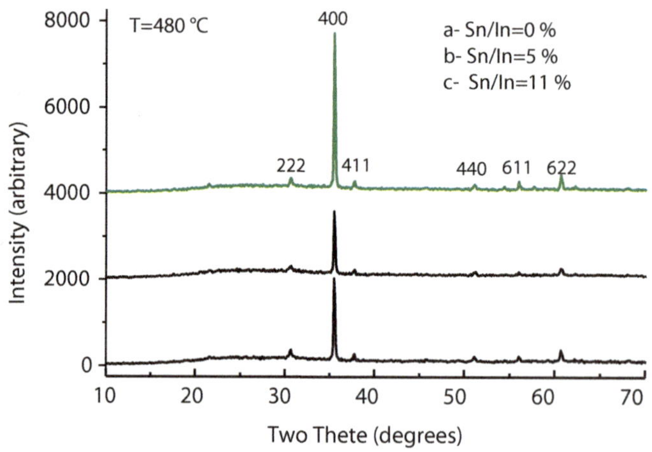

Figure 11. XRD diffraction patterns of the ITO films fabricated using solutions with different Sn/In ratios.

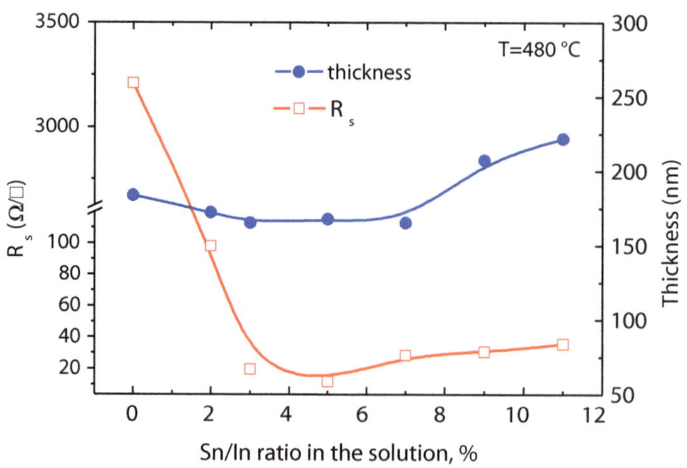

Figure 12. Sheet resistance and thickness of the sprayed ITO films fabricated using solutions with different Sn/In ratios.

The spray films deposited at 480°C, using the solution with a Sn/In = 5% ratio, showed the best electrical parameters, namely, a resistivity of 2×10^{-4} Ω-cm, a carrier concentration of 1.15×10^{21} cm^{-3}, and a mobility above 26 cm^2/V-s.

The optical transmittance (**Figure 14**) of the spray-deposited ITO films exceeded that of the sputtered/annealed films leading to a much higher FOM (**Figure 15**).

Comparison of properties of the ITO films deposited by sputtering and annealed in oxygen at 300°C with the properties of spray deposited films is shown in **Table 2**.

The practical optoelectronic applications (solar cell, hetero-photodiodes, and light emitting diode on silicon) of thin ITO films fabricated by spray pyrolysis are presented in our recently published articles [3–6].

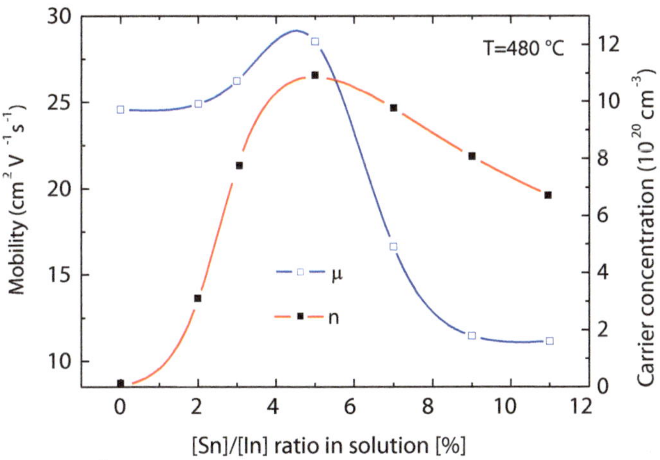

Figure 13. Hall mobility and carrier concentration of the sprayed ITO films fabricated using solutions with different Sn/In ratios.

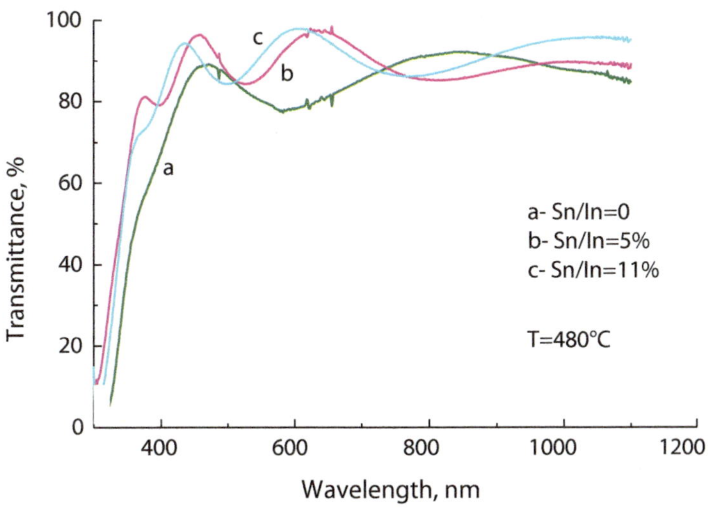

Figure 14. Optical transmittance spectra of the ITO films fabricated using solutions with different Sn/In ratios.

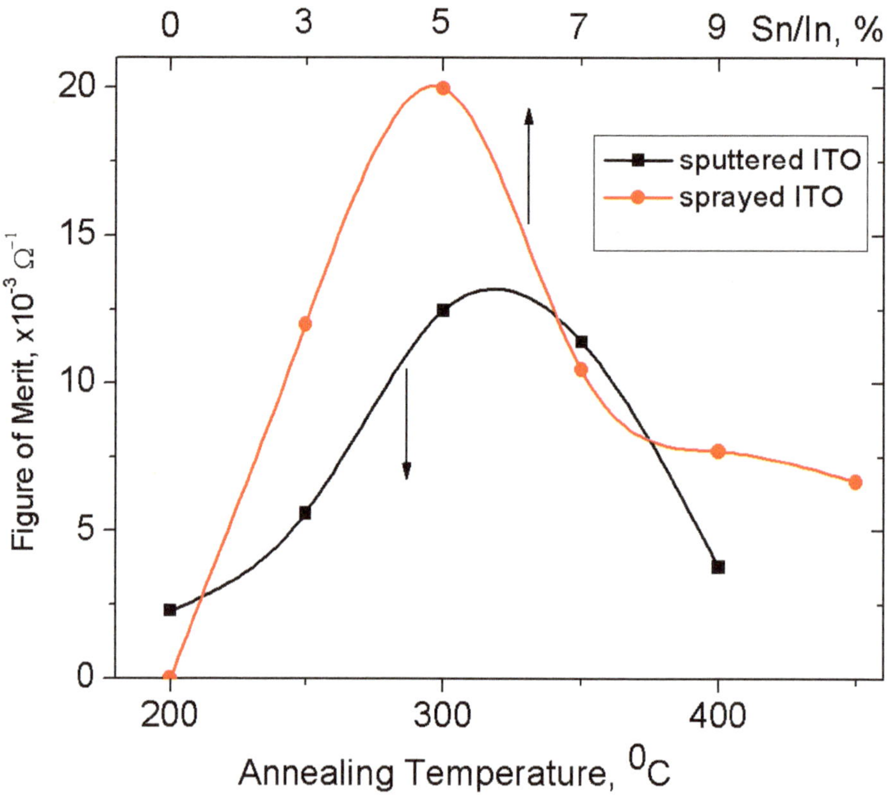

Figure 15. Comparison of the figure-of-merit (FOM) of the ITO films fabricated by sputtering with a subsequent annealing in oxygen at different temperatures, with the FOM of the films fabricated by spray pyrolysis using solutions with different Sn/In ratios.

Parameters	Sputtered and annealed ITO film on SLG	Spray deposited ITO film
Annealing temperature (°C)	300	-
Ratio Sn/In in solution (%)	-	5
Substrate roughness (nm)	2.7	2.7
Ratio XRD peaks (222/400)	1.23	0.032
Grain size (nm)	45.0	160
ITO roughness (nm)	0.43	30
Specific resistance (10^{-4} Ω-cm)	2.8	2
Carrier concentration (10^{20} cm^{-3})	11.8	11.5
Mobility (cm^2 V^{-1} s^{-1})	22.0	28
Integral transparency	0.83	0.85
FOM (10^{-3} Ω^{-1})	12.5	20

Table 2. Comparison of properties of the ITO films deposited by sputtering and annealed in oxygen at 300°C with the properties of spray deposited films.

5. Conclusions

The structural, electrical, and optical properties of sputtered and oxygen-annealed ITO films, deposited on alkali-free glass (AFG) and soda-lime glass (SLG) substrates, were studied. An important result of this work is the evidence that the sputtered ITO films in a pure argon atmosphere on low-cost SLG substrates, after annealing in oxygen at 300–350°C, show high optoelectronics parameters without the need of a diffusion barrier between the substrate and the ITO film. The electrical parameters of ITO films deposited on SLG substrates show a slight difference when compared to those obtained for the films deposited on AFG substrates at annealing temperatures no exceeding 350°C; the films deposited on SLG substrates exhibit better optical parameters (transmittance and value of FOM). Our conclusions are contrary to those reported in Ref. [20], where the ITO films on SLG substrates were annealed in a nitrogen atmosphere at 400°C. Our films were annealed in an oxygen atmosphere that leads to a binding of the diffused alkali ions with oxygen atoms diffused through the ITO films forming a stable oxide or peroxide at the film-substrate interface.

According to our results, we can conclude that SLG substrates are suitable for obtaining ITO films with high electro-optical parameters and small roughness due to a planarization effect. The use of such low-cost substrates is an important factor for a significant reduction in the cost of thin film solar cells and modules for large-scale production.

A comparative study shows that films deposited by spray pyrolysis present better electrical and optical properties than those obtained by sputtering. Hence, such a simple and low-cost deposition method is very promising for the fabrication of high-quality ITO films working as the ohmic contact in thin film optoelectronic devices as well as the active transparent conducting electrode inefficient solar cells and photodetectors with potential surface-barriers.

Author details

Oleksandr Malik* and Francisco Javier de la Hidalga-Wade

*Address all correspondence to: amalik@inaoeo.mx

Electronics Department, Óptica y Electrónica (INAOE), Instituto Nacional de Astrofísica, Puebla, Mexico

References

[1] A. L. Dawar, J.C. Joshi. Review semiconducting transparent thin films: their properties and applications. J. Materials Sci. 1984; 19:1-23.

[2] D. Ginley, H. Hosono, and D. C. Paine Eds. Handbook of Transparent Conductors, Springer, New York: 2011, 534 p.

[3] O. Malik, F.J. De la Hidalga-Wade, C. Zúñiga-Islas, and J. H. Abundis Patiño. UV-sensitive optical sensors based on ITO-GaP heterojunctions. Phys. Stat. Sol. (c). 2010; 7: 1176-1179.

[4] O. Malik, F. J. De la Hidalga-Wade, C. Zúñiga-Islas, and G. Ruíz-T. Efficient ITO-Si solar cells and power modules fabricated with a low temperature technology: results and perspectives. J. Non-Cryst. Solids. 2008; 354: 2472-2477.

[5] O. Malik, A. I. Martinez, and F. J. De la Hidalga-Wade. The physical reason of intense electroluminescence in ITO-Si heterostructures. Thin Solid Films. 2007; 515: 8615-8618.

[6] O. Malik, F.J. De la Hidalga-Wade. Spray deposited thin films of tin-doped indium oxide for optoelectronic applications. Adv. Mater. Res. 2013; 677: 173-178.

[7] H. Enoki, J. Echigoya, and H. Suto. The intermediate compound in the In_2O_3-SnO_2 system. J. Mater. Sci. 1991; 26: 4110.

[8] W. J. Heward, D. J. Swenson. Phase equilibrium in the pseudo-binary In_2O_3-SnO_2 system. J. Mater. Sci. 2007; 42: 7135-7140.

[9] G. B. González, T. O. Mason, J. S. Okasinski, T. Buslaps and V. Honkimäki.Determination of stability of tin in indium oxide in situ and ex situ x-ray diffraction. J. Am. Ceram. Soc. 2012; 95(2): 809-815.

[10] G. Falk. Sintering of transparent conductive oxides: from oxide ceramic powders to advanced optoelectronic materials. Sintering of Ceramics-New Emerging Techniques, A. Lakshmanan, Ed. InTech, Croatia: 2012, 587-610.

[11] T. Minami, H. Sonohara, T. Kakumu, and S. Takata. Physics of very thin ITO conducting films with high transparency prepared by DC magnetron sputtering. Thin Solid Films. 1995; 270(1-2): 37-42.

[12] M. Gheidari, H. H. Dabaghi, D. Kalhor, M. Iraj, M. Kazemzad, and F. Behafarid. Post annealing effects on the properties of sputtered nano-crystallite indium tin oxide thin films on flexible polyimide substrate. Phys. Stat. Sol. (c). 2008; 10: 3338-43.

[13] T. A. Gessert, Y. Yoshida, C. C. Fesenmaier, and T. J. Coutts. Sputtered In_2O_3 and ITO films containing zirconium. J. Appl. Phys. 2009; 105: 083547-53.

[14] M. Gulen, G. Yildirim, S. Bal, A. Varilci, I. Belenli, and M. Oz. Role of annealing temperature on microstructural and electro-optical properties of ITO films produced by sputtering. J. Mater. Sci.: Mater. Electron. 2013; 24: 467-474.

[15] T. A. Gessert, Y. Voshida, C. C. Fesenmaier, and T. J. Coutts. High-transparency sputtered In_2O_3 and ITO films containing zirconium. In: AVS 54th International Symposium & Exhibition, NREL/PR-520-42307, Seattle, Washington, USA: 2007, 14.

[16] C. P. Udawatte, K. Yanagisawa. Hydrothermal preparation of highly sinterable tin doped indium oxide powders: the effect of the processing parameters. In: Proceeding

of 7th International Conference on Ceramic Processing Science Proc., Inuyama City, Japan: 2000, 46.

[17] M. Nisha, S. Anusha, A. Antony, R. Manoj, and M. K. Jayaraj. Effect of substrate temperature on the growth of ITO films. Appl. Surf. Sci. 2005; 252: 1430-1436.

[18] N. Nisha. Growth and characterisation of radio frequency magnetron sputtered indium tin oxide thin films. PhD Thesis, Cochin University, Kerala, India: 2006, 116.

[19] R. Y. Tsai, F. C. Ho and M. Y. Hua. Annealing effects on the properties of indium tin oxide-films coated on soda lime glasses with a barrier layer of TiO_2-SiO_2 composite films. Opt. Eng. 1997; 3: 2335-2340.

[20] J. M. Lee, B. J. Choi, M. J. Ji, Y. T. An, and J. H. Park. Effect of barrier layers on the properties of indium tin oxide thin films on soda lime glass substrates. Thin Solid Films. 2009; 517: 4074-4077.

[21] R. Garkova, G. Volksch, and C. Russel. Precipitation of SnO_2 nano-crystallites from Na_2O/B_2O_3/SnO_2/(Al_2O_3) glasses. J. Non-Cryst. Solids. 2005; 351: 2287-2293.

[22] O. Malik, F. J. De la Hidalga-Wade. Optoelectronics properties of tin-doped indium oxide films fabricated by DC magnetron sputtering in pure argon with post-annealing in oxygen atmosphere. J. Mater. Res. 2015; 30: 18941901.

[23] O. Malik, F. J. De la Hidalga-Wade. Comparison of tin-doped indium oxide films fabricated by spray pyrolysis and magnetron sputtering. Cryst. Res. Technol. 2015; 50: 516-520.

[24] D. Raoufi, J. Alam. Surface characterization and microstructure of ITO thin films at different annealing temperatures. Appl. Surf. Sci. 2007; 253(23): 9085-9090.

[25] A. Salehi. The effects of deposition rate and substrate temperature of ITO thin films on electrical and optical properties. Thin Solid Films. 1998; 324(1-2): 214-218.

[26] J. W. Kim, H. S. Kang, and S. Y. Lee. Effect of deposition rate on the property of ZnO thin films deposited by pulsed laser deposition. J. Electrical Eng. Tech. 2006; 1(1): 98-100.

[27] G. Haacke. New figure of merit for transparent conductors. J. Appl. Phys. 1976; 47(9): 4086-4089.

[28] GEOMATEC for ITO films, thin-films, and coating; www.geomatec-sputtering.com.

[29] Optics Balzers AG; www. opticsbalzers.com.

[30] H. R. Fallah, M. Ghasemi, A. Hassanzadeh, and H. Steki. The effect of annealing on structural, electrical and optical properties of nanostructured ITO films prepared by e-beam evaporation. MRS Bull. 2007; 42 (3): 487-496.

[31] C. Guillén, J. Herrero. Influence of oxygen in the deposition and annealing atmosphere on the characteristics of ITO thin films prepared by sputtering at room temperature. Vacuum. 2006; 80: 615-620.

[32] C. Viespe, I. Nicolae, C. Sima, C. Grigoriu, and R. Medianu. ITO thin films deposited by advanced pulsed laser deposition. Thin Solid Films. 2007; 515: 8771-8775.

[33] M. A. Aouaj, R. Diaz, A. Belayachi, F. Rueda, and M. Abd-Lefdil. Comparative study of ITO and FTO thin films grown by spray pyrolysis. MRS Bull. 2009; 44(7): 1458-1461.

[34] M. J. Alam, D. C. Cameron. Investigation of annealing effects on sol–gel deposited indium tin oxide thin films in different atmospheres. Thin Solid Films. 2002; 420-421: 76-82.

10

Spectral Responses in Quantum Efficiency of Emerging Kesterite Thin-Film Solar Cells

Sanghyun Lee and Kent J. Price

Additional information is available at the end of the chapter

Abstract

The spectral responses in quantum efficiency provide essential information about current generation, recombination, and diffusion mechanisms in a photodetector, photodiode, and photovoltaic devices as the quantum efficiency is a function of the voltage and light biases and the spectral content of the bias light and/or location of the devices. Recently, P-type Kesterite thin-film solar cells are emerging as they have a high absorption coefficient ($>10^4$ cm^{-1}) and ideal direct bandgap (1.4–1.5 eV), which make them a perfect candidate for photovoltaic application. However, a champion device from Zincblende (CdTe) or Chalcopyrite (CIGS) solar cells shows ~21% efficiency (<21.5%, First Solar and <21.7%, ZSW, respectively) while Kesterite devices suffer from severe losses with <12.6% efficiency. Furthermore, the maximum theoretical efficiency based on Shockley-Queisser limit is about 32.2%, which indicates there is much room for the improvement. Consequently, the implication from the current situation highlights the need for a systematic analysis of the loss mechanism in Kesterite devices. In this work, we carried out a systematic study of the efficiency limiting factors based on quantum efficiency to model the quantum efficiency response of current CZTSSe thin-film solar cells. This will provide the guidance for proper interpretation of device behaviors when it is measured by quantum efficiency.

Keywords: solar cells, thin film, spectral response, quantum efficiency, photovoltaics

1. Introduction

Recently, polycrystalline thin-film CdTe and $Cu(In,Ga)Se_2$ (CIGS) solar cells are commercially successful with dramatic conversion efficiency improvements and cost reduction in module mass production as well as the balance of system (BoS), which is the cost except the module production cost. A champion device from CdTe (Zincblende) and CIGS (Chalcopyrite) solar cells shows >21%

efficiency (<21.5%, First Solar and < 21.7%, ZSW, respectively). However, some of these materials used in thin-film solar cells technology are not benign to the environment, which cause toxicity issues and possible scarce in the earth's crust, whereas competing Si solar cell technology uses the second most abundant element, Silicon, in the earth's crust. To fully facilitate the advantage of low manufacturing cost in the polycrystalline thin-film solar cells technology, a research on Kesterite solar cells using earth's abundant materials has gained momentum to pursue low-cost, environment-friendly, and highly efficient Kesterite photovoltaic devices [1–4]. P-type Kesterite solar cells have high absorption coefficient, 10^4 cm^{-1} and ideal direct bandgap (1.0–1.5 eV), which make them a perfect candidate for photovoltaic application [1–6]. Kesterite solar cells are also called CZTSSe (Copper Zinc Tin Sulphur Selenium) solar cells following the crystal structure. However, the record power conversion efficiency is 12.6% with laboratory-level samples [1, 6], which are processed with non-vacuum-based techniques. Since the development is in an early stage, the development of Kesterite solar modules is not considered yet due to incompetence in the market. However, the maximum theoretical efficiency based on Shockley-Queisser limit is about 32.2%, which indicates there is much room for improvement and hence many opportunities [7, 8].

To reduce the discrepancy between actual and maximum possible efficiency, there are major issues that need to be addressed such as absorber quality (electronic band structure, secondary phases, defect concentration, grain boundaries, optical property), CdS or (CdS-less) buffer quality, Molybdenum (Mo) back contact quality, and the quality of interfaces. One of the main challenges is different types of recombination processes at different interfaces such as buffer/window, heterojunction, and absorber/back contact. At these interfaces, recombination active defects are present due to intrinsic lower symmetry effects [9], lattice mismatch, band tails (still open discussion) [10–13], and segregation of impurities aggravated by grain boundaries, which are inevitable for polycrystalline solar cells. The recombination current enhances forward diode current that deteriorates the conversion efficiency, which is inevitable for polycrystalline Kesterite solar cells. The recombination current enhances forward diode current which worsens the conversion efficiency, which is one of the reasons to have lower open circuit voltage (*Voc*). Open circuit voltage results from the forward bias owing to the bias of the solar cell junction with the light-generated current. This is a point where diode current is equivalent to light-generated current (J_L), affected by the saturation current (J_0) of the solar cell. Since *Voc* is expressed through the saturation current as $Voc = (kT/q)\ \ln[(J_L/J_0) + 1]$, *Voc* depends on the recombination in the device, which is strongly dependent on the saturation current. Another factor that influences the diode saturation current is a function of the bandgap, which increases with bandgap.

Despite a recent significant improvement in Kesterite solar cells, one of the most fundamental problems that remain is lower *Voc* [14–17]. Moving *Voc* beyond 750 mV in Kesterite solar cells requires improvements in both built-in voltage and recombination. At the heterojunction of thin-film solar cells, the recombination that contributes to the diode current can occur on different recombination paths. Recombination may occur at the heterojunction interface, in the space charge region (SCR), in the quasi-neutral region (QNR), and at the back contact. In particular, the interface and near-interface recombination (SCR) mainly contribute to the recombination current, which is influenced by a number of recombination active defects and grain boundaries, the number of minority carriers at the junction edge, and diffusion length. One way to characterize the dominant recombination path is the quantum efficiency analysis.

In this chapter, we present the study of photocurrent loss mechanism based on analytical and numerical models of quantum efficiency, which provide insights into the generation and transport of carriers in the CZTSSe solar cell. Numerical models are computational results of solving continuity equation and Poisson equation, whereas analytical models are based on key parameters such as the lifetime of carriers, depletion width, diffusion length and drift length, surface recombination velocity, and thickness of the absorber layer. Key parameters of CZTSSe solar cells are focused to quantitatively describe the spectral responses of quantum efficiency by considering the recombination losses near or at the CdS/CZTSSe interface as well as the CZTSSe absorber layer. The dependency of charge collection efficiencies in the space charge region and CZTSSe absorber is discussed with respect to uncompensated impurity concentration, defect energy state, drift and diffusion components of short circuit current, and recombination velocities.

2. Analytical description of spectral responses in quantum efficiency

For this analytical study, we developed the collection efficiency model for thin-film solar cells and put in the mathematical expression. The external quantum efficiency, $EQE(\lambda, V)$ is the ratio of the number of charge carriers collected by the photovoltaic device to the number of photons of a given wavelength or energy onto the device. When the various wavelengths of light in the spectrum are applied, quantum efficiency corresponds to the spectral responses (amperes per watt) of solar cells per unit area, unit second, unit wavelength (Remember the units in solar spectra radiation).

$$EQE(\lambda, V) = \frac{I_{ph}}{q\Phi_p} = \frac{I_{ph}}{q}\left(\frac{hv}{P_{opt}}\right) \quad (1)$$

where I_{ph} is the photocurrent, Φ_p is the photon flux which is equivalent to P_{opt}/hv, and P_{opt} is the optical power.

The external quantum efficiency is related to transmission, $T(\lambda)$ of all the materials prior to CZTSSe absorber film, which consists of anti-reflection coating (ARC), transparent conducting front contact oxide (TCO), a buffer layer, and an n-type CdS layer. The internal quantum efficiency, $IQE(\lambda, V)$ is associated with the external quantum efficiency as follows.

$$EQE(\lambda, V) = \left(1 - R_{AFC}(\lambda)\right) \cdot T_{TCO}(\lambda) \cdot T_{buffer}(\lambda) \cdot T_{CdS}(\lambda) \cdot IQE(\lambda, V) \quad (2)$$

where T_{TCO} is the transmission of transparent conduction oxide (TCO), T_{buffer} is the transmission of buffer (ZnO), and T_{CdS} is the transmission of CdS.

2.1. Experimental details

In **Figure 1**, the quantum efficiency spectra of the CZTSSe solar cell are measured from 300 to 1400 nm. The cell design is based on the common substrate structure, which consists of soda-lime glass (SLG)/Molybdenum (Mo) layer followed by depositing CZTSSe film (1.8–2.0 um).

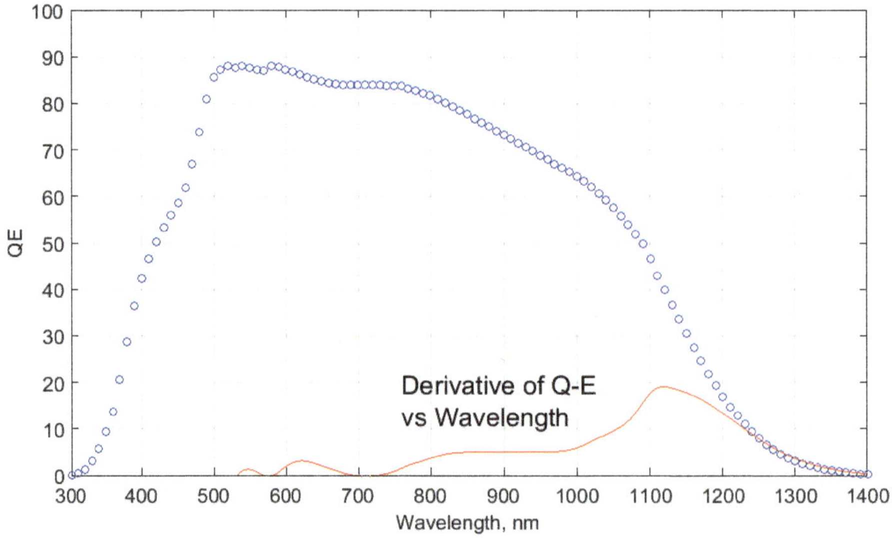

Figure 1. Spectral responses of the external quantum efficiency of a CdS/CZTSSe device measured from 300 to 1400 nm.

The buffer CdS was developed by 0.05 μm and intrinsic ZnO (0.05–0.1 μm) is deposited to prevent damage during ITO deposition at the post-ZnO deposition. Indium tin oxide (ITO) is deposited (350–400 nm) as a transparent conductive window layer; the device is completed with Ni/Al front metal fingers and MgF_2 ARC.

The spectral responses of this sample were characterized in the range of 300–1400 nm with Xenon arc lamp and the spectral distribution of the photon flux at the outlet slit was calibrated with NIST calibrated photodiode G425.

In the wavelength region 300–500 nm, measured spectral responses from the devices demonstrate typical CdS layer response, which agrees with the literature [1–5]. Above 500 nm, the measured spectral response is limited by the band gap of CZTSSe, λ_{Eg} (= $h\nu/Eg$). The bandgap calculated from the derivative of Q-E is 1.11 eV for this device and the total quantum efficiency-calculated short circuit current is 33.59 mA/cm^2. Across wavelength regions, the overall reduction in the quantum efficiency in **Figure 1** is due to the grid shading and front surface reflection as described in the literature [18]. Between 300 and 500 nm, the widely accepted loss mechanisms in quantum efficiency are TCO/ZnO absorption and CdS buffer absorption. Higher the transmission of light through TCO, ZnO, and CdS, lower the quantum efficiency losses. Hence the transmission in Eq. (2) needs to be minimized to maximize the spectral response from the device.

2.2. Energy band diagram

Under the steady-state illuminated condition, we derive the total current density under the applied biased conditions. One of the important parameter sets that determines the spectral responses in photovoltaic devices is the space charge region width, *W*. Although it is widely accepted that CdS does not fundamentally contribute photogenerated current, any changes in effective donor impurity concentration or uncompensated impurity (N_d-N_a) and photoconductivity of CdS buffer layer greatly impact the space charger region due to an asymmetrical p-n heterojunction nature. In fact, there are a number of defects in CdS, which contribute to the increase of effective uncompensated donor concentration. In this case, the depletion layer of n-CdS/p-CZTSSe is located in almost all the p-CZTSSe absorber layer, which is fundamentally similar to an abrupt n+/p junction diode or a Schottky diode. Following the band structure [19], the depletion width (**Figure 2**) in the n-CdS/p-CZTSSe is derived as follows [20].

$$W = \sqrt{\frac{2\epsilon\epsilon_r(\phi_0 - qV)}{q^2(N_a - N_d)}} \tag{3}$$

where ϵ_r is the relative permittivity, ϵ is the permittivity of free space, $\phi_0 = qV_{bi}$ is the barrier height at the absorber side (V_{bi} is the built-in potential), V is the applied voltage, and $N_a - N_d$ is the uncompensated acceptor concentration in the CZTSSe absorber layer.

2.3. Analytical description of bias-dependent quantum efficiency

Under the steady-state illuminated condition, we derive the total current density under widely accepted assumptions, which are no thermal generation current and no photocurrent contribution by n-type CdS layer. Hence, the electron-hole generation rate by absorption is described by the below.

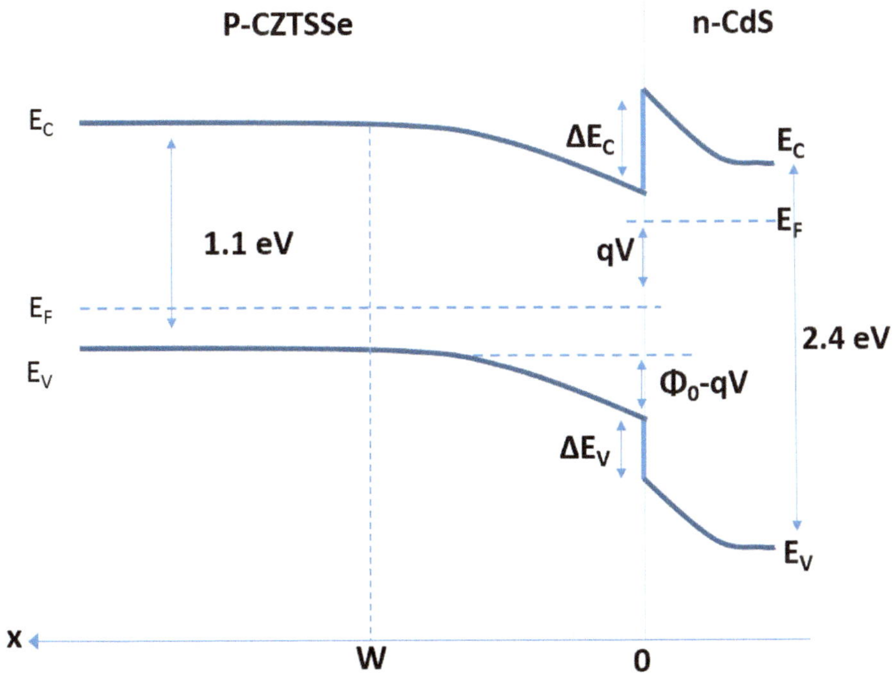

Figure 2. Energy band diagram of n-CdS/p-CZTSSe heterojunction at biased condition (V).

$$G_e(x) = \Phi_p \alpha \exp(-\alpha x) \tag{4}$$

where Φ_p is the incident photon flux per unit area and α is the absorption coefficient of CZTSSe.

To isolate non-transmission terms in the quantum efficiency Eq. (2), we distinguish $EQE(\lambda, V)$ and $IQE(\lambda, V)$ so that we will be able to focus on developing a model for carrier collection function with $IQE(\lambda, V)$. As mathematically derived by Gartner [21] and Sze [20] with boundary conditions, quantum efficiency is the sum of a drift component in the space charge region and a diffusion component in the quasi-neutral region. After simplifying the exact solution of Lavagna's model [22], Kosyachenko derived IQE [23].

$$IQE = \frac{1 + \frac{S}{D_n}\left(\alpha + \frac{2}{W} \cdot \frac{qV_{bi} - qV}{kT}\right)^{-1}}{\frac{S}{D_n}\left(\frac{2}{W} \cdot \frac{qV_{bi} - qV}{kT}\right)^{-1}} - \frac{\exp(-\alpha W)}{1 + \alpha L_n} \tag{5}$$

With the assumption of an abrupt p-n junction, the internal quantum efficiency is described based on the widely accepted Gartner formula [21, 22] as no significant contribution of electron collection by CdS is expected [23, 24].

$$IQE = 1 - \frac{\exp(-\alpha W)}{1 + \alpha L_n} \tag{6}$$

where W is the space charge region width and L_n is the diffusion length of the electron ($\sqrt{D_n \tau_n}$). Photogenerated carriers in the space charge region are assumed to be completely collected as a drift component after integrating the drift current (J_{drift}) [20].

$$J_{drift} = -q\int_0^W \phi_0 \alpha \exp(-\alpha x)dx = q\phi_0\{1 - \exp(-\alpha W)\} \tag{7}$$

$$IQE_{drift} = 1 - \exp(-\alpha W) \tag{8}$$

Photogenerated carriers outside of space charge region are collected by the diffusion process with no electric field [22] as follows.

$$IQE_{diff} = IQE - IQE_{drift} = 1 - \frac{\exp(-\alpha W)}{1 + \alpha L_n} - \{1 - \exp(-\alpha W)\} = \frac{\exp(-\alpha W)}{1 + 1/\alpha L_n} \tag{9}$$

Consequently, the total internal quantum efficiency is the sum of Eqs. (8) and (9). With complete collection of the carrier in the space charge region, EQE is revised after introducing the interface recombination term.

$$EQE(\lambda, V) = \left(1 - R_{ARC}(\lambda)\right) \cdot T_{TCO}(\lambda) \cdot T_{buffer}(\lambda) \cdot T_{CdS}(\lambda) \cdot IQE(\lambda, V) \cdot h(v) \tag{10}$$

$$h(v) = \left(1 + \frac{S_{junction}}{\mu_e E_{junction}}\right)^{-1} \tag{11}$$

where *h(V)* is the interface recombination function due to the lattice mismatch between CdS and CZTSSe absorber, $IQE_{absorber}$ is the internal quantum efficiency of CZTSSe absorber [25], and $S_{junction}$ is the surface recombination velocity, μ_e is electron mobility, and $E_{junction}$ is the electric field at the junction.

However, a complete collection model in the space charge neglects including the space charge region recombination, and a couple of researchers identified that significant uncompensated acceptors and defects cause the recombination in the space charge region [23]. Hardrich revised a collection function (CF_{SCR}) [24] for space charge region with the hypothesis of constant field.

$$CF_{SCR\_CEF} = \exp\left(-\frac{t}{\tau_e}\right) = \exp\left(-\frac{2x}{\mu_e \tau_e E_{max}}\right) = \exp\left(-\frac{x}{L_{drift}}\right) \tag{12}$$

where τ_e is the lifetime of the electron minority carrier in the p-doped CZTSSe absorber, *t* is the time, constant electric field (CEF) is assumed as half of maximum ($E_{max}/2$), and L_{drift} is defined as $\mu_e \tau_e E_{max}/2 = \mu_e \tau_e (V_{bi} - V)/W$. Hence, $IQE_{absorber}$ in the space charge region under constant field yields,

$$IQE_{drift\_CEF} = \int_0^W \alpha exp(-\alpha x) \cdot \exp\left(-\frac{x}{L_{drift}}\right) dx \tag{13}$$

$$IQE_{drift\_CEF} = \frac{L_{drift}}{L_{drift} + (1/\alpha)} \cdot \left\{1 - \exp\left(-W.\left(\alpha + \frac{1}{L_{dirft}}\right)\right)\right\} \tag{14}$$

However, the electric field in the space charge region is not constant. Assuming constant doping at the junction, it linearly decreases along the depth of the CZTSSe absorber. Constant doping does not reflect the exact real situation of the doping profile due to doping concentration fluctuation with grain boundaries, fixed charges, impurities, and so on. We mathematically simplify the case with the linear electric field (LEF) at the junction. Hence, we are able to develop an alternative expression for the linearly decreasing electric field in the space charge region. An alternative expression for the collection in the space charge region begins with the definition of the drift velocity, assuming the one-dimensional charge movement, which limits the lateral movement caused by non-uniformity. To simplify the mathematical processes, we define new x-coordinate, $x' = 0$ at the interface between the space charge region and the quasi-neutral region (x' = W-x). Therefore, the drift time for the electron is as below.

$$t_{drift} = \int_0^{t_{drift}} dt = \int_x^W \frac{1}{v_{x'}} dx' = \int_{x'}^W \frac{1}{\mu_e E(x')} dx' \tag{15}$$

where an electron generated at point x′ is swept to the edge of the space charge region. The electric field inside the space charge region decreases linearly toward the quasi-neutral region of the p-type absorber layer and $E(x')$ can be revised to be $Emax \cdot (x'/W)$.

$$t_{drift} = \int_{x'}^{W} \frac{dx'}{(E_{max}\mu_e/W)\cdot x'}\, dx' = \frac{1}{E_{max}\mu_e/W} ln\left(\frac{W}{x'}\right) \tag{16}$$

$$CF_{SCR\_LEF} = exp\left(-\frac{t_{drift}}{\tau_e}\right) = exp\left(-\frac{W}{\mu_e\tau_e E_{max}} ln\left(\frac{W}{x'}\right)\right) = \left(\frac{x'}{W}\right)^{\frac{W}{\mu_e\tau_e E_{max}}} \tag{17}$$

By defining κ ($= \mu_e\tau_e E_{max}$), the collection function under the linear electric field is

$$CF_{SCR\_LEF} = \left(\frac{x'}{W}\right)^{\frac{W}{\kappa}} \tag{18}$$

And if you integrate throughout x (replacing x' into W-x) and use gamma function (Γ),

$$IQE_{drift\_LEF} = -\int_{W}^{0} \alpha e^{-\alpha(W-x')}\cdot\left(\frac{W-x'}{W}\right)^{\frac{W}{\kappa}} dx' = -\frac{\Gamma(1+\frac{W}{\kappa},\ -\alpha(x-W))}{\alpha^{\frac{W}{\kappa}}W^{\frac{W}{\kappa}}}\Bigg|_{0}^{W} \tag{19}$$

$$IQE_{drift\_LEF} = (\alpha W)^{-\frac{W}{\kappa}}\cdot\Gamma(1+\frac{W}{\kappa},\ \alpha W) \tag{20}$$

For a diffusion component of the photoelectric quantum response, the exact solutions for electron and hole are described [20] for the case of p-layer in a p-n junction, including surface recombination at the back surface CZTSSe absorber. We assume only electron contribution of the absorber layer, considering the hole collection is negligible due to strong compensation with a large number of high p-type compensating defects in CdS [23].

$$IQE_{diff} = \frac{\alpha L_n}{\alpha^2L_n^2-1}\exp(-\alpha W)\cdot \left\{\alpha L_n - \frac{\frac{S_bL_n}{D_n}\left[\cosh\left(\frac{d-W}{L_n}\right)-\exp(-\alpha(d-W))\right]+sinh\left(\frac{d-W}{L_n}\right)+\alpha L_n\exp(\alpha(d-W))}{\frac{S_bL_n}{D_n}\sinh\left(\frac{d-W}{L_n}\right)+\cosh\left(\frac{d-W}{L_n}\right)}\right\} \tag{21}$$

where d is the thickness of the CZTSSe absorber layer and S_b is the recombination velocity at the back surface of the absorber. The sum of drift and diffusion components is the total internal quantum efficiency. The total internal quantum efficiency of the absorber for the constant electric field is the sum of Eqs. (14) and (21), and the IQE for the linear electric field is the sum of Eqs. (20) and (21).

$$IQE_{CEF} = \frac{L_{drift}}{L_{drift}+(1/\alpha)}\cdot\left\{1-\exp\left(-W\cdot\left(\alpha+\frac{1}{L_{dirft}}\right)\right)\right\}+\frac{\alpha L_n}{\alpha^2L_n^2-1}\exp(-\alpha W) \cdot\left\{\alpha L_n - \frac{\frac{S_bL_n}{D_n}\left[\cosh\left(\frac{d-W}{L_n}\right)-\exp(-\alpha(d-W))\right]+sinh\left(\frac{d-W}{L_n}\right)+\alpha L_n\exp(\alpha(d-W))}{\frac{S_bL_n}{D_n}\sinh\left(\frac{d-W}{L_n}\right)+\cosh\left(\frac{d-W}{L_n}\right)}\right\} \tag{22}$$

$$IQE_{LEF} = (\alpha W)^{-\frac{W}{\kappa}} \cdot \Gamma(1+\frac{W}{\kappa}, \alpha W) + \frac{\alpha L_n}{\alpha^2 L_n^2 - 1} \exp(-\alpha W) \cdot$$
$$\left\{ \alpha L_n - \frac{\frac{S_b L_n}{D_n}\left[\cosh\left(\frac{d-W}{L_n}\right) - \exp(-\alpha(d-W))\right] + sinh\left(\frac{d-W}{L_n}\right) + \alpha L_n \exp(\alpha(d-W))}{\frac{S_b L_n}{D_n}\sinh\left(\frac{d-W}{L_n}\right) + \cosh\left(\frac{d-W}{L_n}\right)} \right\} \tag{23}$$

Consequently, external quantum efficiency is the product of the total internal quantum efficiency, which includes recombination in the space charge region (Eqs. (22) and (23)), recombination at the interface (Eq. (1)), and the optical losses prior to CZTSSe absorption.

$$EQE_{CEF} = (1 - R_{ARC}) T_{TCO} T_{Buffer} T_{CdS} \cdot$$
$$\left\{ \frac{L_{drift}}{L_{drift} + \left(\frac{1}{\alpha}\right)} \cdot \left[1 - exp\left(-W \cdot \left(\alpha + \frac{1}{L_{dirft}}\right)\right)\right] + \frac{\alpha L_n}{\alpha^2 L_n^2 - 1} exp(-\alpha W) \cdot \right.$$
$$\left. \left\{ \alpha L_n - \frac{\frac{S_b L_n}{D_n}\left[\cosh\left(\frac{d-W}{L_n}\right) - \exp(-\alpha(d-W))\right] + sinh\left(\frac{d-W}{L_n}\right) + \alpha L_n \exp(\alpha(d-W))}{\frac{S_b L_n}{D_n}\sinh\left(\frac{d-W}{L_n}\right) + \cosh\left(\frac{d-W}{L_n}\right)} \right\} \right\}$$
$$\cdot \left(1 + \frac{S_{junction}}{\mu_e E_{junction}}\right)^{-1} \tag{24}$$

$$EQE_{LEF} = (1 - R_{ARC}) T_{TCO} T_{Buffer} T_{CdS} \cdot \left\{ (\alpha W)^{-\frac{W}{\kappa}} \cdot \Gamma(1+\frac{W}{\kappa}, \alpha W) + \frac{\alpha L_n}{\alpha^2 L_n^2 - 1} \exp(-\alpha W) \cdot \right.$$
$$\left. \left\{ \alpha L_n - \frac{\frac{S_b L_n}{D_n}\left[\cosh\left(\frac{d-W}{L_n}\right) - \exp(-\alpha(d-W))\right] + sinh\left(\frac{d-W}{L_n}\right) + \alpha L_n \exp(\alpha(d-W))}{\frac{S_b L_n}{D_n}\sinh\left(\frac{d-W}{L_n}\right) + \cosh\left(\frac{d-W}{L_n}\right)} \right\} \right\}$$
$$\cdot \left(1 + \frac{S_{junction}}{\mu_e E_{junction}}\right)^{-1} \tag{25}$$

For the non-uniform field, mathematical expressions are developed with an exponential electric field [26] or two different electric field regions [27, 28]. However, it is argued that including the non-uniform electric field with additional parameter sets only adds confusion with no additional insights [29]. Hence, we will explore the quantum efficiency response with a constant electric field model for our analysis for simplification.

2.4. Impurity concentration and electron lifetime by analytical models

Figure 3 shows the computed spectral responses of the external quantum efficiency, $EQE(\lambda)$ with the impact of the uncompensated acceptor impurity at the n-CdS/p-CZTSSe interface. The absorption coefficient, $\alpha(\lambda)$ was taken from Ref. [28] and the front surface recombination velocity (S_f), the default depletion width is 0.3 μm, and the electron lifetime (τ_n) were 10^3 cm/s and 10^{-9} s [29, 30]. In **Figure 3**, the Q-E shape changes disproportionally with significant red light response reduction >600 nm as the uncompensated impurity concentration changes

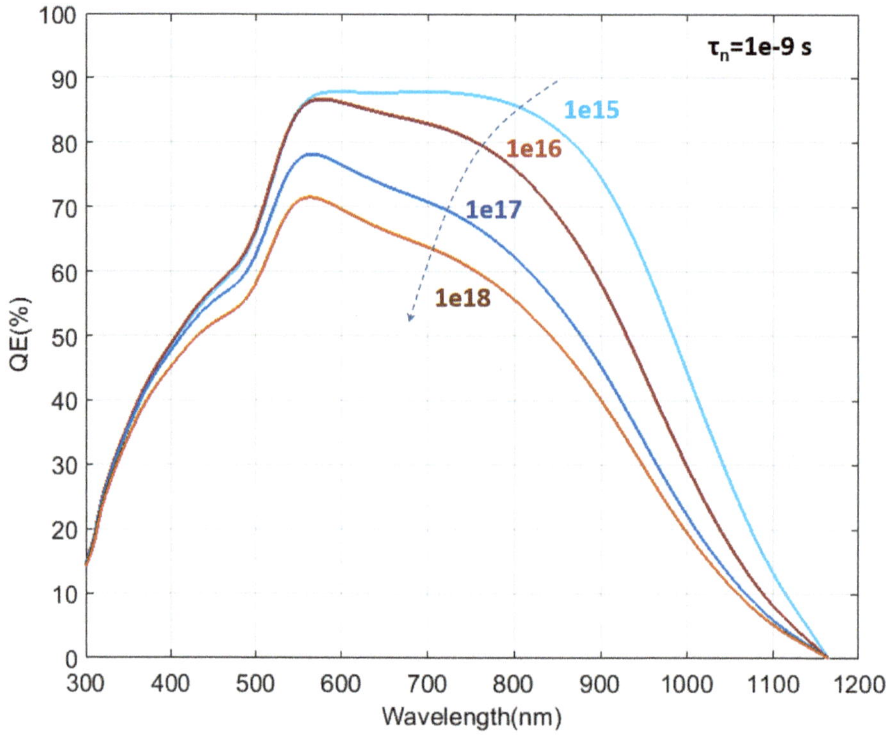

Figure 3. EQE for uncompensated acceptor concentration (N_a–N_d), 10^{15}, 10^{16}, 10^{17}, and 10^{18} cm^{-3}.

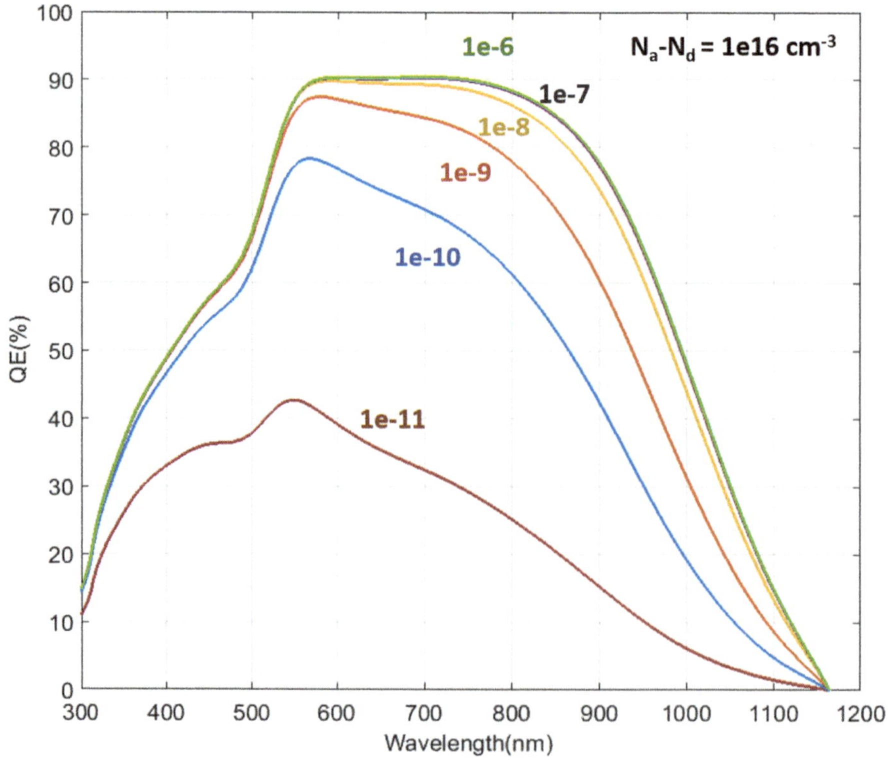

Figure 4. EQE spectra with minority carrier lifetime (τ_n), 10^{-6}, 10^{-7}, 10^{-8}, 10^{-9}, 10^{-10}, and 10^{-11} s.

(N_a–N_d). When the concentration of (N_a–N_d) increases from 10^{15} cm^{-3} to 10^{18} cm^{-3}, the external quantum efficiency decreases. The decrease in EQE(λ) is due to the expansion of the depletion layer, which induces efficient photogenerated minority carrier from the absorber bulk. However, the decrease in photosensitivity between 300 and 500 nm is relatively small, which is mainly influenced by the absorption and recombination of ZnO and CdS. This is also well demonstrated in the Q-E responses with a set of electron lifetime (τ_n) in **Figure 4**, which shows less changes in Q-E between 300 and 500 nm except the extreme case of 10 ps. With the same depletion width by the same uncompensated impurity concentration, electron lifetime that can facilitate an electron to travel throughout the quasi-neutral region to the space charge region can show higher carrier collection and fewer carrier losses in the range from 600 to 900 nm as electron lifetime decreases from 10^{-6} to 10^{-10} s.

2.5. Impact of deletion width by analytical model

In order to provide the losses caused by recombination at the CdS/CZTSSe interface, we now check the spectral response broken into drift and diffusion components in Q-E spectra. **Figure 5** demonstrates the drift component in dashed line and the diffusion component in dash-dotted line as the depletion width increases from 0.1 to 0.9 μm. At the depletion width, 0.9 μm, the photogenerated carrier in the absorber near 600–1000 nm can be efficiently collected mainly via the space charge region by showing the sufficient drift component compared to the weak diffusion component of photogenerated current. However, once the depletion width decreases

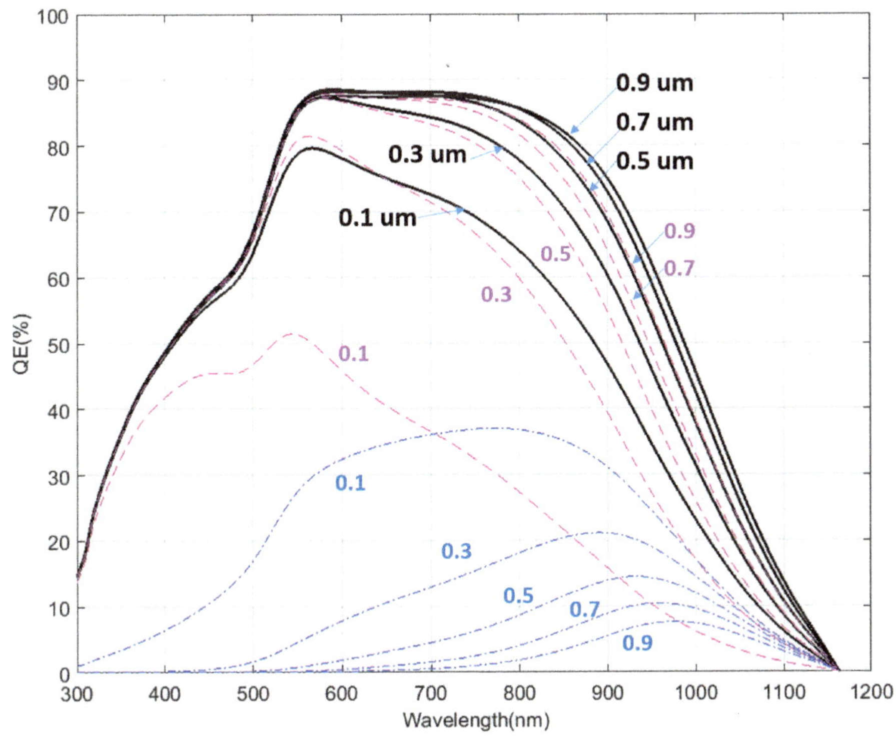

Figure 5. EQE spectra with the depletion width, 0.1, 0.3, 0.5, 0.7, and 0.9 μm broken into the diffusion and drift components. The total EQE (black) is a sum of the diffusion component (blue) and the drift component (magenta).

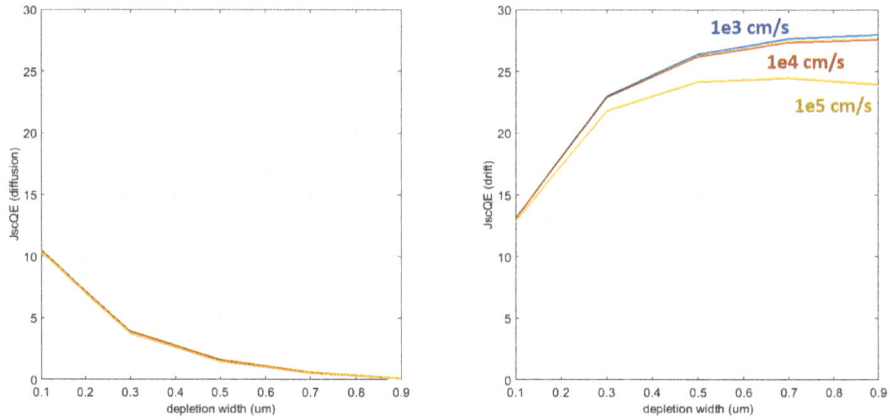

Figure 6. JscQE (drift) and JscQE (diffusion) under the standard AM 1.5 solar radiation (1 sun, 100 mW/cm^2) under different front surface recombination velocities (10^3, 10^4, and 10^5 cm/s).

from 0.9 μm, the drift component begins to weaken in the CZTSSe absorber in 600–1000 nm and the diffusion component increases. At the depletion width, 0.1 μm, the contribution to the carrier collection by drift component is almost equivalent to that of the diffusion component. Based on the calculation of Q-E spectra about each drift and diffusion component, **Figure 6** shows the drift current short circuit current component, JscQE (drift) and the diffusion short circuit current component, JscQE (diffusion) under the standard AM 1.5 solar radiation (1 sun, 100 mW/cm^2) under different front surface recombination velocities (10^3, 10^4, 10^5 cm/s). As *Sf* increases, there are no noticeable changes in the diffusion component whereas the drift component decreases faster.

3. Comparison with modeling and simulation of CZTSSe devices

In this section, we investigate quantum efficiency by utilizing SCAPS simulator (Solar Cell Capacitance Simulator) version 3.2.01 to computationally model the quantum efficiency response of current CZTSSe thin-film solar cells. This will also provide the proper interpretation of device behaviors when it is measured by quantum efficiency. In particular, we focus on the influence of near-interface defect states on quantum efficiency at particular bias conditions with the intention of evaluating peculiar quantum efficiency responses by defect distributions with adjusting 1D numerical parameters in this work.

Each material parameter as an input includes thickness, relative permittivity, electron mobility, hole mobility, acceptor concentration, donor concentration, band gap, and effective density of states. The simulated device structure of CZTSSe thin-film solar cell is with ZnO window layer, CdS buffer, CZTSSe absorber, and Molybdenum (Mo) back contact. In intrinsic defects in the CZTSSe bulk, we updated the parameter sets from the analytical description to reflect the real situation to understand the general trend by interface defect states. Parameters used for this device structures are listed in **Table 1**. Considering the complexity of interface defects at absorber/back contact, absorber/buffer, and buffer/window, we adopted the basic input parameters from the literature, theories, or reasonable estimates [12, 28, 31]. In reality, there is a significant amount of intrinsic bulk defects with different charge states in the CZTSSe

Parameters	Symbol (unit)	ZnO	CdS	CZTSSe
Thickness	d (nm)	200	50	2500
Band gap	Eg (eV)	3.3	2.4	1.2–1.3
Electron affinity	χ (eV)	4.4	4.2	4.1
Relative permittivity	ϵ/ϵ_r	9	10	10
Effective density of state (CB)	N_C (10^{18} cm^{-3})	2.2	2.2	2.2
Effective density of state (VB)	N_V (10^{19} cm^{-3})	1.8	1.8	1.8
Electron thermal velocity	v_n (10^7 cm/s)	1.0	1.0	1.0
Hole thermal velocity	v_p (10^7 cm/s)	1.0	1.0	1.0
Electron mobility	μ_n (cm^2/V-s)	100	100	100
Hole mobility	μ_p (cm^2/V-s)	25	25	25
Donor concentration	N_D (10^{17} cm^{-3})	10.0	1.0	0
Acceptor concentration	N_A (10^{14} cm^{-3})	0	0	2.0
Bulk defect type	a, d, n	d	a	d/a/n
Bulk defect concentration	N_t (10^{14} cm^{-3})	10,000	100	1/1/5

Types of defect are: a, acceptor; d, donor; n, neutral.

Table 1. Parameters set used in this work.

absorber layer as well as ZnO, CdS layers. To introduce different amounts of intrinsic defects in the CZTSSe bulk, we updated the parameter sets from the analytical description to reflect the real situation to understand the general trend by interface defect states. Parameters used for this device structure are listed in **Table 1**. Considering the complexity of interface defects at absorber/back contact, absorber/buffer, and buffer/window, we adopted the basic input parameters from the literature, theories, or reasonable estimates [31–36].

3.1. Interface trap-free case

In the first step, the numerical simulations of both current-voltage (I-V) and quantum efficiency (Q-E) characteristics have been carried out without traps near the heterojunction interface. The default illumination spectrum and operation temperature are set to the standard AM 1.5 condition and 300 K, respectively. A typical result of the I-V curve simulated for a CZTSSe solar cell without heterojunction defects is demonstrated in **Figure 7(a)**. As expected, we observed an ideal steep I-V curve with conversion efficiency, 15.3% with emphasis on higher *Voc* (661 mV) and FF (72.5%) than any reported experimental *Voc* and FF justifying the assumption of an intrinsic defect-free interface.

The simulated I-V curve in **Figure 7** shows the dramatic improvement of *Voc* with nearly zero conductance (<2.6 mS/cm^2) at short circuit condition (in other words, a slope of I-V curve at zero bias) which can be explained by the alleviated recombination and the improved carrier collection with the associated depletion width change. In particular, the reduced recombination can lower the total forward diode current by showing the ideal slope at the short circuit current

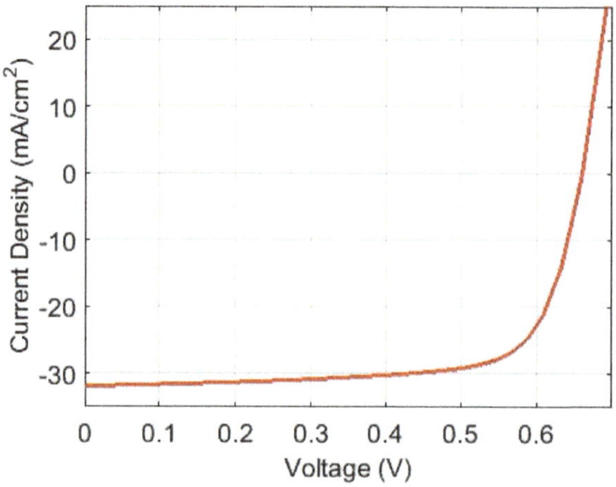

Figure 7. I-V of ideal CZTSSe solar cells without interface defects with efficiency (15.3%), Voc (661 mV), FF (72.5%), and Jsc (32.0 mA/cm^2).

condition [25]. With an ideal device having no near-interface defects in the simulation, an ideal uniform absorber setting keeps the shape and the ratio of the quantum efficiency curves, (H(V)/H(−1 V)) under biases since the shape of the curves in quantum efficiency depends on wavelength and is almost independent of bias voltage [20]. One way to confirm the uniformity of reduced carrier collection at each wavelength region is to compare carrier collection at each bias with full collection by applying negative bias, for example, −1 V [20]. In **Figures 8** and **9**, each quantum efficiency at different biases is normalized by dividing with the quantum efficiency at −1 V. The normalized Q-E (QE(V)/QE(−1 V)) shows two features by comparing each bias-dependent quantum efficiency. First, wavelength-dependent quantum efficiency drop is observed regardless of bias level. Secondly, the shape of each normalized quantum efficiency curve is kept almost the same. In other words, a nearly uniform collapse of quantum efficiency over a wide range of wavelength is demonstrated with the trend of increasing absorber collection loss due to reduced depletion width by applied biases at region 2 (500–800 nm) and region 3 (800–1033 nm).

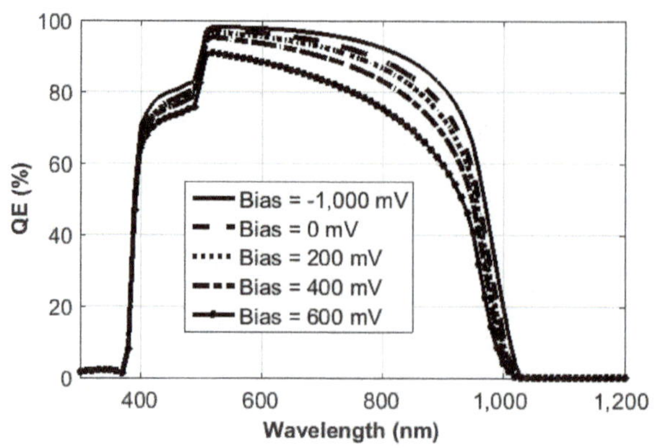

Figure 8. Q-E of ideal CZTSSe solar cells without interface defects at bias conditions (−1.0, 0.0, 0.2, 0.4, and 0.6 V).

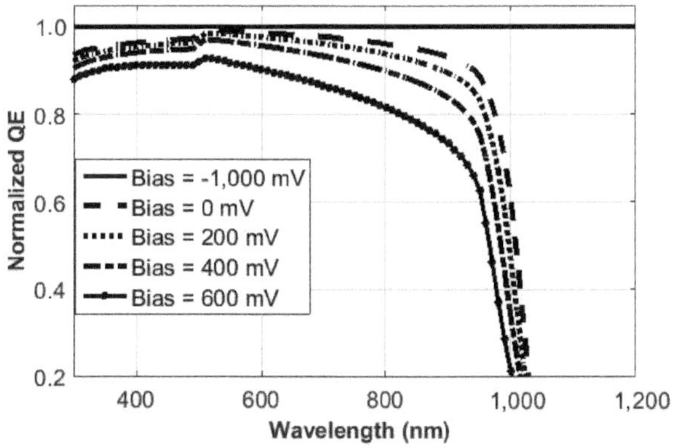

Figure 9. Normalized Q-E of ideal CZTSSe solar cells (QE(V)/QE(−1 V)) at bias conditions (−1.0, 0.0, 0.2, 0.4, and 0.6 V).

3.2. Interface and near-interface defect states case

Before discussing the simulated results, it is worth describing the systematic trends affected by varying the interface defect parameters for the different types of defects. First, we consider acceptor- and donor-type defects near both conduction and valence bands as well as mid-gap. In SCAPS simulation, we can introduce three different parameter sets of interface defects for both acceptor and donor types that significantly affect the performance of the device (capture cross sections, defect energy level: E_t, defect density). However, real CZTSSe cells consist of complex multilayers, which form defects and/or defect compounds at the surface of CZTSSe layer, next to the CdS buffer layer. Furthermore, the direct measurement of this layer's electronic properties is difficult to be separated from other layers' electronic properties which are exacerbated by complicated window structures with CdS, ZnO, and TCO layer interdiffusion. To avoid the complexity of modeling, we focus on numerical modeling of electronic properties, in particular, quantum efficiency by introducing a thin defective interface layer between CdS and absorber.

3.2.1. Trend by defect distribution and concentration

In the first step, numerical simulations have been carried out with a set of defect concentrations within a defective interface layer. To be consistent with the reported literature, theories, or reasonable estimates, capture cross sections are fixed at 1.0×10^{-12} cm^2 for electrons and 1.0×10^{-15} cm^2 for holes at donor-type defect and at 1.0×10^{-15} cm^2 for electrons and 1.0×10^{-12} cm^2 for acceptor-type defect [12, 28–31, 33–37]. For a comparative study of the ideal device in the previous section, the device structure is updated with very thin (5 nm) defective interface layer between CdS and CZTSSe. Input parameters are set to the same as those of CdS buffer except thickness and doping densities which are set to the same values of 1.0×10^{14} cm^{-3} for donor and acceptor assuming the intermixing of n-type dopant (CdS) and p-type dopant (CZTSSe).

Figure 10 shows the simulated quantum efficiency with an inserted defective interface layer between CZTSSe and CdS buffer layers as a function of defect concentration. As we begin with defect density 1.0×10^{15} cm^{-3} in this thin layer as deep acceptor-type bulk defect which is different from donor-like interface trap (as in the conventional Si model), we located this type of defect between 0.9 and 1.5 eV above the top of the valence band around the mid-gap (1.2 eV).

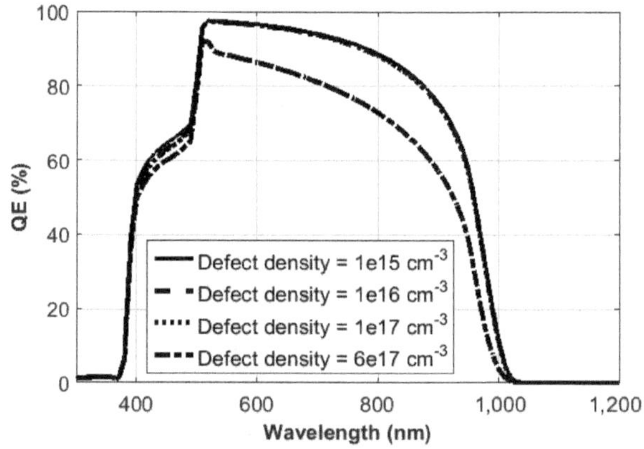

Figure 10. Q-E of CZTSSe device with the inserted interface defective layer between CdS and CZTSSe, varying acceptor-type defect density from 1.0×10^{15} to 6.0×10^{17} cm^{-3}.

As defect concentration increases until 1.0×10^{17} cm^{-3} no specific collection loss at particular regions is observed. However, once defect concentration increases more than 6.0×10^{17} cm^{-3}, serious collection loss at the absorber is demonstrated in regions 2 (500–800 nm) and 3 (800–1033 nm). In region 1 (300–500 nm), there is not much collection loss while slight inflection of collection occurs near 500 nm.

3.2.2. Impact by deep acceptor-type defect distribution

Based on the findings above, defect concentration is set at 6.0×10^{17} cm^{-3} as deep acceptor-type defects, 0.9 eV below and above from the bottom of the conduction band and top of the valence band, respectively. Consequently, the results of the simulated biased quantum efficiency are demonstrated in **Figure 11**. Beginning at −1 V (not shown), quantum efficiency over all wavelength regions shows similar spectral response compared to that of the device without defects in **Figure 8**. Once bias increases up to 0 V, absorber loss slightly increases as mentioned in **Figure 8**. As expected, this reduction in quantum efficiency is caused by the decrease of the depletion layer. However, as bias increases up to 0.2 V, the stronger drop of Q-E spectrum happens at both < 500 nm and > 550 nm disproportionately. Conversely, the reduction of quantum efficiency between 500–550 nm is less severe that the other regions. This peculiar Q-E response occurs with a peak near 520–530 nm (in other words, near blue light region), reported at their measurements in Refs. [12, 37].

Apparently, the reduction in $\lambda < 400$ nm, which is relatively non-sensitive, is due to absorption by the CdS layer and the transparent conducting oxide/substrate. In the spectral region, < 500 nm, the absorption by CdS at 0.2 V is slightly decreased due to the smaller photocurrent contribution and adjustment of space charge region of window layer under the presence of strongest electric field at the heterojunction. On the other hand, the intensified quantum efficiency drop in <500 nm and distinctively >550 nm looks interesting. When it comes to describing the bias-dependent quantum efficiency, the Gartner model describes the photogenerated carrier collection with the bias-dependent depletion width (Eq. (6)) and diffusion length, which describes the exact expression (Eq. (21)) as we discussed in the previous section [20, 36].

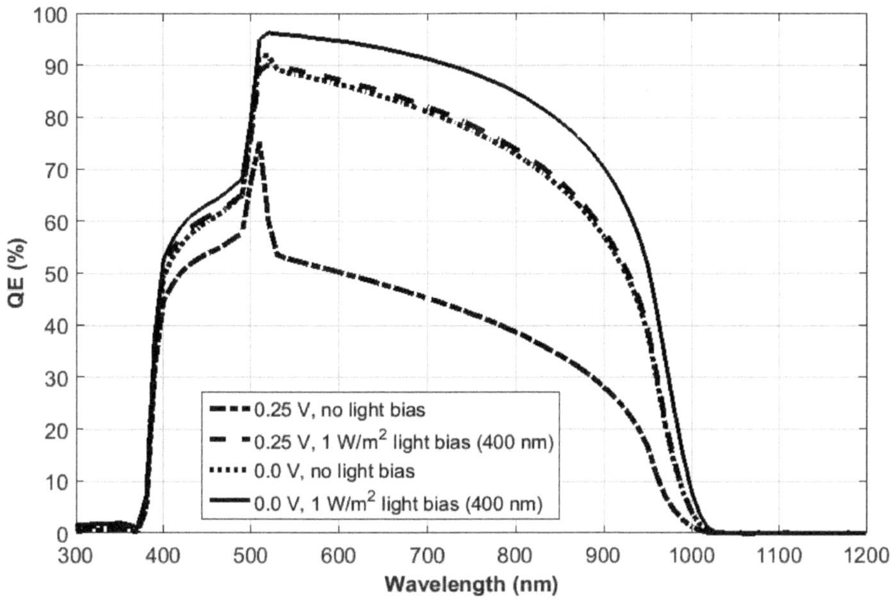

Figure 11. Light-biased (1 mW/cm^2, 0.01 sun at 400 nm) Q-E of CZTSSe solar cell with the defective interface layer at 0 and 0.25 V.

With the given depletion width (W_d), diffusion length (L_n), absorber thickness (d), and surface recombination velocity (S_n), the exact expression predicts the trend of spectral response in thin-film solar cells under the presence of voltage and, in particular, quantum efficiency response for an ideal device. However, the exact model or the Hecht-like equation in Ref. [29] is solely dependent on the depletion width (or applied bias) and diffusion length. Under applied voltage biases compared to the full collection, the analytical quantum efficiency analysis is unable to demonstrate the intensified reduction at both <500 nm and >550 nm but it rather provides partial interpretation of the quantum efficiency with the above equations by keeping a similar shape of curves. Hence, analytical expressions do not represent the peculiar Q-E response near 520–530 nm over other wavelength regions from our simulation.

However, the intensified drop in both <500 nm and distinctively >550 nm can be explained by assuming deep acceptor-type defects near the heterojunction interface layer. In this interface layer, a higher concentration of deep ionized-acceptor (negative charge) type defects 0.3 eV above and below the mid-gap impacts the electric field and hence the space charge region of absorber toward the back contact. These deep acceptor-type defects induce electronic doping at room temperature by trapping holes from the valence band, which eventually lower the band bending of the CdS/CZTSSe interface layer upon additional blue light. In other words, higher electric field near the heterojunction interface increases absorption by a part of the absorber toward CdS and near-CdS layer, whereas it reduces absorption by the majority of CZTSSe absorber toward the back contact. This reduced depletion width by higher interface defects can be deteriorated by increased biases such as 0.2 and 0.25 V, which furthers the depletion width decrease. One way to confirm depletion width reduction induced by CdS/absorber interface defects is to apply weak blue light bias (400 nm) and observe spectral response at >550 nm in quantum efficiency. Using this model, weak light bias (1 mW/cm^2, 0.01 sun) is applied to the same device at 0 and 0.25 V in **Figure 11**. Weak blue light bias onto window and buffer layers almost fully recovers CZTSSe absorber carrier collection even at 0.25 V. After filtering out noises in the regions of <400 and >1000 nm, light-biased Q-E is normalized by dividing with the standard no-light-biased Q-E in **Figure 12**. This clearly indicates that weak blue light bias intensifies the carrier collection of absorber in regions 2 and 3 (>550 nm) at both 0 and 0.25 V, whereas minimal increases are shown in region 1 (<500 nm). At 0.25 V, the depletion width of CZTSSe absorber without light bias is narrower than that with weak blue light bias. Hence, weak light bias activates hole trapping through deep acceptor-type defects and near-interface defects between 0.9 and 1.5 eV and recovers the depletion width.

The presence of ionized deep acceptor-type bulk defects (0.9–1.5 eV) in the defective interface layer can effectively decrease the depletion width near the heterojunction interface by increasing negative charge density (**Figure 13**), which could only be activated by weak bias light via hole charge trapping. The band diagram and carrier density of holes and electrons indicate

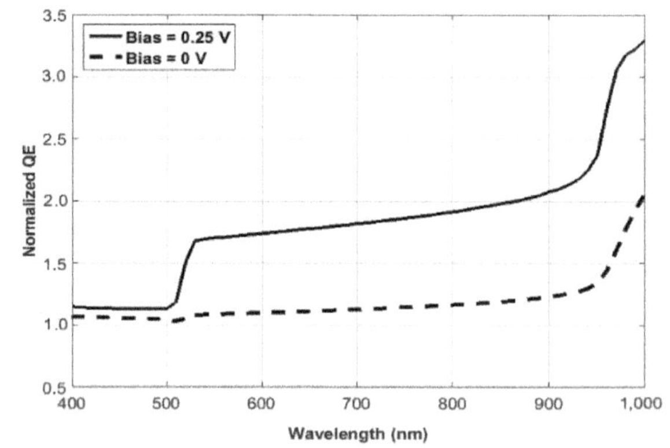

Figure 12. Light bias-dependent normalization of Q-E (Q-E at blue light bias divided by Q-E at no light bias) illustrating weak blue light bias (400 nm).

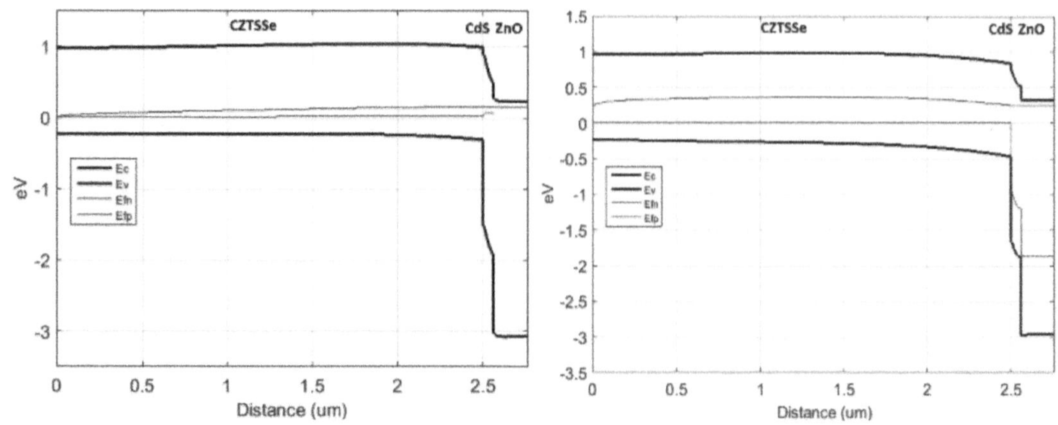

Figure 13. Band diagram without light bias at 0.25 V (left) and with light bias at 0.25 V (right), showing larger depletion width.

higher electric field with less hole concentration at the interface, resulting in smaller depletion width in the presence of the defective interface layer in **Figure 13**.

4. Summary

The spectral response of CZTSSe thin-film solar cells has been modeled analytically and numerically with simulation study under a set of different parameters and biased conditions about efficiency-limiting factors through quantum efficiency. The analytical model describes the drift and diffusion photocurrent components to reflect the recombination losses in the space charge region and near the CdS/CZTSSe heterojunction interface. The trend that we found about the uncompensated impurity concentration, minority carrier lifetime, surface recombination velocity, and the depletion width shows qualitatively good agreement with the reported data in the literature. Furthermore, the simulations address the role of interface and near-interface defects for the spectral response of thin-film solar cells. For this simulation study, it is assumed that deep acceptor-type near-heterojunction interface defects from 0.9 to 1.5 eV above the top of the valence band are difficult to increase the depletion width, which provide additional electronic doping in space charge region. These ionized acceptor-type defects at room temperature can trap holes from the valence band only activated by additional bias light and reduce the effective doping concentration at the heterojunction interface. Consequently, these deep acceptor-type defects cramp the depletion width and hence, weaker spectral responses near red and IR regions. This trend is intensified with biased quantum efficiency conditions. The findings of this work give further insight into the issues and provide some requirements on the parameter sets of the CZTSSe absorber layer as well as the CdS/CZTSSe heterojunction interface layer, which shows fairly good agreement with numerical calculation and reported experimental results.

Author details

Sanghyun Lee[1]* and Kent J. Price[2]

*Address all correspondence to: sanghyun.lee0304@gmail.com

1 School of Engineering and Information Systems, Morehead State University, Morehead, KY, USA

2 Department of Physics, Morehead State University, Morehead, KY, USA

References

[1] W. Wang, M.T. Winkler, O. Gunawan, T. Gokmen, T.K. Todorov, Y. Zhu, D.B. Mitzi. Device characteristics of CZTSSe thin-film solar cells with 12.6 % efficiency. Adv. Energy Mater. 2014;**4**:1301465.

[2] D. Mitzi, T. Todorov, K. Wang, S. Gunha. The path towards a high-performance solution-processed kesterite solar cell. Sol. Energy Mater. Sol. Cells. 2011;**95**:1421–1436.

[3] Y. Kumar, G. Babu, P. Bhaskar, V. Raja. Preparation and characterization of spray-deposited Cu_2ZnSnS_4 thin films. Sol. Energy Mater. Sol. Cells. 2009;**93**:1230–1237.

[4] C.P. Chan, H. Lam, C. Surya. Preparation of Cu_2ZnSnS_2 films by electrodeposition using ionic liquids. Sol. Energy Mater. Sol. Cells. 2010;**94**:207–211.

[5] H. Katagiri, K. Jimbo, S. Yamada, T. Kamimura, W.S. Maw, T. Fukano, T. Ito, T. Motohiro. Detailed balance limit of efficiency of p-n junction solar cells. Appl. Phys. Express. 2008;**1**:41201.

[6] M. Paranthaman, W. Wong-Ng, R. Bhattacharya, editors. Semiconductor materials for solar photovoltaic cells (1st ed.). Springer International Publishing, Switzerland; 2016. doi:10.1007/978-3-319-20331-7_2.

[7] P. Jackson, D. Hariskos, R. Wuerz, W. Wischmann, M. Powalla. Compositional investigation of potassium doped $CuInGaSe_2$ solar cells with efficiencies up to 20.8 %. Phys. Stat. Sol. RRL. 2014;**8**:219–222.

[8] W. Shockley, H.K. Queisser. Detailed balance limit of efficiency of p-n junction solar cells. Jpn. J. Appl. Phys. 1961;**32**:510–519.

[9] G. Lucovsky, S. Lee, J.P. Long, H. Seo, J. Luning. Interfacial transition regions at germanium/ Hf oxide based dielectric interfaces: qualitative differences between non-crystalline Hf Si oxynitride and nanocrystalline HfO2 gate stacks. Microelectron. Eng. 2009;**86**(3):224–234.

[10] J. Moore, C. Hages, R. Agrawal, M. Lundstrom. The importance of band tail recombination on current collection and open-circuit voltage in CZTSSe solar cells. Appl. Phys. Lett. 2016;**109**:021102.

[11] T. Gokmen, O. Gunawan, T. Todorov, D. Mitzi. Band tailing and efficiency limitation in kesterite solar cells. Appl. Phys. Lett. 2013;**103**:103506.

[12] T. Gokmen, O. Gunawan, D.B. Mitzi. Semi-empirical device model for $Cu_2ZnSn(S,Se)_4$. Appl. Phys. Lett. 2014;**105**:033903.

[13] J. Moore, C. Hages, N. Carter, R. Agrawal, J. Gray, M. Lundstrom. Current–voltage analysis of band tail effects in CZTSSe through numerical simulation. 2015 IEEE 42nd Photovoltaic Specialist Conference (PVSC). 2015.

[14] D. Mitzi, O. Gunawan, T. Todorov, D. Barkhouse. Preparation of Cu_2ZnSnS_4 films by electrodeposition using ionic liquids. Phil. Trans. R. Soc. A. 2013;**371**(20110432):1–22.

[15] O. Simya, A. Mahaboobbatcha, K. Balachander. A comparative study on the performance of Kesterite based thin film solar cells using SCAPS simulation program. Superlatt. Microstruct. 2015;**82**:248–261.

[16] T. Jager, Y. Romanyuk, B. Bissig, F. Pianezzi, S. Nishiwaki, P. Reinhard, J. Steinhauser, J. Schwenk, A. Tiwari. Improved open-circuit voltage in $Cu(In,Ga)Se_2$ solar cells with high work function transparent electrodes. J. Appl. Phys. 2015;**117**:225303.

[17] F.-J. Haug, R. Biron, G. Kratzer, F. Leresche, J. Besuchet, C. Ballif, M. Dissel, S. Kretschmer, W. Soppe, P. Lippens, K. Leitner. Improvement of the open circuit voltage by modifying the transparent indium-tin oxide front electrode in amorphous n-i-p solar cells. Prog. Photo. 2012;**20**(6):727–734.

[18] S. Hegedus, W. Shafarman. Thin-film solar cells: device measurements. Prog. Photovolt. Res. Appl. 2004;**12**(2–3):155–176. doi:10.1002/pip.518.

[19] C. Hages, N. Carter, R. Agrawal, T. Unold. Generalized current–voltage analysis and efficiency limitations in non-ideal solar cells: case of $Cu_2ZnSn(S_xSe_{1-x})_4$ and $Cu_2Zn(Sn_yGe_{1-y})(S_xSe_{1-x})_4$. J. Appl. Phys. 2014;**115**:234504.

[20] S. Sze, K. Ng. Physics of semiconductor devices (3rd.). New York: John Wiley & Sons; 2007.

[21] W. Gartner. Depletion-layer photoeffects in semiconductors. Phys. Rev. 1959;**116**:84–87.

[22] M. Lavagna, J.P. Pique, Y. Marfaing. Theoretical analysis of the quantum photoelectric yield in Schottky diode. Solid State Electron. 1977;**20**:235–240.

[23] L.A. Kosyachenko. Problems of efficiency of photoelectric conversion in thin-film CdS/CdTe solar cells. Semiconductors. 2006;**40**(6):710–727.

[24] M. Hadrich, H. Metzner, U. Reislohner, C. Kraft. Modelling the quantum efficiency of cadmium telluride solar cells. Sol. Energy Mater. Sol. Cells. 2011;**95**:887–893.

[25] A.L. Fahrenbruch, R.H. Bube. Fundamentals of solar cells: photovoltaic solar energy conversion (1st ed.). Academic Press, New York; 1983.

[26] A. Nakane, H. Tampo, M. Tamakoshi, S. Fujimoto, K.M. Kim, S. Kim, H. Shibata, S. Niki, H. Fukiwara. Quantitative determination of optical and recombination losses in thin-film photovoltaic devices based on external quantum efficiency analysis. J. Appl. Phys. 2016;**120**(6):064505. doi:10.1063/1.4960698.

[27] K. Misiakos, F. Lindholm. Analytical and numerical modelling of amorphous silicon p-i-n solar cells. J. Appl. Phys. 1988;**64**:383.

[28] V. Chu, J. Conde, D. Shen, S. Wagner. Photocurrent collection in a Schottky barrier on an amorphous silicon-germanium alloy structure with 1.23 eV optical gap. Appl. Phys. Lett. 1989;**55**:262.

[29] S. Hegedus, D. Desai, C. Thompson. Voltage dependent photocurrent collection in CdTe/CdS solar cells. Prog. Photovolt. Res. Appl. 2007;**15**:587–602.

[30] A. Pu, F. Ma, C. Yan, J. Huang, K. Sun, M. Green, X. Hao. Sentaurus modelling of 6.9% Cu_2ZnSnS_4 device based on comprehensive electrical & optical characterization. Sol. Energy Mater. Sol. Cells. 2017;**160**:372–381. doi:10.1016/j.solmat.2016.10.053.

[31] O. Gunawan, T.K. Todorov, D.B. Mitzi. Loss mechanisms in hydrazine-processed $Cu_2ZnSn(Se,S)_4$ solar cells. Appl. Phys. Lett. 2010;**97**:233506.

[32] P. Viktorovich, G. Moddel, J. Blake, W. Paul. Carrier-collection efficiencies in amorphous hydrogenated silicon Schottky-barrier solar cells. J. Appl. Phys. 1981;**52**:6203.

[33] W. Zhao, W. Zhou, X. Miao. Numerical simulation of CZTS thin film solar cell. In 7th IEEE International Conference on Nano/Micro Engineered and Molecular Systems, 2012, pp. 502–505.

[34] T. Gokmen, O. Gunawan, D.B. Mitzi. Modelling polycrystalline semiconductor solar cells. Thin Solid Films. 2000;**361–362**(033903):527–532.

[35] A. Polizzott, I.L. Repins, R. Noufi, S-Hi We, D.B. Mitzi. The state and future prospects of kesterite photovoltaics. Energy Environ. Sci. 2013;**6**:3171–3182.

[36] R.D. Rugescu, L. Kosyachenko. Solar energy (1st ed.). InTech, Croatia; 2010, pp. 105–130.

[37] G. Brammertz, S. Oueslati, M. Buffiere, J. Bekaert, H.E. Anzeery, K.B. Messaoud, S. Sahayarj, T. Nuytten, C. Koble, M. Meuris, J. Poortmans. Investigation of properties limiting efficiency in $Cu_2ZnSn(S,Se)_4$-based solar cells. IEEE J. Photovolt. 2015;**527–532** (5):649–655.

Permissions

All chapters in this book were first published in OPTOELECTRONICS, by InTech Open; hereby published with permission under the Creative Commons Attribution License or equivalent. Every chapter published in this book has been scrutinized by our experts. Their significance has been extensively debated. The topics covered herein carry significant findings which will fuel the growth of the discipline. They may even be implemented as practical applications or may be referred to as a beginning point for another development.

The contributors of this book come from diverse backgrounds, making this book a truly international effort. This book will bring forth new frontiers with its revolutionizing research information and detailed analysis of the nascent developments around the world.

We would like to thank all the contributing authors for lending their expertise to make the book truly unique. They have played a crucial role in the development of this book. Without their invaluable contributions this book wouldn't have been possible. They have made vital efforts to compile up to date information on the varied aspects of this subject to make this book a valuable addition to the collection of many professionals and students.

This book was conceptualized with the vision of imparting up-to-date information and advanced data in this field. To ensure the same, a matchless editorial board was set up. Every individual on the board went through rigorous rounds of assessment to prove their worth. After which they invested a large part of their time researching and compiling the most relevant data for our readers.

The editorial board has been involved in producing this book since its inception. They have spent rigorous hours researching and exploring the diverse topics which have resulted in the successful publishing of this book. They have passed on their knowledge of decades through this book. To expedite this challenging task, the publisher supported the team at every step. A small team of assistant editors was also appointed to further simplify the editing procedure and attain best results for the readers.

Apart from the editorial board, the designing team has also invested a significant amount of their time in understanding the subject and creating the most relevant covers. They scrutinized every image to scout for the most suitable representation of the subject and create an appropriate cover for the book.

The publishing team has been an ardent support to the editorial, designing and production team. Their endless efforts to recruit the best for this project, has resulted in the accomplishment of this book. They are a veteran in the field of academics and their pool of knowledge is as vast as their experience in printing. Their expertise and guidance has proved useful at every step. Their uncompromising quality standards have made this book an exceptional effort. Their encouragement from time to time has been an inspiration for everyone.

The publisher and the editorial board hope that this book will prove to be a valuable piece of knowledge for researchers, students, practitioners and scholars across the globe.

List of Contributors

Salah Ibrahim and Ryo Takahashi
NTT Device Technology Laboratories, NTT Corporation, Kanagawa, Japan

Andrey Kosarev, Svetlana Mansurova, Antonio J. Olivares and Hiram E. Martinez
National Institute for Astrophysics, Optics and Electronics, Optics and Electronics (INAOE), Puebla, Mexico

Ismael Cosme
CONACyT-INAOE, Mexico

Chien-Hung Yeh
Department of Photonics, Feng Cha University, Taichung, Taiwan

Chi-Wai Chow
Department of Photonics, National Chiao Tung University, Hsinchu, Taiwan

Binhao Wang
Hewlett Packard Labs, Hewlett Packard Enterprise, USA

Sergei L. Pyshkin
Institute of Applied Physics, Academy of Sciences of Moldova, Kishinev, Moldova, Republic of Moldova Clemson University, South Carolina, USA

John Ballato
Clemson University, South Carolina, USA

Erinn van Wynsberghe and Ayse Turak
McMaster University, Hamilton, Ontario, Canada

Xiaolu Zhu and Yifei Yang
College of Mechanical and Electrical Engineering, Hohai University, China

Oleksandr Malik and Francisco Javier De la Hidalga-Wade
Electronics Department, Instituto Nacional de Astrofísica, Óptica y Electrónica (INAOE), Puebla, Mexico

Oleksandr Malik and Francisco Javier de la Hidalga-Wade
Electronics Department, Óptica y Electrónica (INAOE), Instituto Nacional de Astrofísica, Puebla, Mexico

Index